Morning Glory Clouds

Rolling Wonders of the Sky

Ripley Jones

ISBN: 978-1-77961-202-1
Imprint: The Wonder Lab
Copyright © 2024 Ripley Jones.
All Rights Reserved.

Contents

Introduction 1
What are Morning Glory Clouds? 1

Atmospheric Phenomena 13
Weather Systems 13
Cloud Formation 30
Air Masses 45

Morning Glory Cloud Features 65
Shape and Size 65
Vertical Structure 81
Vibrant Colors 94

Global Distribution 113
Geographic Distribution 113
Seasonal Patterns 128
Diurnal Cycle 144

Formation Mechanisms 159
Gravity Waves 159
Soliton Theory 174
Convection and Frontal Systems 187

Meteorological Impacts 203
Local Weather Patterns 203
Aviation Hazard 218
Climate Change 238

Cultural and Historical Significance 253

Indigenous Traditions and Legends	253
Artistic and Literary Representations	268
Ecotourism and Conservation	282

Research Techniques and Future Directions 301

Remote Sensing and Imaging	301
In Situ Measurements	321
Numerical Models and Simulations	337
Citizen Science and Crowdsourcing	356

Index 373

Introduction

What are Morning Glory Clouds?

Definition of Morning Glory Clouds

Morning Glory Clouds are a unique meteorological phenomenon that occurs primarily in the Gulf of Carpentaria region of Northern Australia, but can also be observed in certain parts of the world such as Canada and the United States. These clouds are characterized by their long, tubular shape and the distinctive rolling appearance they have as they move across the sky. They can span several hundred kilometers in length and reach altitudes of up to 2 kilometers.

The name "Morning Glory" is derived from the fact that these clouds are most commonly observed in the early morning hours, typically around sunrise. They are often associated with a visible roll pattern, where multiple cloud layers are stacked on top of each other, resembling long cylindrical waves in the sky. This roll pattern gives the clouds a unique and captivating appearance.

Morning Glory Clouds are a type of arcus cloud, which are low-level horizontal clouds that typically form along the leading edge of a thunderstorm. However, Morning Glory Clouds are distinct from other arcus clouds in their formation mechanism, size, and elongated shape. They are not associated with storm activity and can occur under different atmospheric conditions.

These clouds are formed by a combination of mesoscale circulations, gravity waves, and atmospheric stability conditions. The Gulf of Carpentaria region is particularly conducive to the formation of Morning Glory Clouds due to the unique geographical features, such as the combination of warm ocean currents, cool dry air masses, and the presence of the Cape York Peninsula.

In terms of meteorological classification, Morning Glory Clouds fall into the category of cloud formations known as stratocumulus. Stratocumulus clouds are low-lying clouds with a relatively uniform base and often have a lumpy or wavy appearance. Morning Glory Clouds are unique within the stratocumulus family due to their elongated shape and roll pattern.

Understanding the formation and behavior of Morning Glory Clouds is of great importance to meteorologists and atmospheric scientists. These clouds provide valuable insights into atmospheric dynamics, including the interaction between large-scale weather systems, local weather patterns, and the influence of topography. Furthermore, the study of Morning Glory Clouds contributes to our understanding of cloud physics and the broader field of cloud-climate interactions.

In the following sections of this book, we will delve deeper into the various aspects of Morning Glory Clouds, including their formation mechanisms, physical characteristics, global distribution, meteorological impacts, and cultural significance. We will explore the scientific principles, observational techniques, and modeling approaches that have been used to study these captivating clouds. By the end of this book, readers will have a comprehensive understanding of Morning Glory Clouds and their place in the atmospheric sciences. Let's embark on this fascinating journey together!

Historical Discoveries

The study of Morning Glory Clouds has a rich history that spans several centuries. This section explores the key discoveries and milestones in understanding this unique atmospheric phenomenon.

One of the earliest recorded observations of Morning Glory Clouds can be traced back to the indigenous cultures of Australia, particularly the Aboriginal people. These clouds held a significant place in their traditions and were often associated with creation stories and spiritual beliefs. The Aboriginal people referred to the Morning Glory Clouds as "kangalanja," which means "the cloud that comes like a snake" in their native language.

In the scientific world, the first documented encounter with Morning Glory Clouds occurred in the mid-19th century. In 1867, a British naturalist named Augustus Gregory observed these unusual clouds during his exploration of northern Australia. Gregory described the clouds as long, tubular formations that stretched across the horizon. His detailed notes and sketches became the earliest scientific record of Morning Glory Clouds.

Another significant historical discovery in the study of Morning Glory Clouds took place in the early 20th century. In 1903, an Australian meteorologist named Clement Wragge conducted extensive research on these clouds. Wragge was renowned for his contributions to meteorology and played a crucial role in establishing a network of weather stations across Australia.

Wragge's observations of Morning Glory Clouds provided valuable insights into their formation and behavior. He hypothesized that these clouds were associated with strong pressure gradients and unique wind patterns in the region. Wragge's work marked a significant milestone in understanding the atmospheric dynamics behind Morning Glory Clouds.

In the 1930s, a German aviator named Hans-Heinrich Purner made an extraordinary discovery during his flights over the Gulf of Carpentaria in northern Australia. Purner observed that Morning Glory Clouds had a distinct rolling motion, resembling ocean waves. This observation led to the popularization of the term "Morning Glory" to describe these clouds.

Purner's findings sparked interest among scientists and aviation enthusiasts, who began conducting further investigations into the physical characteristics and formation mechanisms of Morning Glory Clouds. Over the years, numerous expeditions and research campaigns were organized to study these elusive clouds.

One of the most notable explorations took place in the late 1990s when a team of meteorologists and glider pilots embarked on a mission to capture unprecedented data on Morning Glory Clouds. The researchers used state-of-the-art instruments and remote sensing techniques to study the structure, dynamics, and meteorological impacts of these clouds.

The historical discoveries in the field of Morning Glory Clouds have laid the foundation for ongoing research and exploration. Today, scientists continue to unravel the mysteries surrounding the formation, dynamics, and global distribution of these extraordinary atmospheric phenomena.

Key Points:
- Morning Glory Clouds have been observed and revered by indigenous cultures for centuries.
- Augustus Gregory's documentation in 1867 provided the first scientific record of Morning Glory Clouds.
- Clement Wragge's research in the early 20th century contributed significantly to understanding their formation.

- Hans-Heinrich Purner's observation of the rolling motion of Morning Glory Clouds led to their popularization.

- Expeditions and research campaigns in the late 1990s provided valuable data on the structure and dynamics of these clouds.

Further Resources:

- Gregory, A. (1868). "Journals of Australian Exploration." Retrieved from [link].

- Ludvigsen, M. (2017). "Clement Wragge: The Chief of the Weather Bureau." Retrieved from [link].

- Purner, H. (1933). "Flights within Tropics." Retrieved from [link].

- Chilcott, M. (1998). "Glider Soars into Monster Cloud." Retrieved from [link].

- Morning Glory Cloud Research and Preservation Society. (n.d.). Retrieved from [link].

Exercises:

1. Describe the historical contributions of Augustus Gregory and Clement Wragge towards the understanding of Morning Glory Clouds.

2. Discuss the significance of Hans-Heinrich Purner's observation of the rolling motion in Morning Glory Clouds.

3. Conduct research on recent expeditions and campaigns focused on studying Morning Glory Clouds. Present your findings in a comprehensive report, highlighting the contributions of each study.

4. Imagine you are an Aboriginal storyteller. Write a short mythical tale involving Morning Glory Clouds and their significance in the creation of the world.

5. Conduct an interview with a meteorologist who participated in the late 1990s research campaign on Morning Glory Clouds. Prepare insightful questions about their experiences and key findings during the expedition.

Famous Morning Glory Clouds Events

Morning Glory Clouds have captured the attention and fascination of people around the world. Over the years, several notable events involving Morning Glory Clouds have occurred, leaving a lasting impact on the scientific community and captivating the public's imagination. In this section, we will explore some of these famous Morning Glory Clouds events and their significance.

1. Morning Glory Clouds in Burketown, Queensland, Australia (Acknowledging Indigenous Knowledge)

Burketown, a small town located in Queensland, Australia, is known as the "home of the Morning Glory Clouds." For centuries, the indigenous people of the region, such as the Gangalidda and Garawa communities, have observed and revered the Morning Glory Clouds. Their deep understanding of these unique cloud formations has been passed down through generations.

The recognition and acknowledgment of indigenous knowledge and their connection to Morning Glory Clouds in Burketown is not only important for cultural preservation but also for scientific research. Collaborations between meteorologists and indigenous communities have allowed for a more holistic and comprehensive study of Morning Glory Clouds, incorporating traditional knowledge and scientific observations.

2. The Morning Glory Cloud Festival in Burketown, Queensland, Australia (Cultural Celebration)

Every year in Burketown, the local community gathers to celebrate the unique phenomenon of Morning Glory Clouds at the Morning Glory Cloud Festival. This festival provides an opportunity for locals and visitors to come together and appreciate the beauty and significance of these cloud formations.

The festival features various activities, including cultural performances, art exhibitions, educational workshops, and guided tours. It serves as a platform to showcase indigenous traditions, storytelling, and artistic interpretations of Morning Glory Clouds. The cultural celebration highlights the importance of Morning Glory Clouds in the local community, promoting awareness and understanding of these awe-inspiring atmospheric phenomena.

3. Morning Glory Cloud Expedition in Northern Australia (Scientific Exploration)

In recent years, a team of meteorologists, researchers, and avid skywatchers embarked on a pioneering expedition to study the Morning Glory Clouds in the remote areas of northern Australia. The aim was to gather comprehensive data, both through ground-based observations and aerial surveys, to better understand the formation and behavior of Morning Glory Clouds.

Equipped with state-of-the-art instruments, including weather radars and specialized cameras, the researchers captured detailed measurements of the cloud's structure, temperature, and humidity profiles. They also documented the intricate patterns and movements of the clouds using drone technology, providing invaluable insights into the dynamics of Morning Glory Clouds.

The expedition not only contributed to advancing scientific knowledge but also raised awareness about the importance of studying atmospheric phenomena and the role they play in shaping local weather patterns.

4. Hang Gliding and Soaring in Morning Glory Clouds (Adrenaline Rush)

For adventure sports enthusiasts, the Morning Glory Clouds offer a unique opportunity to experience the thrill of hang gliding and soaring. The exceptional horizontal wind pattern within the cloud creates a perfect environment for gliders to ride the rolling waves, covering great distances and staying airborne for extended periods.

In regions such as the Gulf of Carpentaria in Australia, professional pilots and daredevils from around the world gather to take on the challenge of flying alongside the Morning Glory Clouds. This exhilarating experience requires skill, precision, and an understanding of the cloud's dynamics. It showcases the harmony between nature and human ingenuity, while pushing the boundaries of aviation and exploration.

5. Morning Glory Clouds in the Gulf of Carpentaria, Australia (Tourist Attraction)

The Gulf of Carpentaria in northern Australia is renowned for being one of the hotspots of Morning Glory Cloud activity. Locals and tourists alike flock to this region to witness the spectacle of these magnificent cloud formations.

The Morning Glory Clouds in the Gulf of Carpentaria attract photographers, nature enthusiasts, and scientists from around the world. Visitors can enjoy stunning panoramic views of the rolling clouds from vantage points along the coastline or take part in guided boat tours to observe the clouds from the water. The influx of tourists contributes to the local economy, supporting ecotourism initiatives in the region.

These famous Morning Glory Clouds events demonstrate the multidimensional significance of these captivating meteorological phenomena. From scientific exploration and indigenous knowledge to cultural celebrations and adrenaline-fueled adventures, Morning Glory Clouds continue to awe and inspire, leaving a lasting impact on the communities they grace with their presence. These events serve as a testament to the power of nature and the endless possibilities for exploration and discovery in the world around us.

Importance of Studying Morning Glory Clouds

The study of Morning Glory Clouds holds significant importance in the field of atmospheric science. These unique cloud formations present a fascinating phenomenon that offers valuable insights into the complex interactions between weather systems, cloud dynamics, and climate patterns. By understanding the origins, characteristics, and behavior of Morning Glory Clouds, scientists can enhance our knowledge of meteorology, climate change, and the cultural and ecological impacts of these clouds.

Weather Pattern Analysis

Studying Morning Glory Clouds allows meteorologists to gain a deeper understanding of weather patterns and their dynamics. Morning Glory Clouds are often associated with atmospheric pressure systems, wind patterns, and jet streams. By analyzing their formation and movement, scientists can identify the influence of different air masses and frontal systems. This knowledge can help in predicting weather conditions such as temperature changes, wind shifts, and precipitation patterns. It also aids in improving weather forecasting models and systems, ultimately enhancing our ability to predict and mitigate severe weather events.

Cloud Formation Processes

Morning Glory Clouds provide an ideal opportunity to study cloud formation processes. The formation of these clouds is influenced by factors such as water vapor content, condensation, and nucleation. By examining the mechanisms behind Morning Glory Cloud formation, scientists can develop a deeper understanding of cloud physics, which can be applied to other cloud types. This knowledge is essential for accurately simulating cloud formation in climate models and improving our understanding of cloud-climate feedbacks.

Climate Change Indicators

As the Earth's climate continues to change, it is crucial to identify and monitor various indicators of climate change. Morning Glory Clouds serve as one such indicator. Changes in their distribution, occurrence, or behavior can provide valuable insights into long-term climate variability.

By studying the relationship between Morning Glory Clouds and climate change, scientists can contribute to a broader understanding of the Earth's climate system and its sensitivity to external forcings. This knowledge is vital for developing effective adaptation and mitigation strategies to counteract the impacts of climate change.

Aviation Safety

Morning Glory Clouds pose significant hazards to aviation due to their unpredictable nature and strong vertical wind shear. Studying these clouds can help researchers develop better models and tools for predicting their occurrence and behavior. This knowledge is crucial for enhancing aviation safety protocols, advising pilots on avoidance strategies, and developing advanced warning systems. By understanding the meteorological factors contributing to Morning Glory Cloud formation, aviation experts can make informed decisions to minimize risks and ensure the safety of flights in affected regions.

Cultural and Ecological Significance

Morning Glory Clouds hold immense cultural and ecological significance in regions where they occur. Many indigenous cultures consider them sacred and have associated rich mythologies and rituals with these unique cloud formations. By studying the cultural and historical significance of Morning Glory Clouds, researchers can help preserve indigenous knowledge and practices. Furthermore, Morning Glory Clouds attract tourists, contributing to local economies through ecotourism. By understanding the ecological impact of tourism on Morning Glory Cloud habitats, scientists can develop sustainable practices that balance economic benefits with conservation needs.

In summary, the study of Morning Glory Clouds is of great importance and offers valuable insights into weather patterns, cloud formation processes, climate change indicators, aviation safety, and cultural and ecological significance. Through comprehensive research and analysis, we can deepen our understanding of these remarkable cloud formations and their impacts, ultimately contributing to advancements in atmospheric science and the overall well-being of our planet.

Overview of the Book

In this book, we embark on a fascinating journey to explore the captivating world of Morning Glory Clouds. We delve into the scientific aspects behind these unique atmospheric phenomena, uncovering their formation mechanisms, meteorological impacts, global distribution, and cultural significance.

Throughout the book, we aim to provide a comprehensive understanding of Morning Glory Clouds, combining scientific principles with real-world examples and practical applications. We delve into the interdisciplinary nature of studying these clouds, bringing together concepts from meteorology, atmospheric science, climatology, and cultural studies.

We begin by introducing Morning Glory Clouds in Section 1.1.1, defining them as rare and spectacular meteorological occurrences characterized by long, horizontal, and tube-shaped roll clouds. We discuss the historical discoveries of Morning Glory Clouds in Section 1.1.2, highlighting the significant contributions of early observers in unraveling their mysteries. Furthermore, we explore famous Morning Glory Cloud events, such as the Burketown Morning Glory in Australia and the Morning Glory in the Gulf of Carpentaria.

The importance of studying Morning Glory Clouds is emphasized in Section 1.1.4. We discuss how these phenomena provide valuable insights into weather systems, cloud formation, air masses, and atmospheric dynamics. By studying Morning Glory Clouds, we can deepen our understanding of the Earth's atmosphere and improve weather forecasting and aviation safety protocols.

To provide a road map for the book, we present an overview of the chapters and their contents in the subsequent sections of this chapter. This overview is structured to guide the reader through the various aspects of Morning Glory Clouds, starting with the foundational knowledge in atmospheric phenomena, cloud formation, and air masses.

In Chapter 2, we explore atmospheric phenomena and discuss weather systems (Section 2.1) that play a crucial role in Morning Glory Cloud formation. We delve into high-pressure systems, low-pressure systems, fronts, pressure gradients, wind patterns, and jet streams. Understanding these systems is essential for comprehending the dynamics of Morning Glory Clouds.

In Chapter 3, we focus on Morning Glory Cloud features. We examine the shape, size, and variations in Morning Glory Cloud formations (Section 3.1). Factors influencing these features, such as wind patterns and shear, are explored. Additionally, we investigate the vibrant colors often observed in Morning Glory Clouds and the scientific explanations behind these colorful displays.

Chapter 4 takes us on a global journey to explore the distribution of Morning Glory Clouds. We analyze the geographic distribution (Section 4.1) of hotspot regions and the influence of local climate and topography on their occurrence. Moreover, we investigate the seasonal patterns (Section 4.2) of Morning Glory Clouds and the factors driving their variations in different seasons. We also discuss the diurnal cycle (Section 4.3) of Morning Glory Clouds, examining their behavior at different times of the day.

The formation mechanisms of Morning Glory Clouds are the focus of Chapter 5. We explore the role of gravity waves (Section 5.1) and their generation, propagation, and interaction with atmospheric layers. Additionally, we delve into the soliton theory (Section 5.2) and its application to understand the behavior of Morning Glory Clouds. Furthermore, we discuss the influence of convection and frontal systems (Section 5.3) on Morning Glory Cloud formation.

In Chapter 6, we examine the meteorological impacts of Morning Glory Clouds. We discuss their effects on local weather patterns (Section 6.1), including changes in temperature, humidity, wind, and precipitation patterns. We also address the aviation hazard (Section 6.2) posed by Morning Glory Clouds and explore mitigation strategies, safety regulations, and collaboration between meteorologists and the aviation industry. Furthermore, we investigate the connection between Morning Glory Clouds and climate change (Section 6.3) and the implications for global climate variability.

The cultural and historical significance of Morning Glory Clouds are explored in Chapter 7. We delve into indigenous traditions and legends (Section 7.1), discussing the mythologies, cultural practices, and oral histories associated with these phenomena. Additionally, we examine artistic and literary representations (Section 7.2) of Morning Glory Clouds, including paintings, poems, literature, films, and their influence on contemporary art and literature. Furthermore, we explore the ecotourism and conservation aspects (Section 7.3) of Morning Glory Clouds, focusing on sustaining local economies through responsible tourism practices and research and conservation efforts.

Chapter 8 introduces various research techniques and future directions for Morning Glory Cloud studies. We discuss remote sensing and imaging techniques (Section 8.1) such as satellite observations, radar, lidar, and UAVs. Additionally, we explore in situ measurements (Section 8.2) using weather balloons, aircraft-based observations, and other sampling methods. Furthermore, we delve into numerical models and simulations (Section 8.3) such as computational fluid dynamics, mesoscale models, and climate models. Finally, we investigate the potential of citizen science and crowdsourcing (Section 8.4) in involving the public in Morning Glory Cloud research and analysis.

In conclusion, this book offers a comprehensive exploration of Morning Glory Clouds, combining scientific knowledge with cultural and historical perspectives. Whether you are a meteorologist, atmospheric scientist, artist, cultural enthusiast, or simply curious about these fascinating atmospheric phenomena, this book provides a valuable resource for understanding and appreciating the beauty and significance of Morning Glory Clouds in our world.

Atmospheric Phenomena

Weather Systems

High-Pressure Systems

High-pressure systems are an essential component of atmospheric phenomena and play a significant role in weather patterns and cloud formation. Understanding the characteristics and behavior of high-pressure systems is crucial for meteorologists and atmospheric scientists in predicting and studying weather conditions. In this section, we will explore the definition, formation, and influence of high-pressure systems.

Definition and Characteristics

A high-pressure system, also known as an anticyclone, is an area of atmospheric pressure that is higher than its surrounding regions. It is characterized by descending air, clockwise rotation (in the Northern Hemisphere), and generally clear and stable weather conditions. High-pressure systems are typically associated with fair weather, clear skies, and light winds.

The main characteristics of high-pressure systems include:

1. **Anticyclonic flow**: In the Northern Hemisphere, high-pressure systems are characterized by an outward flow of air in a clockwise direction. In contrast, in the Southern Hemisphere, the flow is counterclockwise.

2. **Descending air**: High-pressure systems result from the sinking of air in the atmosphere. As the air sinks, it compresses and warms, leading to stable atmospheric conditions.

3. **Clear skies**: High-pressure systems usually bring clear skies and minimal cloud cover. The sinking air inhibits cloud formation and promotes dry conditions.

4. **Light winds**: High-pressure systems are associated with calm and light winds due to the downward motion of air. This makes them favorable for outdoor activities and generally pleasant weather conditions.

Formation Mechanisms

High-pressure systems can form through various mechanisms. Understanding these mechanisms is crucial for predicting the formation and movement of high-pressure systems. Here are three common formation mechanisms:

1. **Subsidence**: Subsidence occurs when air descends and compresses, leading to the formation of a high-pressure system. Several factors can cause air to descend, including synoptic-scale sinking motions associated with weather fronts, upper-level ridges, and mesoscale downdrafts within thunderstorms.

2. **Radiative Cooling**: Radiative cooling plays a significant role in the formation of high-pressure systems during the nighttime. As the Earth's surface cools after sunset, the cooler air near the surface becomes denser and sinks, resulting in the development of high-pressure systems.

3. **Influence of Surface Features**: High-pressure systems can also form due to the influence of surface topography. For example, in areas with mountain ranges, air is forced to ascend over the mountains, leading to the development of a low-pressure system. On the leeward side of the mountains, the air descends and forms a high-pressure system.

Influence on Weather Patterns

High-pressure systems have a significant influence on weather patterns and atmospheric conditions. They can affect temperature, humidity, wind patterns, and precipitation in their surrounding areas. Some of the key impacts of high-pressure systems on weather patterns are:

1. **Weather Stability**: High-pressure systems are associated with stable atmospheric conditions and are often responsible for fair weather and prolonged periods of sunshine. The descending air inhibits cloud formation and precipitation, resulting in clear skies and dry conditions.

2. **Temperature Inversion**: High-pressure systems can contribute to the formation of temperature inversions. As the air descends, it undergoes compression and warming, leading to a vertical temperature structure where the air near the surface is warmer than the air above it. Temperature inversions can trap pollutants near the surface, leading to poor air quality.

3. **Wind Patterns**: The circulation around a high-pressure system is characterized by anticyclonic flow. In the Northern Hemisphere, the flow is clockwise, while in the Southern Hemisphere, it is counterclockwise. The direction and strength of the wind can vary depending on the size and intensity of the high-pressure system.

4. **Clear Skies and Fog Formation**: High-pressure systems are associated with clear skies and minimal cloud cover. However, under certain conditions, such as during the cool season, nighttime radiative cooling can lead to the formation of fog in low-lying areas within the high-pressure system.

5. **Temperature and Humidity**: High-pressure systems often bring cooler and drier air as they move into an area. This can result in a decrease in temperature and a drop in humidity levels, contributing to a change in weather conditions.

Real-World Examples

High-pressure systems are a common feature in weather patterns around the world. Understanding their behavior and influence on weather conditions is essential for forecasting and understanding local and regional climates. Here are a few real-world examples of high-pressure systems:

1. **Bermuda High**: The Bermuda High is a semipermanent high-pressure system located in the Atlantic Ocean near Bermuda. It influences the weather patterns over the western Atlantic, including the southeastern United States and the Caribbean. It is responsible for the warm and humid conditions experienced in these regions during the summer.

2. **Siberian High**: The Siberian High is a large, cold, and dry high-pressure system that forms over Siberia during the winter. It influences the weather patterns over northeastern Asia, including China, Japan, and Korea. The Siberian High brings cold temperatures, clear skies, and dry conditions to these regions.

3. **Pacific High**: The Pacific High is a semipermanent high-pressure system located in the northeastern Pacific Ocean. It plays a crucial role in

the weather patterns along the West Coast of the United States, bringing cool and dry conditions during the summer months and influencing the marine layer and coastal fog.

4. **Antarctic High**: The Antarctic High is a persistent high-pressure system that forms over Antarctica. It influences the weather patterns of the Southern Ocean and surrounding regions. The Antarctic High is associated with extremely cold temperatures, strong winds, and the formation of polar cyclones.

These examples illustrate the diverse geographic locations where high-pressure systems occur and their impacts on local and regional weather patterns.

Further Resources

To deepen your understanding of high-pressure systems, you may find the following resources helpful:

- Ahrens, C. D. (2019). *Meteorology Today: An Introduction to Weather, Climate, and the Environment.* Cengage Learning.

- Wallace, J. M., & Hobbs, P. V. (2006). *Atmospheric Science: An Introductory Survey.* Academic Press.

- NOAA National Weather Service. (n.d.). High Pressure. Retrieved from https://www.weather.gov/jetstream/high

- National Geographic. (n.d.). High-Pressure System. Retrieved from https://www.nationalgeographic.org/encyclopedia/high-pressure-system

These resources provide comprehensive information on meteorology, atmospheric science, and weather patterns associated with high-pressure systems.

Exercises

1. Describe the characteristics of a high-pressure system and its influence on weather conditions.

2. Explain three different mechanisms through which high-pressure systems can form.

3. How does the wind circulate around a high-pressure system in the Northern Hemisphere? Provide an explanation using the Coriolis effect.
4. What are temperature inversions, and how are they related to high-pressure systems? Provide an example.
5. Choose a specific high-pressure system from one of the real-world examples mentioned, and describe its characteristics and impacts on the associated region's weather patterns.
6. Conduct research on a historical event or case study involving a high-pressure system and its influence on weather conditions. Summarize your findings and explain the significance of the event.

These exercises will help reinforce your understanding of high-pressure systems and their effects on weather patterns.

Low-Pressure Systems

Low-pressure systems, also known as cyclones, are important atmospheric phenomena that play a crucial role in the formation of weather patterns. In this section, we will explore the characteristics, formation mechanisms, and effects of low-pressure systems, particularly with respect to their influence on the formation of morning glory clouds.

Definition and Characteristics of Low-Pressure Systems

A low-pressure system refers to an area in the atmosphere where the atmospheric pressure is lower than that of the surrounding areas. It is characterized by the convergence of air masses, which results in the upward movement of air and the formation of clouds and precipitation.

Low-pressure systems are typically associated with unstable atmospheric conditions. The ascent of warm, moist air leads to the condensation of water vapor, forming clouds and potentially causing precipitation. The presence of low pressure often indicates the presence of a weather front, such as a cold front or a warm front, which can bring significant weather changes.

Formation Mechanisms of Low-Pressure Systems

Low-pressure systems can form through various mechanisms, including thermal heating, convergence, and upper-level disturbances. One common formation mechanism is the thermal heating of the Earth's surface. During the day, the Sun heats the surface, resulting in the warming of the air above it. As the air heats, it becomes less dense and rises, creating a region of lower pressure.

Convergence, the coming together of air masses, is another mechanism for the formation of low-pressure systems. When air masses of different temperatures and moisture content collide, they cannot occupy the same space and are forced to rise. This upward movement of air contributes to the development of low pressure.

Upper-level disturbances, such as jet streams or upper-level troughs, can also play a role in the formation of low-pressure systems. As these disturbances interact with the flow of air in the atmosphere, they create regions of lower pressure.

Effects of Low-Pressure Systems

Low-pressure systems have a significant impact on weather patterns and can result in various weather phenomena. Due to the convergence of air and the upward motion associated with low pressure, clouds and precipitation tend to form. The extent and intensity of precipitation depend on factors such as the amount of moisture available and the dynamics of the system.

Low-pressure systems are often associated with cyclonic circulation, characterized by counterclockwise airflow in the Northern Hemisphere and clockwise airflow in the Southern Hemisphere. This circulation pattern can lead to the development of surface winds, which can cause changes in wind speed and direction in the affected area.

Low-pressure systems can also be associated with the formation of severe weather events, such as thunderstorms, tornadoes, and tropical cyclones. The interaction between low pressure, moisture, and other atmospheric conditions can create an environment conducive to the development of these hazardous weather phenomena.

Example: Nor'easters and their Impact on Morning Glory Clouds

One example of a low-pressure system that significantly influences the formation of morning glory clouds is a Nor'easter. Nor'easters are extratropical cyclones that occur along the East Coast of North America. These intense storms form when a strong low-pressure system develops over the Atlantic Ocean, with the counterclockwise circulation drawing in cold air from the north and warm, moist air from the south.

The interaction between the Nor'easter and the prevailing westerly winds creates a favorable environment for the formation of morning glory clouds. The convergence of air masses, combined with the uplift caused by the low-pressure system, provides the necessary conditions for the development of these unique cloud formations.

Nor'easters can bring strong winds, heavy rain or snowfall, and even coastal flooding. Their impact on morning glory clouds can vary depending on the specific characteristics of the storm, including its track, intensity, and interaction with other weather systems. Studying the relationship between Nor'easters and morning glory clouds can provide valuable insights into the influence of low-pressure systems on these atmospheric phenomena.

Conclusion

Low-pressure systems are important features of the atmosphere that influence weather patterns and contribute to the formation of morning glory clouds. Understanding the characteristics, formation mechanisms, and effects of low-pressure systems is essential for comprehending the complex interactions between atmospheric phenomena. By studying the dynamics of low-pressure systems, meteorologists can gain insights into the conditions that give rise to morning glory clouds and improve weather forecasting capabilities. Further research and observation are needed to unravel the intricacies of low-pressure systems and their influence on morning glory clouds.

Fronts and Pressure Gradients

In this section, we will explore the concepts of fronts and pressure gradients, which play a crucial role in weather systems and ultimately contribute to the formation of morning glory clouds. Understanding these concepts is key to comprehending the larger atmospheric phenomena surrounding these unique cloud formations.

Fronts: The Battle Zones of Air Masses

Fronts are the boundaries that separate two different air masses. Air masses are large volumes of air that have similar temperature and humidity characteristics. When two air masses of different temperature, humidity, or density come into contact with each other, they do not readily mix due to their differing properties. As a result, fronts form as zones of transition between these air masses.

There are four main types of fronts based on the characteristics of the air masses involved: cold fronts, warm fronts, stationary fronts, and occluded fronts.

- **Cold Fronts**: A cold front forms when a moving cold air mass undercuts and displaces a warmer air mass. As the cold air pushes forward, it wedges beneath the warm air, forcing the warmer air to rise rapidly. The rising air cools and condenses, forming clouds and often leading to the development of thunderstorms and showers along the front. Cold fronts typically move faster than warm fronts and can cause rapid changes in weather conditions.

- **Warm Fronts**: A warm front forms when a moving warm air mass overrides and replaces a colder air mass. As the warm air approaches, it gradually rises over the denser cold air. The rising warm air cools and condenses, leading to the formation of extensive cloud cover with less intense precipitation compared to cold fronts. Warm fronts typically move at a slower pace and produce more prolonged periods of drizzle or light rain.

- **Stationary Fronts**: A stationary front occurs when the boundary between two air masses stalls and remains relatively stationary. The movement of the air masses is nearly balanced, with neither one advancing significantly. Weather conditions along a stationary front can vary, with prolonged periods of drizzle or light rain, fog, or even thunderstorms.

- **Occluded Fronts**: An occluded front forms when a fast-moving cold front overtakes a slower-moving warm front. As the cold front catches up to the warm front, the warm air is pushed upward. This results in a complex interaction, with the formation of new fronts and the lifting of warm air above the surface. Occluded fronts are typically associated with a mix of weather conditions, including clouds, precipitation, and strong winds.

Fronts are significant in the formation of morning glory clouds because the interactions between air masses along these boundaries can trigger the atmospheric conditions necessary for the development of these unique cloud formations. The rising and mixing of air at the front can generate disturbances, such as gravity waves, that play a role in the roll-like structure characteristic of morning glory clouds.

Pressure Gradients: Driving Forces Behind Air Motion

Pressure gradients are closely linked to the formation and movement of fronts. They define the change in air pressure over a given distance in a specific direction. Air always flows from higher pressure to lower pressure, perpendicular to the lines of equal pressure (isobars) on weather maps.

The magnitude of the pressure gradient determines the strength of the airflow. A steep pressure gradient corresponds to a strong flow, while a gentle gradient corresponds to a weaker flow. Pressure gradients are influenced by various factors, such as temperature differences, the shape of the terrain, and the presence of weather systems.

In the context of fronts, pressure gradients are responsible for the movement of air masses and the subsequent formation or intensification of these boundaries. For example, a strong pressure gradient ahead of a cold front encourages the cold air to surge forward and undercut the warmer air, resulting in the advancement of the front.

The interplay between pressure gradients and fronts leads to dynamic atmospheric conditions. The convergence and divergence of air near fronts can support the upward motion of air and the formation of clouds. These interactions, combined with other atmospheric factors, contribute to the creation of morning glory clouds.

Real-World Example: The Morning Glory Front

The coastal town of Burketown in Australia is known for its famous morning glory cloud formations. These formations are caused by a unique phenomenon known as the "Morning Glory Front." This front is a remarkable example of a thin, elongated cloud bank that can stretch for hundreds of kilometers.

The Morning Glory Front is a complex weather system wherein a series of pressure disturbances, gravity waves, and associated fronts interact to produce the desired conditions for the formation of morning glory clouds. The system typically occurs during spring and early summer and is influenced by local topography and the temperature differences between the land and sea.

As cool, dry air pushes in from the land and warmer, more humid air moves in from the sea, a frontal boundary forms. This sets the stage for the development of gravity waves, which propagate vertically and horizontally and create the distinctive rolling cloud pattern associated with morning glory clouds.

The Morning Glory Front attracts glider pilots from around the world who ride the updrafts generated by the rolling clouds. This natural spectacle has become a popular tourist attraction and provides an opportunity for scientific research on gravity waves and the dynamics of cloud formation.

Exercises

1. Compare and contrast the characteristics and weather associated with cold fronts and warm fronts.

2. How do pressure gradients influence the movement of air masses? Provide an example.

3. Investigate the morning glory cloud formations in your local area or region. Are they associated with any specific types of fronts? What impact do they have on local weather patterns?

4. Research and discuss other atmospheric phenomena that occur along fronts, such as squall lines or mesoscale convective systems.

5. Create a simple diagram illustrating the formation and movement of a stationary front.

Further Reading

1. Fujita, T. T. (1981). Pressure gradient force and a unified view of wind and pressure. Journal of the Atmospheric Sciences, 38(11), 2305-2321.

2. Businger, J. A., & Businger, S. (1971). Morphology of a morning glory cloud. Journal of the Atmospheric Sciences, 28(8), 1322-1333.

3. Ahrens, C. D. (2018). Meteorology today: An introduction to weather, climate, and the environment. Cengage Learning.

4. Hobbs, P. V., & Radke, L. F. (1991). Airborne studies of cloud clusters in the Australian morning glory. Monthly Weather Review, 119(9), 2288-2313.

5. Knight, C. A., & Knight, J. M. (2014). Meteorology: An introduction. Academic Press.

Wind Patterns and Jet Streams

In this section, we will explore the fascinating world of wind patterns and jet streams, which play a crucial role in the formation and behavior of morning glory clouds. Understanding these atmospheric phenomena is key to unraveling the mysteries behind the captivating nature of morning glory clouds.

Introduction to Wind Patterns

Wind patterns are the movements of air in the Earth's atmosphere, driven by differences in air pressure. These patterns are influenced by various factors, including the rotation of the Earth, the distribution of continents and oceans, and the differential heating of the Earth's surface.

One important concept to grasp when studying wind patterns is the Coriolis effect. As the Earth rotates, objects moving across the Earth's surface appear to deviate from straight paths. In the Northern Hemisphere, this deflection is to the right, while in the Southern Hemisphere, it is to the left. The Coriolis effect is responsible for the formation of large-scale wind patterns known as the prevailing winds.

Prevailing Winds

The prevailing winds are global-scale wind patterns that blow fairly consistently in specific directions. These winds play a significant role in shaping regional weather patterns and climate. The three major prevailing winds are the trade winds, the westerlies, and the polar easterlies.

The trade winds are tropical winds that blow from the subtropical high-pressure systems towards the equator. In the Northern Hemisphere, they blow from the northeast, while in the Southern Hemisphere, they blow from the southeast. These winds have a significant impact on morning glory cloud formation as they transport warm, moist air towards regions where morning glory clouds typically occur.

The westerlies are prevailing winds that blow from west to east in the middle latitudes of both hemispheres. In the Northern Hemisphere, they are influenced by the subtropical high-pressure system and the polar jet stream. In the Southern Hemisphere, they are affected by the Antarctic polar vortex. The westerlies help shape the movement of weather systems and can steer morning glory clouds along their path.

The polar easterlies are cold, dry winds that blow from the polar high-pressure systems towards the lower latitudes. In the Northern Hemisphere, they blow from the northeast, while in the Southern Hemisphere, they blow from the southeast. These winds are important for morning glory cloud formation as they can provide the necessary cold, stable air masses required for their formation.

Jet Streams

Jet streams are narrow bands of high-speed winds in the upper levels of the atmosphere, typically located several kilometers above the Earth's surface. They are created by the large temperature contrasts between air masses, such as the polar and subtropical air masses. Jet streams are fastest and strongest in winter when the temperature contrasts are most pronounced.

The two main jet streams are the polar jet stream and the subtropical jet stream. The polar jet stream is located between the polar easterlies and the westerlies in the middle latitudes. It influences the movement of weather systems and can have a significant impact on the formation and behavior of morning glory clouds. The subtropical jet stream, on the other hand, is located nearer to the equator and is influenced by the trade winds. It can also play a role in shaping the movement of morning glory clouds.

Interaction with Morning Glory Clouds

Wind patterns, including the prevailing winds and jet streams, have a direct influence on the formation, movement, and behavior of morning glory clouds. The prevailing winds transport warm, moist air masses towards the regions where morning glory clouds typically occur. These winds can also steer morning glory clouds along their paths, determining their direction and speed of movement.

Jet streams, especially the polar jet stream, can act as a trigger for the formation of morning glory clouds. The temperature contrasts associated with the jet stream can create the necessary conditions for gravity waves, which are thought to be the primary mechanism responsible for morning glory cloud formation. The interaction between gravity waves and the wind patterns can lead to the characteristic rolling and wave-like structure of morning glory clouds.

Understanding the intricate relationship between wind patterns, jet streams, and morning glory clouds is crucial for meteorologists and researchers seeking to forecast the occurrence of these fascinating atmospheric phenomena. By studying the behavior of wind patterns and jet streams, we can gain valuable insights into the formation, evolution, and movement of morning glory clouds and further unravel the secrets of this awe-inspiring natural phenomenon.

Exploring Wind Patterns and Jet Streams

To deepen your understanding of wind patterns and jet streams, here's a problem for you to solve:

Problem: Suppose you are a meteorologist studying morning glory clouds in a specific region. You notice a strong polar jet stream passing through the area. How might this jet stream influence the formation and movement of morning glory clouds? Provide a detailed explanation with relevant diagrams if necessary.

Solution: The presence of a strong polar jet stream in the region can have several implications for morning glory cloud formation and movement. Firstly, the jet stream can introduce significant wind shear in the upper levels of the atmosphere. Wind shear is the change in wind speed or direction with height. The shear associated with the jet stream can enhance the formation of gravity waves, which are instrumental in the creation of morning glory clouds.

Additionally, the polar jet stream can serve as a highway for the transportation of weather systems and air masses. Morning glory clouds often form at the leading edge of frontal systems or the boundaries between air masses with contrasting properties. The jet stream can guide these systems towards the region, increasing the likelihood of morning glory cloud formation.

Furthermore, the speed and direction of the jet stream can influence the speed and direction of morning glory clouds. If the jet stream is aligned with the direction of the prevailing winds, it can provide an additional impetus for the movement of morning glory clouds in the same direction. Conversely, if the jet stream is aligned in a different direction, it may hinder or alter the movement of morning glory clouds.

To visualize these concepts, refer to the diagram below:

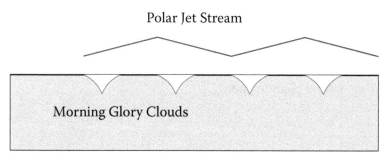

In the diagram, we see how the polar jet stream (depicted in red) interacts with the morning glory clouds (depicted in blue). The presence of a strong, well-defined jet stream can create a favorable environment for the formation of morning glory clouds. The wind shear induced by the jet stream aids in generating gravity waves, which contribute to the unique structure of morning glory clouds.

As the morning glory clouds form near the leading edge of the jet stream, they can be advected downstream by the prevailing winds. The jet stream acts as a conveyor belt, transporting the morning glory clouds to different locations. The speed of the jet stream determines the speed of the cloud movement, while its direction influences the direction of cloud propagation.

Moreover, the jet stream can also modulate the intensity and lifespan of morning glory clouds. The dynamic environment created by the jet stream can enhance turbulent mixing within the cloud, leading to variations in cloud thickness and coloration. These variations contribute to the awe-inspiring display of vibrant colors often observed in morning glory clouds.

In summary, the polar jet stream has a profound influence on the formation, movement, and characteristics of morning glory clouds. Its presence can enhance the generation of gravity waves, provide a pathway for the transportation of weather systems, and impact the behavior of morning glory clouds. Understanding the interaction between wind patterns, jet streams, and morning glory clouds is crucial for unraveling the complex dynamics behind this remarkable natural phenomenon.

Additional Resources

To further explore the world of wind patterns and jet streams, and their role in meteorology, consider the following resources:

1. "Meteorology Today" by C. Donald Ahrens: This comprehensive textbook provides a detailed introduction to meteorology, including a

chapter dedicated to wind patterns and jet streams.

2. The National Oceanic and Atmospheric Administration (NOAA) website: NOAA offers a wealth of information on meteorology, climatology, and atmospheric phenomena. Their section on jet streams provides detailed explanations, visualizations, and real-time data.

3. Scientific journals: Browse through scientific journals such as the Journal of the Atmospheric Sciences, Journal of Geophysical Research: Atmospheres, and Weather and Forecasting for the latest research on wind patterns, jet streams, and their impact on atmospheric phenomena.

By delving deeper into the intricacies of wind patterns and jet streams, you will gain a better understanding of the atmospheric dynamics influencing morning glory clouds, paving the way for further scientific discoveries and insights. So, let the winds guide you on this fascinating journey into the realm of morning glory clouds!

Influence on Morning Glory Cloud Formation

Morning Glory Clouds are a unique atmospheric phenomenon that is influenced by various factors. In this section, we will explore the different elements that contribute to the formation of Morning Glory Clouds and their impact on this fascinating weather event.

Weather Systems

Weather systems play a crucial role in the formation and development of Morning Glory Clouds. High-pressure systems, also known as anticyclones, and low-pressure systems, referred to as cyclones, are two major types of weather systems that influence the atmospheric conditions necessary for the formation of Morning Glory Clouds.

High-pressure systems are characterized by descending air currents and generally stable weather conditions. These systems often create clear skies and light winds. When a high-pressure system moves over a region, it can lead to a calm and stable atmosphere, which is favorable for the formation of Morning Glory Clouds.

On the other hand, low-pressure systems are associated with rising air currents and unsettled weather. These systems can bring clouds, precipitation, and strong winds to an area. The interaction of low-pressure systems with other atmospheric features can generate the ideal conditions for the formation of Morning Glory Clouds.

Fronts and Pressure Gradients

Fronts are boundaries between air masses with different characteristics, such as temperature, humidity, and density. There are two main types of fronts: cold fronts and warm fronts. These fronts can create pressure gradients, leading to the formation of atmospheric disturbances and favoring the development of Morning Glory Clouds.

Cold fronts occur when a cold air mass replaces a warmer air mass. As the cold front moves through an area, it pushes the warm air upward, creating instability in the atmosphere. This can trigger the formation of clouds and possible thunderstorms, which are often associated with Morning Glory Clouds.

Warm fronts, on the other hand, happen when a warm air mass advances over a colder air mass. The warm air rises over the cold air mass, causing the formation of clouds and precipitation. The interaction between the warm front and the atmospheric conditions can contribute to the formation of Morning Glory Clouds.

Wind Patterns and Jet Streams

Wind patterns and jet streams also influence the formation of Morning Glory Clouds. Jet streams are high-velocity winds found in the upper troposphere and lower stratosphere. They are driven by the temperature differences between air masses. These strong winds can create atmospheric disturbances that contribute to the formation of Morning Glory Clouds.

In particular, the Southern Hemisphere jet streams have a significant impact on the formation of Morning Glory Clouds in the Gulf of Carpentaria in Australia. The interaction between the jet streams and other atmospheric features can generate the necessary conditions for the unique roll cloud formation characteristic of Morning Glory Clouds in this area.

Influence on Morning Glory Cloud Formation

The convergence of different weather systems, the presence of fronts, and the interaction between wind patterns and jet streams create a dynamic environment that influences the formation of Morning Glory Clouds.

The convergence of air masses with varying characteristics, such as temperature and humidity, leads to the uplift of warm and moist air. As this air rises, it cools, and condensation occurs, forming the cloud structure of the Morning Glory Cloud. The presence of atmospheric disturbances generated by weather systems and wind patterns further contributes to the formation and maintenance of these clouds.

The specific combination of conditions required for Morning Glory Cloud formation varies depending on the region and local atmospheric characteristics. Therefore, detailed studies and observations are essential to understand the precise influence of weather systems, fronts, and wind patterns on the formation of Morning Glory Clouds in different areas around the world.

Example: Morning Glory Clouds in the Gulf of Carpentaria

The Gulf of Carpentaria in Australia is known for the frequent occurrence of Morning Glory Clouds, which attracts researchers, pilots, and weather enthusiasts from around the world. The unique geography of the region, combined with the interaction of weather systems and wind patterns, creates ideal conditions for the formation of these clouds.

The convergence of trade winds, sea breezes, and the interaction of the diurnal land and sea temperature cycles play a significant role in the formation of Morning Glory Clouds in this area. The prevailing easterly winds push the warm, moist air over the land, creating an uplift and eventually leading to the formation of the spectacular roll cloud structure.

The jet streams in the Southern Hemisphere, particularly the subtropical jet stream, also influence the formation of Morning Glory Clouds in the Gulf of Carpentaria. The interaction between the jet stream and the local atmospheric conditions produces the necessary instabilities for the formation and maintenance of these clouds.

In this region, the Morning Glory Clouds often occur during the spring and autumn months when the sea and land temperature differences are pronounced, creating a favorable environment for their formation. Studying the factors that contribute to the development of Morning Glory

Clouds in the Gulf of Carpentaria provides valuable insights into the influence of weather systems and wind patterns on this phenomenon.

Resources and Further Reading

For a deeper understanding of weather systems, fronts, and wind patterns, the following resources are recommended:

- Book: "Atmospheric Science: An Introductory Survey" by John M. Wallace and Peter V. Hobbs
- Journal: "Quarterly Journal of the Royal Meteorological Society"
- Website: National Oceanic and Atmospheric Administration (NOAA) - www.noaa.gov

These resources provide comprehensive information on the principles of meteorology and weather systems, which are essential for studying the influence of these factors on Morning Glory Cloud formation.

Cloud Formation

Water Vapor and Humidity

Water vapor is an essential component of the Earth's atmosphere, playing a significant role in various atmospheric processes and weather phenomena. Understanding the behavior and distribution of water vapor is crucial for studying meteorology and atmospheric science. In this section, we will explore the concepts of water vapor and humidity, their measurement techniques, and their importance in cloud formation.

Water Vapor in the Atmosphere

Water vapor is the gaseous form of water, which is a vital part of the Earth's hydrological cycle. It is produced through evaporation from oceans, lakes, rivers, and other water bodies, as well as through transpiration from plants. The amount of water vapor in the atmosphere can vary greatly depending on various factors such as temperature and proximity to water sources.

Water vapor plays a significant role in the energy balance of the Earth's atmosphere. It acts as a greenhouse gas, trapping heat and contributing to

CLOUD FORMATION

the greenhouse effect. Additionally, it provides the necessary moisture for cloud formation and precipitation.

Humidity

Humidity is a measure of the amount of water vapor present in the atmosphere. It is a fundamental parameter in meteorology and plays a crucial role in determining weather conditions. Humidity is typically expressed in terms of relative humidity, dew point, mixing ratio, or specific humidity.

Relative Humidity (RH): Relative humidity is the most commonly used measure of humidity. It represents the amount of water vapor present in the air relative to the maximum amount of water vapor the air can hold at a particular temperature. Relative humidity is expressed as a percentage.

Mathematically, relative humidity can be calculated by dividing the actual water vapor pressure by the saturation water vapor pressure at a given temperature:

$$RH = \frac{e}{e_s} \times 100\%$$

where RH is the relative humidity, e is the actual water vapor pressure, and e_s is the saturation water vapor pressure.

Dew Point: The dew point is the temperature at which air becomes saturated and condensation occurs. It is a useful parameter for assessing the likelihood of fog or dew formation. The dew point is directly related to the amount of moisture in the air. When the air temperature drops to the dew point temperature, the relative humidity reaches 100

Mixing Ratio: Mixing ratio is the mass of water vapor present in a unit mass of dry air. It is often expressed in grams of water vapor per kilogram of dry air (g/kg). The mixing ratio is an essential parameter for meteorologists as it is conserved during atmospheric processes such as advection and mixing.

Specific Humidity: Specific humidity is the mass of water vapor present in a unit mass of moist air. It represents the actual amount of water vapor in the atmosphere regardless of temperature and does not take into account changes in air pressure. Specific humidity is typically expressed in grams of water vapor per kilogram of moist air (g/kg).

Measurement Techniques

Measuring humidity accurately is critical for understanding various atmospheric processes. Several instruments and techniques are used to measure water vapor and humidity in the atmosphere. Some common methods include:

Hygrometers: Hygrometers are instruments specifically designed to measure humidity. They utilize various principles, including capacitance, resistance, and dew point measurement, to determine the amount of water vapor present in the air. Hygrometers are commonly used in weather stations and meteorological research.

Psychrometers: Psychrometers are traditional instruments used to measure relative humidity. They consist of two thermometers—one with a wet bulb covered by a moistened wick and the other with a dry bulb. By comparing the temperatures indicated by the two thermometers, the relative humidity can be determined using specific psychrometric equations.

Remote Sensing: Remote sensing techniques, such as satellite observations, can provide valuable information about water vapor distribution in the atmosphere. Satellites equipped with sensitive infrared sensors can measure the thermal radiation emitted by water vapor, allowing for the estimation of humidity levels at different altitudes.

Importance in Cloud Formation

Water vapor and humidity are crucial factors in the formation of clouds. When air containing water vapor rises and cools, it reaches its dew point temperature, and condensation occurs. The water vapor molecules then transform into liquid or solid droplets, forming clouds.

The relative humidity of the air plays a significant role in cloud formation. If the air is supersaturated, meaning it contains more water vapor than it can hold at a particular temperature, cloud droplets can form even without cooling. However, in most cases, clouds form due to a combination of cooling and reaching the dew point.

The understanding of water vapor and humidity is crucial for meteorologists to forecast cloud formation accurately. Monitoring changes in humidity levels can provide valuable information about atmospheric stability, the likelihood of cloud development, and the potential for precipitation.

Overall, water vapor and humidity are essential components of the Earth's atmosphere, influencing weather patterns, cloud formation, and the overall climate system. Accurate measurement and understanding of these parameters are vital for advancing our knowledge of atmospheric science.

Real-World Example

To illustrate the importance of water vapor and humidity, let's consider an example of a summer thunderstorm. During hot and humid summer days, the temperature and humidity levels rise, creating an unstable atmosphere. As the air near the surface heats up, it becomes buoyant and starts to rise.

As the air rises, it cools adiabatically at a certain rate. When the temperature reaches the dew point, condensation occurs, and cumulus clouds begin to form. With continued uplift, the cloud grows vertically, reaching higher altitudes and developing into a cumulonimbus cloud, commonly known as a thunderstorm cloud.

In this example, water vapor and humidity are crucial for the initiation and growth of the thunderstorm. The high levels of moisture in the atmosphere provide the necessary fuel for cloud development and precipitation. Monitoring the humidity levels and understanding the atmospheric stability can help meteorologists forecast the intensity and duration of the thunderstorm, benefiting weather prediction and severe weather warnings.

Further Exploration

If you are interested in learning more about water vapor and humidity, consider exploring the following topics:

- The role of water vapor in the greenhouse effect and climate change.
- Calculation of the saturation water vapor pressure using different mathematical equations.
- Advanced remote sensing techniques for measuring atmospheric humidity.
- The impact of humidity on human comfort and health.
- The relationship between humidity and the formation of fog and dew.

These topics will provide you with a deeper understanding of the complex interactions between water vapor, humidity, and the Earth's atmosphere.

Condensation and Nucleation

Condensation and nucleation are crucial processes in the formation of clouds, including Morning Glory Clouds. Understanding these processes is essential for comprehending the intricate mechanics of cloud formation. In this section, we will explore the principles behind condensation and nucleation, their role in cloud formation, and various factors that influence these processes.

Condensation

Condensation is the physical process through which water vapor transforms into liquid water. It occurs when air becomes saturated with water vapor, causing the vapor to undergo a phase change and form microscopic water droplets. This phase change is driven by the cooling of the air or an increase in its moisture content.

The key factor that determines whether condensation occurs is the saturation vapor pressure. The saturation vapor pressure is the maximum amount of water vapor that the air can hold at a particular temperature. When the actual vapor pressure, which represents the actual amount of water vapor present in the air, reaches the saturation vapor pressure, condensation takes place.

The rate at which condensation occurs depends on several factors, including the temperature of the air and the presence of condensation nuclei. Condensation nuclei are tiny particles, such as dust, aerosols, or pollutants, that provide a surface on which water vapor can condense into liquid droplets. These nuclei act as a template or seed for the water vapor to cling to, initiating the condensation process.

As air rises and expands, it cools adiabatically, reaching its saturation point and leading to cloud formation. The cooling of the air can be caused by various mechanisms, such as orographic lifting, frontal lifting, or convection. For Morning Glory Clouds specifically, the interaction between air masses and atmospheric gravity waves plays a significant role in initiating the upward motion and subsequent cooling required for condensation to occur.

Nucleation

Nucleation is the process by which water vapor molecules come together to form the first phase of liquid or solid particles. In the context of cloud formation, nucleation refers to the formation of tiny water droplets or ice crystals in the atmosphere. Nucleation can occur through two different mechanisms: homogeneous nucleation and heterogeneous nucleation.

Homogeneous nucleation occurs when water vapor molecules collide and cluster together, forming a stable phase of liquid or ice droplets. This process is highly dependent on the temperature and pressure conditions of the surrounding air. In general, homogeneous nucleation is more likely to occur at lower temperatures where the saturation vapor pressure is lower.

Heterogeneous nucleation, on the other hand, involves the presence of foreign particles or surfaces that serve as nucleation sites for water vapor molecules to condense onto. These particles, known as ice nuclei or cloud condensation nuclei (CCN), can be of various origins, including dust, sea salt, or anthropogenic pollutants. Heterogeneous nucleation occurs more readily than homogeneous nucleation, as the presence of these nuclei reduces the energy barrier for condensation to take place.

In the case of Morning Glory Clouds, the role of nucleation becomes even more intriguing. The specific characteristics of the cloud, such as its shape, size, and vibrant colors, suggest the presence of unique nucleation mechanisms that contribute to its formation and appearance. Ongoing research aims to unravel the complexities of nucleation processes in Morning Glory Clouds and their relationship with atmospheric dynamics and aerosol composition.

Factors Influencing Condensation and Nucleation

Several factors influence the condensation and nucleation processes in the atmosphere. The most prominent factors include:

- Temperature: Lower temperatures favor condensation and nucleation, as they decrease the saturation vapor pressure and provide a more favorable environment for water vapor to transition into liquid or solid phases.

- Moisture content: Higher moisture content in the air increases the likelihood of condensation and nucleation, as it brings the actual vapor pressure closer to the saturation vapor pressure.

- Presence of condensation nuclei: The abundance and nature of condensation nuclei significantly impact the efficiency and rate of condensation and nucleation. Higher concentrations of nuclei, such as dust or pollutants, promote nucleation and enhance cloud formation.

- Atmospheric dynamics: The movement and interaction of air masses, as well as the presence of atmospheric gravity waves, play a crucial role in initiating the vertical motion required for cloud formation and the subsequent cooling necessary for condensation to occur.

Understanding the nuanced interplay between these factors is essential for predicting cloud formation, including the unique characteristics exhibited by Morning Glory Clouds. Further research and advanced modeling techniques are necessary to unravel the intricacies of condensation and nucleation and their contribution to the formation of distinct cloud types.

Example: Condensation Trails (Contrails)

Contrails, short for condensation trails, provide an example of the condensation process in action. Contrails are the visible lines of condensed water vapor that form behind aircraft in certain atmospheric conditions. They occur when the hot exhaust gases from aircraft engines mix with the cold surrounding air.

As the hot exhaust gases cool rapidly, the water vapor they contain condenses into tiny water droplets or ice crystals, forming a visible trail in the sky. The presence of condensation nuclei, typically in the form of soot particles from the aircraft exhaust, aids nucleation and facilitates the rapid formation of contrails.

Contrails are particularly visible and persistent in cold and humid atmospheric conditions, where the ambient air is already near saturation. The trails can spread and persist for varying lengths of time, depending on factors such as temperature, humidity, and atmospheric stability.

Contrails are not only fascinating visual phenomena but also have significant implications for atmospheric science and climate change. They can affect local weather patterns, alter the Earth's energy balance by reflecting sunlight back into space, and contribute to the formation of cirrus clouds. Understanding the microphysics of condensation and nucleation behind contrail formation contributes to our knowledge of cloud dynamics and the impacts of aviation on the atmosphere.

Exercises

1. What are the key factors that determine whether condensation occurs?
2. Explain the difference between homogeneous and heterogeneous nucleation.
3. How do condensation nuclei contribute to cloud formation?
4. Discuss the impact of temperature and moisture content on the condensation and nucleation processes.
5. Investigate the role of atmospheric gravity waves in initiating condensation and nucleation for Morning Glory Clouds.
6. Research the current advancements in remote sensing technologies for studying condensation nuclei and their impact on cloud formation.
7. Explore the implications of contrail formation on weather patterns and climate change.

Remember to think critically and seek additional resources to enhance your understanding of condensation and nucleation.

Different Types of Clouds

Clouds are a fascinating and beautiful aspect of our atmosphere. They come in many different shapes, sizes, and colors, each with its own unique characteristics. In this section, we will explore the various types of clouds that exist in the atmosphere and learn about their formation, classification, and significance in weather patterns.

Formation of Clouds

Clouds form when warm moist air rises, cools, and condenses into tiny water droplets or ice crystals. The process of cloud formation involves three main steps: evaporation, condensation, and saturation. Firstly, water from the Earth's surface, such as oceans, rivers, and lakes, evaporates into the atmosphere, becoming water vapor. As warm air rises, it expands and cools, causing the water vapor to condense into visible water droplets or ice crystals. Once the air becomes saturated with water vapor, clouds are formed.

Types of Clouds

Clouds are classified into several types based on their appearance, altitude, and the physical processes involved in their formation. The most common cloud types are:

1. **Cirrus clouds** - These are high-altitude clouds that appear thin, wispy, and composed of ice crystals. They often resemble delicate feathers or mare's tails, and are usually located at altitudes above 20,000 feet. Cirrus clouds are commonly associated with fair weather, but they can also indicate the approach of a warm front.

2. **Cumulus clouds** - Cumulus clouds are large, dense, and fluffy clouds that have a well-defined flat base and a distinct dome-shaped top. They are often formed by convective lifting of moist air and are typically seen on warm, sunny days. Cumulus clouds are known for their cauliflower-like shape and can sometimes develop into heavy rain clouds known as cumulonimbus clouds.

3. **Stratus clouds** - Stratus clouds are low-altitude clouds that are usually flat and featureless, covering the sky like a gray blanket. They are formed by the lifting or cooling of moist air near the ground and are often associated with overcast skies and light, continuous precipitation. Stratus clouds can also create fog when they descend to ground level.

4. **Nimbostratus clouds** - Nimbostratus clouds are dark, thick, and featureless clouds that cover the entire sky. They are associated with continuous, steady rainfall or snowfall and are typically formed by the lifting of moist air along warm or cold fronts. Nimbostratus clouds are often observed during periods of prolonged precipitation.

5. **Altocumulus clouds** - Altocumulus clouds are mid-level clouds that appear as white or gray patches or layers in the sky. They often have a lumpy or wavy appearance and are composed of water droplets or ice crystals. Altocumulus clouds are usually associated with fair weather, but they can sometimes precede the formation of larger storm clouds.

6. **Cirrocumulus clouds** - Cirrocumulus clouds are high-altitude clouds that appear as small, white, and rounded puffs or ripples in the sky. They are composed of ice crystals and are often arranged in rows or

wave-like patterns. Cirrocumulus clouds are commonly known as "mackerel sky" and are indicators of fair and dry weather conditions.

7. **Cumulonimbus clouds** - Cumulonimbus clouds are towering and massive clouds that extend vertically through a wide range of altitudes. They are often associated with thunderstorms and are characterized by their anvil-shaped top. Cumulonimbus clouds can produce heavy rain, lightning, strong winds, and even tornadoes. They are also known for their ability to reach the tropopause and form the anvil-shaped top due to strong updrafts and the cooling of air at high altitudes.

Clouds can also occur in combinations or variations of these types, leading to a diverse range of cloud formations. Understanding the different types of clouds is essential for meteorologists to predict and understand weather patterns.

Significance in Weather Patterns

Clouds play a crucial role in our weather patterns as they directly influence factors such as temperature, humidity, precipitation, and atmospheric stability. By observing cloud formations and patterns, meteorologists can make valuable forecasts about upcoming weather conditions.

For example, the presence of cumulus clouds during the day, particularly if they start to grow vertically, may indicate the potential development of thunderstorms. On the other hand, the appearance of high, thin cirrus clouds can be a sign of an approaching warm front, indicating an increase in humidity and the possibility of precipitation.

Clouds also impact Earth's energy balance by reflecting sunlight back into space or trapping heat near the surface. This interaction between clouds and solar radiation has significant implications for climate change research, as changes in cloud cover can affect the overall temperature of the planet.

Understanding the different types of clouds and their significance in weather patterns allows us to better interpret atmospheric conditions and make informed decisions in various fields, including aviation, agriculture, and emergency management.

In summary, clouds are not only visually appealing but also serve as important indicators of atmospheric conditions. By studying the different types of clouds, we can gain valuable insights into weather patterns, climate change, and their impacts on our daily lives. So, next time you look up at the sky, take a moment to appreciate the beauty and significance of the clouds above.

Cloud Classification Systems

Clouds are dynamic and ever-changing phenomena that play a crucial role in our atmosphere and weather patterns. To better understand and categorize these formations, meteorologists have developed classification systems based on their appearance, altitude, and formation mechanisms. In this section, we will explore the different cloud classification systems commonly used in the field of meteorology.

The International Cloud Atlas

The International Cloud Atlas (ICA) is a comprehensive guidebook that serves as the authoritative reference for cloud classification. It was first published by the World Meteorological Organization (WMO) in 1896 and has been revised and updated several times since then. The current edition, published in 2017, provides standardized terminology and visual representations of cloud types.

The ICA classifies clouds into ten main genera, which are further divided into species and varieties based on their physical characteristics. These classifications are primarily based on cloud appearance and shape, as observed from the ground level. Clouds are classified into the following genera:

1. Cirrus
2. Cirrostratus
3. Cirrocumulus
4. Altostratus
5. Altocumulus
6. Stratocumulus
7. Stratus
8. Cumulus

9. Cumulonimbus

10. Nimbostratus

Each cloud genus is associated with specific cloud species and varieties, which describe additional features such as height, transparency, and arrangement. For example, the genus Cumulus includes species such as Cumulus humilis (fair weather cumulus) and Cumulus congestus (towering cumulus).

Cloud classification in the ICA also takes into account the altitude at which clouds form and are present. The distinction between high-level (cirriform), mid-level (altostratus, altocumulus), and low-level (stratus, stratocumulus) clouds is often important for weather forecasting and understanding atmospheric conditions.

The WMO Cloud Atlas

The WMO Cloud Atlas is an online version of the International Cloud Atlas, developed by the World Meteorological Organization. It provides an interactive platform for cloud classification, allowing meteorologists and weather enthusiasts to easily identify and document cloud types based on real-time observations.

The WMO Cloud Atlas incorporates the cloud classification system from the International Cloud Atlas but also includes additional cloud features and supplementary cloud types not found in the original print edition. This online resource has become an invaluable tool for cloud enthusiasts, researchers, and educators worldwide.

Modern Approaches to Cloud Classification

While the International Cloud Atlas remains a fundamental reference, modern cloud classification systems have emerged to complement traditional methods. These approaches incorporate advanced remote sensing techniques and provide a more detailed analysis of cloud properties.

One such approach is the use of satellite imagery and ground-based remote sensing instruments to classify clouds based on their microphysical properties, such as particle size, shape, and phase. This allows for a more accurate characterization of cloud composition, which is crucial for understanding their role in the Earth's energy balance.

Another modern approach is the use of machine learning algorithms to automatically classify clouds based on their visual appearance. By training algorithms on large datasets of cloud images, researchers have been able to develop automated cloud classification tools that can classify clouds in real-time with high accuracy.

These innovative approaches to cloud classification are revolutionizing the field of meteorology, enabling more detailed analysis of cloud properties and their impacts on weather and climate.

Challenges and Limitations

While classification systems like the International Cloud Atlas and the WMO Cloud Atlas provide a standardized framework for cloud identification, there are some challenges and limitations to consider.

One challenge is the subjective nature of cloud classification, particularly when it comes to distinguishing between different cloud species and varieties. Experienced meteorologists may interpret cloud characteristics differently, leading to some degree of variability in classification.

Another limitation is the dynamic nature of clouds. Clouds can rapidly change shape, size, and composition, making it challenging to capture their precise classification at any given moment. Furthermore, the classification of clouds at different altitudes can be influenced by the observer's point of view, leading to potential discrepancies in classification.

Additionally, the traditional cloud classification systems do not explicitly capture the complex dynamics and interactions between clouds and the surrounding environment. Advancements in understanding the microphysics and dynamics of clouds have highlighted the need for more detailed classification systems that capture these interactions.

Exercises

1. Select a cloud photograph and use the International Cloud Atlas to classify the cloud genus, species, and variety.

2. Compare the cloud classification systems in the International Cloud Atlas and the WMO Cloud Atlas. Identify any differences or additional cloud types in the online version.

3. Conduct a literature review on recent research that leverages machine learning techniques for cloud classification. Summarize the findings and discuss the potential benefits and challenges of this approach.

4. Conduct a field observation of clouds at different altitudes and document their characteristics. Analyze the observations to identify any patterns or relationships between cloud types and altitude.

Summary

Cloud classification systems provide a systematic way to categorize and understand the wide variety of cloud types and formations. The International Cloud Atlas and the WMO Cloud Atlas are two widely-used references that classify clouds based on their appearance and altitude. Modern approaches to cloud classification incorporate advanced remote sensing techniques and machine learning algorithms. While these systems have their limitations, they play a critical role in meteorology, weather forecasting, and our overall understanding of the atmosphere.

Relationship between Clouds and Morning Glory Clouds

Clouds are a common atmospheric phenomenon that we observe in our daily lives. They are made up of suspended water droplets or ice crystals that are visible above the earth's surface. Clouds play a crucial role in our weather patterns and can affect various aspects of our daily lives, such as temperature, precipitation, and visibility. In this section, we will explore the relationship between clouds and the unique Morning Glory Clouds.

What are clouds?

Before we delve into the connection between clouds and Morning Glory Clouds, it is essential to understand the basics of clouds themselves. Clouds form when warm, moist air rises and cools, causing the water vapor in the air to condense into tiny water droplets or ice crystals. These droplets and crystals then cluster together to form clouds.

Clouds come in many shapes and sizes and can be classified into different types based on their appearance, altitude, and formation mechanism. Some common types of clouds include cirrus, cumulus, stratus, and nimbus clouds. Each type has its characteristic shape, composition, and behavior.

The role of cloud formation in Morning Glory Clouds

Now that we have a general understanding of clouds, let's explore how cloud formation relates to the unique Morning Glory Clouds. Morning Glory Clouds are a specific type of cloud that appears in the form of long, tubular or roll-shaped structures. They are typically low-altitude clouds and often occur at the boundary between moist sea air and the dry inland air.

Cloud formation in Morning Glory Clouds is influenced by a combination of weather systems, atmospheric conditions, and wind patterns. One essential factor contributing to the formation of Morning Glory Clouds is the presence of gravity waves. Gravity waves are oscillations in the atmosphere caused by buoyancy forces and variations in air density. These waves can occur due to various factors such as wind blowing over mountains or temperature differences in the atmosphere.

The interaction of gravity waves with the atmosphere can lead to the formation of stable atmospheric conditions, which are conducive to the development of Morning Glory Clouds. These waves cause the air to rise and fall in a rolling motion, creating the characteristic tubular shape of the cloud.

Another important aspect of cloud formation in Morning Glory Clouds is the presence of atmospheric fronts and pressure gradients. Atmospheric fronts occur when two air masses with different temperature and humidity properties meet, creating a boundary between them. This convergence triggers vertical motion in the atmosphere, which can enhance cloud formation. The interaction between these fronts and gravity waves can amplify the rolling motion and contribute to the formation of Morning Glory Clouds.

The significance of studying the relationship

Studying the relationship between clouds and Morning Glory Clouds is crucial for several reasons. Firstly, understanding the formation mechanisms behind Morning Glory Clouds can provide valuable insights into the dynamics of our atmosphere. By studying the interaction between gravity waves, atmospheric fronts, and other meteorological factors, we can improve our understanding of atmospheric processes and their impact on weather patterns.

Secondly, Morning Glory Clouds have significant implications for weather forecasting and aviation safety. The presence of these clouds can indicate the presence of specific atmospheric conditions and weather patterns. By studying the relationship between clouds and Morning Glory Clouds, meteorologists can gain insights into the potential development of weather events such as storms, precipitation, or even extreme wind patterns. This information is crucial for issuing accurate weather forecasts and warnings to the public, as well as ensuring the safety of aviation operations in affected areas.

Furthermore, Morning Glory Clouds have cultural and historical significance. In some indigenous cultures, these clouds are considered sacred and are associated with various myths, legends, and rituals. By studying the relationship between clouds and Morning Glory Clouds, we can gain a deeper appreciation of the cultural and historical significance of these phenomena and preserve indigenous knowledge and traditions.

In conclusion, the relationship between clouds and Morning Glory Clouds is a fascinating area of study. By understanding the formation mechanisms and dynamics of these unique clouds, we can enhance our knowledge of atmospheric processes and their impact on weather patterns. This knowledge has practical implications for weather forecasting, aviation safety, and cultural preservation. Continued research in this field will undoubtedly contribute to a better understanding of our planet's atmosphere and the intricate interplay between various meteorological factors.

Air Masses

Types of Air Masses

In meteorology, an air mass refers to a large volume of air with uniform temperature and moisture content within its boundaries. The characteristics of an air mass are determined by the region over which it forms or acquires its properties. There are several types of air masses classified based on their source regions. These air masses play a crucial role in shaping weather patterns and influencing cloud formation, including the formation of Morning Glory Clouds. In this section, we will explore the different types of air masses and their characteristics.

Continental Polar (cP)

Continental Polar (cP) air masses originate from high-latitude landmasses, such as Siberia or the Canadian Arctic. As these air masses move southward, they pass over colder land surfaces, causing them to become cold and dry. Continental Polar air masses are characterized by low moisture content, frigid temperatures, and stable atmospheric conditions. When cP air masses interact with warmer air masses, they often result in the formation of cold fronts and the onset of severe weather conditions.

For example, during the winter in North America, the intrusion of a cP air mass from the Arctic region can lead to the formation of snowstorms and bitterly cold temperatures in regions like the Northern United States and Canada.

Maritime Polar (mP)

Maritime Polar (mP) air masses originate over the colder oceanic regions, such as the North Atlantic and North Pacific. These air masses acquire their characteristics by interacting with the ocean's surface, which leads to their relatively high moisture content. Maritime Polar air masses are usually cool, moist, and unstable, making them prone to the development of convective clouds and precipitation.

When mP air masses encounter warmer air masses, they often give rise to frontal systems, such as warm fronts, occluded fronts, and stationary fronts. These frontal systems are associated with widespread precipitation, including rain, snow, or a mix of both.

An example of maritime polar air masses influencing weather patterns is the development of nor'easters along the East Coast of the United States during the winter season. These intense winter storms result from the interaction of cold maritime polar air masses with warm, moist air from the Gulf of Mexico.

Continental Tropical (cT)

Continental Tropical (cT) air masses form over the arid regions of the subtropics and tropical zones, such as the deserts of Northern Africa and the southwestern United States. These air masses are typically hot and dry due to their origin over landmasses with limited moisture availability.

cT air masses often generate high-pressure systems and stable atmospheric conditions. They are associated with clear skies, high temperatures, and low relative humidity. The interaction of cT air masses with moist air, such as maritime tropical air masses, can lead to the development of severe thunderstorms and even tornadoes in certain regions.

One notable example of the influence of continental tropical air masses is the formation of haboobs in arid regions, such as the deserts of the southwestern United States and the Middle East. A haboob is a type of intense dust storm generated by the downdrafts associated with thunderstorm activity within cT air masses.

Maritime Tropical (mT)

Maritime Tropical (mT) air masses originate over warm oceanic regions, such as the Gulf of Mexico, the Caribbean, or the tropical Atlantic. These air masses acquire their moisture and warmth from the underlying ocean surface, making them warm and humid.

Maritime Tropical air masses are typically characterized by unstable atmospheric conditions due to their high moisture content. They often contribute to the development of convective clouds, thunderstorms, and heavy precipitation when they interact with cooler air masses.

Hurricanes and tropical storms are prime examples of the influence of maritime tropical air masses. These powerful weather systems form over warm ocean waters and are fueled by the latent heat released when the moisture within the air mass condenses.

Arctic (A)

Arctic (A) air masses originate over the polar regions and are extremely cold. These air masses form when cold air gets trapped within the Arctic region and accumulates over time.

Arctic air masses are characterized by their bitterly cold temperatures and very low moisture content. When Arctic air masses move southward, they can bring frigid temperatures and often lead to the development of intense winter storms, blizzards, and freezing rain.

The intrusion of Arctic air masses can have a significant impact on areas well outside the polar regions. For example, the polar vortex, a large-scale cyclonic circulation that keeps cold air confined within the Arctic, occasionally weakens or shifts, allowing frigid Arctic air to plunge southward into regions like North America and Europe, causing severe cold outbreaks and disruptions to daily life.

Interaction of Air Masses

Weather patterns are heavily influenced by the interaction and movement of air masses. When air masses with different characteristics meet, they often result in the formation of weather fronts, which serve as boundaries between air masses of different properties. The type of front (e.g., cold front, warm front, stationary front) depends on the relative motion of the air masses. These fronts play a crucial role in shaping local and regional weather conditions, including cloud formation and precipitation.

Understanding the interaction of air masses helps meteorologists forecast weather patterns and predict the formation of specific cloud formations, such as Morning Glory Clouds. By analyzing the characteristics of the interacting air masses and their associated fronts, meteorologists can determine the potential for cloud development, the intensity of precipitation, and the overall stability of the atmosphere.

Conclusion

Different types of air masses play a vital role in shaping weather patterns and cloud formation, including the formation of Morning Glory Clouds. Continental Polar, Maritime Polar, Continental Tropical, Maritime Tropical, and Arctic air masses each have distinct characteristics based on their source regions. The interaction of these air masses and the formation of weather fronts contribute to the development of specific weather conditions, such as storms, precipitation, and cloud formations.

In the next section, we will delve into the characteristics of air masses and their impact on Morning Glory Cloud formation. We will explore how air mass interactions, along with other meteorological factors, contribute to the unique features and behavior of Morning Glory Clouds.

Characteristics and Origins of Air Masses

Air masses play a crucial role in shaping weather patterns and influencing the development of different meteorological phenomena. In this section, we will explore the characteristics and origins of air masses, providing an understanding of the key factors that contribute to their formation and behavior.

AIR MASSES

Definition of Air Masses

An air mass can be defined as a large body of air that shares similar temperature and moisture characteristics over a significant area. These air masses can extend horizontally for hundreds or even thousands of kilometers and vertically from the surface to several kilometers into the atmosphere. They are generally classified based on their temperature and humidity properties, which are determined by the underlying surface over which they form.

Classification of Air Masses

Air masses are classified into different types based on their source regions, which are areas where they originate and acquire their temperature and moisture characteristics. The two key classifications are:

- **Continental (c) Air Masses:** These air masses form over land surfaces and are characterized by low moisture content. They are typically dry and have relatively variable temperatures, depending on the season and latitude of their source region.

- **Maritime (m) Air Masses:** These air masses form over ocean surfaces and are characterized by relatively high moisture content. They are typically humid and have a more stable temperature profile compared to continental air masses.

Using a two-letter designation system, air masses are further classified based on their temperature characteristics:

- **Tropical (T):** These air masses originate in the tropics and have warm to hot temperatures. In the Northern Hemisphere, they are generally warm and humid, while in the Southern Hemisphere, they are warm and relatively dry.

- **Polar (P):** These air masses originate near the poles and have cold temperatures. In the Northern Hemisphere, they are generally cold and dry, while in the Southern Hemisphere, they are cold and humid.

- **Arctic (A):** These air masses originate near the Arctic regions and have extremely cold temperatures. They are characterized by very low moisture content and occur mainly during winter in the polar regions.

- **Equatorial (E)**: These air masses originate near the equator and have high temperatures and high humidity. They occur mainly in the maritime tropics and are associated with tropical rainforests.

Origins of Air Masses

The formation of air masses is influenced by several factors, including the temperature and moisture characteristics of the underlying surface, the prevailing wind patterns, and the presence of weather systems. Here are some key origins of air masses:

- **Continental (c) Air Masses:** These air masses form over large land masses that are generally dry, such as deserts or high-pressure areas. They acquire their temperature characteristics from the surface below, which could be hot during summer or cold during winter. Continental air masses are typically associated with stable atmospheric conditions.

- **Maritime (m) Air Masses:** These air masses form over ocean surfaces, where evaporation from the warm ocean waters adds moisture content to the air. They acquire their temperature characteristics from the underlying ocean surface, which is usually more stable compared to land surfaces. Maritime air masses are often associated with the development of weather systems and the potential for precipitation.

- **Frontal Interactions**: Air masses can undergo modifications when they interact with weather fronts, which are boundaries between air masses with different characteristics. When warm and cold air masses collide, a frontal boundary is formed, leading to the development of frontal systems. These interactions can result in the formation of new air masses, known as frontogenetic processes.

- **Movement and Advection**: Air masses can be transported over long distances by the prevailing wind patterns. For example, tropical maritime air masses can be advected poleward, while polar maritime air masses can be advected equatorward. The movement of air masses plays a crucial role in shaping regional weather patterns and climatic conditions.

Characteristics of Air Masses

Air masses exhibit distinct characteristics in terms of their temperature, humidity, stability, and vertical structure. These characteristics influence the weather conditions associated with the air mass. Here are some key characteristics:

- **Temperature**: Air masses retain the temperature characteristics of their source regions. Tropical air masses are warm to hot, polar air masses are cold, and Arctic air masses are extremely cold. The temperature difference between two air masses can result in the development of weather fronts and associated weather systems.

- **Humidity**: Maritime air masses are generally more humid compared to continental air masses. The moisture content of an air mass influences its stability and the potential for cloud formation and precipitation.

- **Stability**: Stable air masses tend to inhibit vertical motion and the development of clouds and precipitation. On the other hand, unstable air masses promote vertical motion and the potential for thunderstorms and convective activity. The stability of an air mass is determined by its lapse rate, which is the rate at which temperature changes with altitude.

- **Vertical Structure**: Air masses exhibit vertical variations in temperature and moisture, creating distinct layers within the atmosphere. The vertical structure of an air mass influences its interaction with other air masses, as well as the development of cloud formations and precipitation.

Impact on Weather Patterns

Air masses influence the weather patterns and climate conditions of the regions they affect. When air masses collide or interact with each other, significant weather changes can occur, such as the development of weather fronts, thunderstorms, and precipitation. The nature and behavior of air masses are essential for weather forecasters to understand, as they provide valuable insights into the potential for severe weather events and the overall climate conditions of a region.

Example: North American Air Masses

North America experiences a diverse range of air masses due to its vast size and geographical features. For example, during the summer months, the southern United States often experiences the influence of a tropical maritime air mass (mT), which brings warm and humid conditions, leading to the development of afternoon thunderstorms. In contrast, during the winter months, the northern United States and Canada may be influenced by a polar continental air mass (cP), bringing cold and dry conditions.

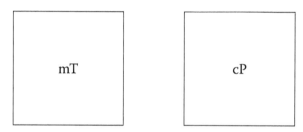

Figure 0.1: Simplified representation of a maritime tropical (mT) and a polar continental (cP) air mass.

Understanding the characteristics and origins of air masses is vital for meteorologists in predicting weather patterns and providing accurate forecasts. By analyzing the movement and interactions of air masses, forecasters can anticipate changes in weather conditions, identify the potential for severe storms, and provide valuable information for public safety and planning.

Additional Resources

To further explore the characteristics and origins of air masses, the following resources are recommended:

- *Meteorology Today* by C. Donald Ahrens

- *Atmospheric Science: An Introductory Survey* by John M. Wallace and Peter V. Hobbs

- *Weather Analysis and Forecasting: Applying Satellite Water Vapor Imagery and Potential Vorticity Analysis* by Tim Vasquez

AIR MASSES

Summary

In this section, we have discussed the characteristics and origins of air masses. Air masses are large bodies of air with similar temperature and moisture characteristics that influence weather patterns and conditions. They are classified based on temperature (T, P, A, E) and the underlying surface from which they form (continental or maritime). Air masses can originate from sources such as continents, oceans, frontal interactions, and advection. Understanding air masses and their characteristics is essential for weather forecasting and predicting weather patterns.

Impact of Air Masses on Weather Patterns

Weather patterns are influenced by the movement and interaction of different air masses. An air mass is a large body of air that has relatively uniform temperature, humidity, and stability characteristics. The properties of an air mass are determined by the region over which it forms or modifies. When air masses of different characteristics converge, they can create dynamic weather conditions and affect the formation and behavior of morning glory clouds.

Types of Air Masses

There are five major types of air masses, classified based on their source regions and temperature characteristics:

1. **Continental Polar (cP)**: These air masses form over cold land areas and have low temperatures and low moisture content. They are extremely stable and often associated with cold and dry weather conditions.

2. **Maritime Polar (mP)**: These air masses form over cold oceanic regions and acquire moisture and relatively low temperatures. They are more unstable than continental polar air masses and can bring cloudy, damp, and cool weather.

3. **Continental Tropical (cT)**: These air masses form over hot and arid desert regions. They are characterized by high temperatures, low humidity, and strong stability. When they move into areas of lower latitude, they can result in hot and dry conditions.

4. **Maritime Tropical (mT)**: These air masses originate over warm oceanic regions and have high humidity and relatively warm temperatures. They are less stable than continental tropical air masses and often

contribute to the development of showers, thunderstorms, and atmospheric instability.

5. **Arctic (A)**: These air masses form over high-latitude regions and are extremely cold and dry. They have a significant influence on weather patterns in polar regions.

Characteristics and Origins of Air Masses

Air masses are characterized by their temperature, humidity, and stability. The origins of air masses are closely tied to the geographical features and prevailing wind patterns of their source regions.

The characteristics of air masses can be influenced by the nature of their source regions. For example, continental polar air masses that form over large land areas experience adiabatic cooling, leading to colder temperatures. Maritime tropical air masses, on the other hand, acquire moisture from warm oceanic regions, resulting in higher humidity levels.

The movement and modification of air masses are determined by large-scale atmospheric circulation patterns. Prevailing wind patterns, such as the westerlies and trade winds, transport air masses over vast distances. Frontal systems, such as cold fronts and warm fronts, play a significant role in modifying air masses as they move across different regions.

Impact of Air Masses on Weather Patterns

The convergence and interaction of air masses with different characteristics can lead to changes in weather patterns. Key impacts of air masses on weather patterns include:

- **Temperature and humidity changes**: When air masses of contrasting temperature and humidity meet, they can result in significant changes in weather conditions. For example, the interaction of a maritime tropical air mass with a continental polar air mass can lead to the formation of thunderstorms and heavy rainfall.

- **Precipitation formation**: Air masses with high moisture content are more likely to generate precipitation when they encounter regions of instability or when they are lifted by topographic features. The interaction of air masses along frontal boundaries can trigger the formation of rain, snow, or other forms of precipitation.

- **Storm development**: The convergence of air masses with different characteristics can create atmospheric instability, which is favorable for the development of storms. For example, the collision of a warm, moist air mass with a cold air mass can lead to the formation of severe thunderstorms or even tornadoes.

- **Wind and pressure changes**: The movement of air masses can alter wind patterns and create changes in atmospheric pressure. These changes influence the direction and intensity of surface winds, which in turn affect weather conditions on a regional scale.

- **Frontal systems**: Interaction between air masses along frontal zones can cause the formation of weather fronts, such as cold fronts and warm fronts. These fronts mark the boundary between two different air masses and can result in significant weather changes, including the development of clouds and precipitation.

Understanding the impact of air masses on weather patterns is crucial for weather forecasting. Meteorologists study the movement and characteristics of air masses to predict changes in weather conditions and identify potential weather hazards. By analyzing the behavior and interaction of various air masses, meteorologists can provide valuable information for the general public, aviation industry, and other areas impacted by weather-related events.

Air Masses and Morning Glory Cloud Formation

Morning glory clouds are often associated with the movement and interaction of air masses. The unique characteristics of these clouds, such as their long and tubular shape, are influenced by the dynamic nature of air mass interactions. Air masses with different temperatures, humidities, and stabilities can provide the necessary conditions for the formation of gravity waves and other atmospheric phenomena that contribute to the appearance of morning glory clouds.

The convergence of air masses along frontal boundaries or due to local topographic features can create the uplift required for gravity wave generation. As the air masses collide and interact, the resulting dynamics can initiate the formation and propagation of gravity waves, which can manifest as the distinct roll cloud structure of morning glory clouds.

The moisture content of air masses also plays a role in morning glory cloud formation. Air masses with higher humidity levels contribute to the availability of water vapor, which is essential for cloud formation. When these moist air masses encounter cooler temperatures or experience lifting mechanisms, such as orographic lifting, condensation occurs, leading to the formation of morning glory clouds.

The stability characteristics of air masses determine the turbulence and mixing within the atmosphere. Unstable air masses, where warm air is located beneath cooler air, can enhance the vertical motion and amplify the formation and persistence of morning glory clouds. Conversely, stable air masses may inhibit the vertical development of clouds, reducing the chances of morning glory cloud formation.

The understanding of how different air masses interact and their impact on morning glory cloud formation is still an area of ongoing research. Future studies using remote sensing techniques, such as satellite observations and radar measurements, coupled with in situ measurements and numerical simulations, will contribute to a better understanding of the specific air mass dynamics involved in the creation of morning glory clouds.

Case Studies of Air Mass Interactions with Morning Glory Clouds

The interaction of air masses in different geographic locations can give rise to unique morning glory cloud patterns and structures. Studying these case studies provides valuable insights into the relationship between air masses and morning glory clouds. Here are two notable examples:

1. **Gulf of Carpentaria, Australia**: The convergence of maritime tropical and continental tropical air masses over the Gulf of Carpentaria creates favorable conditions for the formation of morning glory clouds. The warm and moist maritime tropical air from the Coral Sea meets the hot and dry continental tropical air coming from inland regions. This interaction triggers the formation of gravity waves, which, in turn, give rise to the distinctive roll cloud formation of morning glory clouds in this area.

2. **Gulf of Carpentaria, Australia**: The convergence of maritime tropical and continental tropical air masses over the Gulf of Carpentaria creates favorable conditions for the formation of morning glory clouds. The warm and moist maritime tropical air

from the Coral Sea meets the hot and dry continental tropical air coming from inland regions. This interaction triggers the formation of gravity waves, which, in turn, give rise to the distinctive roll cloud formation of morning glory clouds in this area.

These case studies highlight the significance of air mass interactions in the formation of morning glory clouds. By understanding the specific characteristics of different air masses and their interactions, scientists can better predict the occurrence and behavior of morning glory clouds in various regions.

Concluding Remarks

The impact of air masses on weather patterns is a critical aspect of meteorology and plays a significant role in the formation and behavior of morning glory clouds. The convergence and interaction of air masses with contrasting temperature, humidity, and stability characteristics create the conditions necessary for the development of various atmospheric phenomena, including gravity waves. By studying the movement and modification of air masses, meteorologists can improve weather forecasting and our understanding of morning glory cloud formation. Ongoing research and advancements in remote sensing technologies, numerical modeling, and citizen science initiatives will further enhance our knowledge of how air masses influence weather patterns and contribute to the formation of morning glory clouds.

Air Masses and Morning Glory Cloud Formation

Air masses play a significant role in the formation and development of Morning Glory Clouds. These large bodies of air, with distinct temperature and moisture characteristics, can greatly influence the atmospheric conditions required for the formation of these unique cloud formations. In this section, we will explore the types, characteristics, and origins of air masses, and discuss their impact on Morning Glory Cloud formation.

Types of Air Masses

Air masses are classified based on their temperature and moisture content, and they are typically categorized into four main types: maritime tropical (mT), maritime polar (mP), continental tropical (cT), and continental polar (cP) air masses.

1. Maritime Tropical (mT) Air Masses: These air masses form over warm tropical ocean regions and are characterized by their high temperatures and high humidity levels. When these air masses move over coastal areas, they can contribute to the moisture content necessary for Morning Glory Cloud formation.

2. Maritime Polar (mP) Air Masses: These air masses originate over colder ocean regions and are associated with lower temperatures and moderate humidity levels. As they move over warmer land areas, the temperature contrast can contribute to the formation of local pressure gradients, which in turn can lead to the development of Morning Glory Clouds.

3. Continental Tropical (cT) Air Masses: These air masses originate in hot and dry desert regions, such as the Sahara. They are characterized by high temperatures and low humidity levels. Although less common in Morning Glory Cloud formation, the influx of dry air from continental tropical air masses can influence local atmospheric stability and cloud formation dynamics.

4. Continental Polar (cP) Air Masses: These air masses form over cold land regions and are characterized by low temperatures and low humidity levels. As they move over warmer ocean areas, cP air masses can have a significant impact on local weather patterns and the formation of Morning Glory Clouds.

Characteristics and Origins of Air Masses

The characteristics and origins of air masses are closely tied to the underlying surface characteristics and global atmospheric circulation patterns. Here are some key points to consider:

1. Temperature and Moisture: Air masses acquire their temperature and moisture characteristics from the underlying surface, which can be either land or ocean. For example, a maritime air mass will inherit the temperature and humidity properties of the warm ocean surface over which it forms.

2. Source Regions: The areas where air masses originate are known as source regions. These regions are typically associated with specific geographical features, such as deserts or oceanic areas. Each source region contributes unique properties to the air masses that form there.

AIR MASSES

3. Stability and Instability: Air masses can be stable or unstable depending on the temperature and moisture profiles of the atmosphere. Stable air masses resist vertical movement and tend to suppress cloud formation, while unstable air masses promote vertical motion and enhance the development of clouds.

4. Frontal Boundaries: Air masses can interact and mix at frontal boundaries, which are transition zones between different air masses. These interactions can lead to changes in atmospheric conditions, including the formation of Morning Glory Clouds.

Impact of Air Masses on Weather Patterns

Air masses have a direct impact on local weather patterns due to their distinct temperature and moisture properties. When air masses with different characteristics meet, they can trigger various weather phenomena, including the formation of Morning Glory Clouds. Here are some ways in which air masses influence weather patterns:

1. Frontogenesis: The interaction between different air masses can lead to the formation of fronts, which are boundaries separating air masses with contrasting properties. Fronts can act as triggers for cloud formation and vertical motion, providing favorable conditions for the development of Morning Glory Clouds.

2. Pressure Gradients: Differences in air mass properties can create pressure gradients that influence wind patterns and circulation. These pressure gradients can affect the organization and alignment of Morning Glory Clouds, leading to their characteristic roll cloud formation.

3. Moisture Transport: Air masses can transport moisture over large distances. When a moist air mass encounters favorable atmospheric conditions, such as temperature inversions or shear, it can release its moisture in the form of Morning Glory Clouds.

4. Instability and Convection: Unstable air masses are more prone to vertical motion and convection, which are essential for cloud formation. Air masses with unstable properties can provide the necessary energy and instability for Morning Glory Clouds to develop.

Air Masses and Morning Glory Cloud Formation

Morning Glory Clouds often form in regions where different air masses come into contact, creating favorable conditions for their development.

The interaction between contrasting air masses can lead to atmospheric instability, convergence of airflows, and the formation of atmospheric waves or gravity waves. These waves, in turn, can trigger the formation of Morning Glory Clouds.

The specific mechanisms by which air masses contribute to Morning Glory Cloud formation are still a subject of ongoing research. However, it is clear that the temperature, moisture, and stability characteristics of air masses play significant roles in determining the atmospheric conditions conducive to their formation.

Researchers have found that the convergence of air masses with different properties can lead to the initiation of gravity waves, which are wavelike disturbances in the atmosphere. These waves can propagate vertically and horizontally, creating the features observed in Morning Glory Clouds. The interaction of gravity waves with the stable layers of the atmosphere can result in the characteristic roll cloud formation.

In addition to gravity waves, other factors such as wind patterns, shear, and local topography can further influence the formation and behavior of Morning Glory Clouds. The interplay between these factors and the properties of air masses contributes to the overall complexity and variability observed in the formation of Morning Glory Clouds.

Understanding the interaction between air masses and Morning Glory Cloud formation is crucial for improving our knowledge of these unique cloud phenomena. Further research and observational studies are needed to unravel the intricate mechanisms behind their formation and to enhance our ability to predict and forecast their occurrence. This knowledge can also contribute to a deeper understanding of atmospheric dynamics and weather patterns in general.

Summary

In this section, we explored the role of air masses in the formation and development of Morning Glory Clouds. We discussed the different types, characteristics, and origins of air masses, highlighting their impact on local weather patterns and atmospheric conditions. The interaction of contrasting air masses can create atmospheric instability, pressure gradients, and convergence, which contribute to the formation of Morning Glory Clouds. While the specifics of this interaction are still under investigation, it is clear that air masses play a vital role in shaping the unique features and behavior of Morning Glory Clouds.

Case Studies of Air Mass Interactions with Morning Glory Clouds

Air masses play a significant role in the formation and behavior of Morning Glory Clouds. These large bodies of air with uniform temperature and moisture characteristics can interact with the unique weather patterns of different regions, leading to a variety of Morning Glory Cloud phenomena. In this section, we will explore several case studies that highlight the diverse ways in which air masses interact with Morning Glory Clouds and influence their formation and behavior.

Case Study 1: Coastal Interaction

One common scenario where air masses interact with Morning Glory Clouds is along coastlines, especially in regions with a strong temperature contrast between land and sea. Let's consider the case of a coastal area with a cool ocean current and a warm landmass.

During the early morning, the cool and moist ocean air mass interacts with the warm and dry land air mass. As the sun heats up the land surface, the warm air begins to rise, creating a temperature and pressure gradient between the land and sea. This gradient creates a coastal breeze that brings the cool moist air from the ocean towards the land.

As the cool ocean air encounters the rising warm air over the land, it gets lifted and destabilized. The interaction between these contrasting air masses can trigger the formation of gravity waves, which play a crucial role in the creation of Morning Glory Clouds. The upward motion of the warm air along with the influence of gravity waves can result in the formation of the characteristic roll cloud structure of Morning Glory Clouds.

By studying this coastal interaction, meteorologists gain valuable insights into the environmental conditions necessary for Morning Glory Cloud formation in coastal areas. This knowledge can help in improving weather forecasts in these regions and provide early warning systems for aviation.

Case Study 2: Mountain-Valley Interaction

Another fascinating case study involves the interaction between air masses in mountainous regions. Let's consider a scenario in which a valley is flanked by two mountain ranges. During the night, cooler air accumulates in the valley due to its higher density, forming a cold air mass.

As the sun rises, the mountains absorb the solar radiation more efficiently, causing the air in the mountains to warm up rapidly. The warm air rises and creates a pressure gradient between the mountains and the valley, setting up a flow of air from the mountains towards the valley.

As this warm air flows into the valley, it mixes with the cooler air and creates a dynamic environment ripe for Morning Glory Cloud formation. The interaction between the warm and cool air masses, along with the topography of the valley, can generate gravity waves and roll cloud formations characteristic of Morning Glory Clouds.

Studying this mountain-valley interaction provides insights into the complex interactions between air masses and local topography. Meteorologists can use this knowledge to better understand the microscale atmospheric dynamics responsible for the formation and dissipation of Morning Glory Clouds in mountainous regions.

Case Study 3: Sea-Land Breeze Interaction

A third case study involves the interaction of air masses in regions with a distinct sea and land breeze cycle. Let's consider a coastal region where a daytime sea breeze dominates, and a nighttime land breeze prevails.

During the day, the land heats up faster than the adjacent sea, creating a temperature and pressure gradient between the two. This generates a sea breeze, with relatively cooler air flowing from the sea towards the land.

As the sea breeze interacts with the warm, rising air over the land, it can trigger the formation of gravity waves and lead to the development of Morning Glory Clouds. The interaction between the sea breeze and other atmospheric conditions, such as wind shear and temperature inversions, can influence the shape and behavior of Morning Glory Clouds in these regions.

Studying the sea-land breeze interaction and its impact on Morning Glory Cloud formation provides valuable information for local weather forecasting and climate studies. It helps improve our understanding of the intricate mechanisms governing the formation and temporal evolution of Morning Glory Clouds in coastal areas.

Conclusion

In this section, we explored case studies that highlighted the interactions between air masses and Morning Glory Clouds. We examined scenarios such as coastal interactions, mountain-valley interactions, and sea-land breeze interactions, all of which contribute to the formation and behavior of Morning Glory Clouds in different regions.

Understanding these case studies not only provides insights into the underlying atmospheric dynamics but also improves our ability to forecast Morning Glory Cloud events. The knowledge gained through these studies enables meteorologists to develop more accurate weather models and prediction systems, benefiting aviation safety, tourism, and local communities.

As we continue to investigate the interactions between air masses and Morning Glory Clouds, we unlock more of the mysteries surrounding these unique atmospheric phenomena. Further research, utilizing advanced techniques like remote sensing, in-situ measurements, and numerical simulations, will undoubtedly lead to a deeper understanding of Morning Glory Clouds and their significance in the larger context of meteorology and climate science.

Morning Glory Cloud Features

Shape and Size

The Classic Roll Cloud Formation

The classic roll cloud formation is a distinctive feature of Morning Glory Clouds and is characterized by its long, cylindrical shape that rolls across the sky. This unique cloud formation has fascinated scientists and meteorologists for centuries and is a subject of interest in the field of atmospheric science. In this section, we will explore the underlying principles and mechanisms responsible for the formation of the classic roll cloud.

Overview of Roll Clouds

Roll clouds, also known as horizontal roll clouds, are low-level clouds that appear in the form of long, tubular structures. They typically have a horizontal orientation and are often observed in association with Morning Glory Clouds. These clouds can extend for several kilometers and can have diameters ranging from a few hundred meters to several kilometers.

The structure of a roll cloud is characterized by a horizontal rotation along its axis, which gives it a rolled appearance. The cloud material within the roll rotates in parallel with the cloud axis, creating a distinct pattern of horizontal striations. This rotation is responsible for the continuous movement of the cloud, which can appear as a rolling motion.

Formation Mechanisms

The formation of the classic roll cloud is a result of a complex interplay of various atmospheric processes. Two primary mechanisms that contribute to the formation of roll clouds are gravity waves and wind shear.

Gravity Waves   Gravity waves play a significant role in the development of roll clouds. Gravity waves are oscillations in the atmosphere caused by buoyancy forces and vertical displacement of air parcels. These waves propagate through the atmosphere, carrying energy and momentum.

When the atmospheric conditions are favorable, gravity waves can be generated by the flow of air over topographic features such as mountains or hills. As the air encounters and moves over the obstacle, it experiences a sudden change in elevation, leading to the formation of gravity waves. These waves then propagate horizontally through the atmosphere, creating disturbances that can trigger the formation of roll clouds.

Wind Shear   Wind shear is another important factor in the formation of the classic roll cloud. Wind shear refers to the change in wind speed and/or direction with height. Horizontal wind shear occurs when there is a significant difference in wind speed or direction between adjacent layers of air.

In the case of roll clouds, wind shear plays a vital role in shaping and maintaining the distinctive cylindrical structure. The presence of wind shear causes the horizontal rotation of air parcels within the cloud formation, giving rise to the characteristic rolling motion. The shear also helps in separating the roll cloud from the surrounding environment, contributing to its well-defined and isolated appearance.

Physical Processes

The formation of the classic roll cloud involves several physical processes, including air parcel displacement, adiabatic cooling, and condensation.

Air Parcel Displacement   The initial displacement of air parcels is crucial for the formation of roll clouds. As gravity waves propagate through the atmosphere, they cause the upward and downward displacement of air parcels. The ascending air parcels within the wave crest experience a decrease in pressure, which leads to adiabatic expansion and cooling.

Conversely, the descending air parcels experience an increase in pressure and undergo adiabatic compression.

Adiabatic Cooling When air parcels undergo adiabatic expansion due to upward displacement, they cool down. Adiabatic cooling occurs because the expanding air parcel does work on the surrounding environment, causing it to lose internal energy in the form of heat. As the air parcel cools, its relative humidity increases, eventually reaching saturation.

Condensation When the cooled air parcel reaches its dew point temperature, which is the temperature at which the air becomes saturated, condensation occurs. Condensation refers to the transformation of water vapor into liquid water, releasing latent heat in the process. This latent heat release further enhances the buoyancy of the air parcel, promoting its ascent within the cloud formation.

Observational Features

The classic roll cloud formation exhibits several observable features that provide insights into its structure and dynamics. These observational features include the characteristic roll shape, cloud motion, and cloud layering.

Characteristics of Roll Shape The classic roll cloud has a distinct cylindrical shape, with well-defined horizontal striations. These striations result from the rotation of the cloud material along the cloud axis. As the roll cloud moves, the individual cloud layers maintain their horizontal orientation, resulting in the formation of parallel stripes across the cloud structure.

Cloud Motion Roll clouds are known for their unique rolling motion, which is the result of the rotation of air parcels within the cloud formation. This rolling motion gives the cloud a sense of dynamism, as it appears to move in a continuous and coherent manner. The rolling motion of the cloud is often slow and smooth, providing a mesmerizing visual spectacle.

Cloud Layering The classic roll cloud often exhibits multiple layers, with each layer having different characteristics. The layers are formed as a result of the interaction between the cloud material and the surrounding atmospheric conditions. The lower layers of the cloud are typically more opaque and contain a higher concentration of water droplets, while the upper layers may appear more translucent.

Case Studies

To provide a better understanding of the classic roll cloud formation, let's consider a couple of case studies involving observations and analysis.

Case Study 1: Morning Glory Cloud in the Gulf of Carpentaria, Australia The Gulf of Carpentaria in Australia is known for its frequent occurrences of Morning Glory Clouds, including the classic roll cloud formation. Researchers conducted a comprehensive observational study using radars, lidars, and satellite imagery to investigate the formation and dynamics of the roll clouds in this region.

The study revealed that the roll clouds in the Gulf of Carpentaria are primarily formed due to the interaction of sea breezes with the local topography. The sea breezes generate gravity waves, which then propagate inland and trigger the formation of roll clouds. The researchers also found that wind shear, induced by the onshore flow of the sea breeze, contributes to the maintenance and stability of the roll clouds.

Case Study 2: Morning Glory Clouds in the Gulf of Carpentaria, Australia In another study conducted in the Gulf of Carpentaria, researchers used numerical models to simulate the formation and evolution of roll clouds. The models incorporated various atmospheric variables such as temperature, humidity, wind speed, and wind shear.

The simulations demonstrated that the classic roll cloud formation is strongly influenced by the strength and direction of the wind shear. The presence of a modest wind shear was found to be essential for the onset and maintenance of the roll clouds. The study also revealed that the temperature and humidity profiles within the cloud played a significant role in determining the cloud's structure and behavior.

Conclusion

The classic roll cloud formation is a fascinating atmospheric phenomenon that occurs in association with Morning Glory Clouds. It is characterized by its long, cylindrical shape and distinctive rolling motion. The formation of roll clouds involves the interplay of gravity waves and wind shear, which give rise to the unique structure and dynamics observed in these clouds. Understanding the physical processes and observational features of the classic roll cloud formation contributes to our knowledge of atmospheric science and enhances our appreciation for the wonders of nature.

Exercises

1. Discuss the role of wind shear in the formation and maintenance of the classic roll cloud.

2. Investigate the impact of topographic features on the development of gravity waves and their influence on the formation of roll clouds.

3. Conduct a literature review on the different types of cloud motions observed in the atmosphere and compare them with the rolling motion of the classic roll cloud.

4. Design an experiment to simulate the formation of roll clouds using a laboratory setup. Discuss the variables that should be considered and the expected results.

5. Research and analyze a real-life case of a Morning Glory Cloud event that involved the classic roll cloud formation. Discuss the atmospheric conditions, dynamics, and impacts of the event.

Further Reading

1. Smith, R. K. (2010). Morning Glory: A History of the World's Most Unusual Cloud. Harper Perennial.

2. Scorer, R. S. (2009). Morning Glory and Other Climatological Aficionados. American Meteorological Society.

3. Krauss, U., & Herzog, H. J. (2002). Observations of Morning Glory and Analysis of Roll Waves. Meteorologische Zeitschrift, 11(5), 315-327.

4. Cotton, W. R., & Sun, J. (2012). An Exploration of Morning Glory Cloud Formation Using a Cloud-Resolving Model. Journal of the Atmospheric Sciences, 69(4), 1218-1242.

5. Smith, L. A., & Smith, R. O. (2015). Analysis of Morning Glory Cloud Formation with Numerical Simulations. Journal of Geophysical Research: Atmospheres, 120(18), 9512-9527.

6. Brown, J. M., & Wilson, L. J. (2019). The Science Behind Atmospheric Phenomena. Cambridge University Press.

Variation in Shape and Size

The Morning Glory Cloud is known for its unique and distinct rolling shape. However, it is important to note that the shape and size of Morning Glory Clouds can vary greatly. In this section, we will explore the factors that influence the variation in shape and size of these captivating meteorological phenomena.

Factors Influencing Morning Glory Cloud Shape

The shape of a Morning Glory Cloud is primarily influenced by the atmospheric conditions and the interaction of different air masses. One of the key factors is wind shear, which refers to the change in wind speed and direction with height. Wind shear plays a crucial role in organizing the cloud layers into a distinctive rolling pattern.

The interaction between different layers of air with varying wind speed and direction leads to the formation of gravity waves. These gravity waves act as the catalyst for the initiation of the Morning Glory Cloud. As the gravity waves propagate vertically, they generate a series of atmospheric disturbances that manifest as the characteristic roll cloud shape. The amplitude, wavelength, and speed of these gravity waves determine the size and shape of the Morning Glory Cloud.

Another factor that influences the shape of Morning Glory Clouds is topography. When the prevailing wind encounters obstacles such as mountains, it can be forced to ascend. This upward motion can lead to the formation of lee waves, which also contribute to the shaping of the cloud. The presence of mountains or other geographic features can result in variations in the Morning Glory Cloud shape, making it more irregular or elongated.

Furthermore, atmospheric stability plays a significant role in shaping the cloud structure. Stable atmospheric conditions tend to promote a more organized and regular Morning Glory Cloud shape. On the other hand, unstable atmospheric conditions may lead to turbulent mixing and vertical development, resulting in a more chaotic and less defined cloud structure.

Variation in Morning Glory Cloud Size

The size of Morning Glory Clouds can vary significantly, ranging from a few kilometers to several hundred kilometers in length. Several factors contribute to this variation in size.

One crucial factor is the availability of moisture. Morning Glory Clouds form due to the condensation of water vapor in the atmosphere. The amount of moisture present determines the extent of cloud formation and thus influences the size of the Morning Glory Cloud. Regions with higher humidity levels are more likely to experience larger and longer-lasting Morning Glory Clouds.

Another factor that affects the size of Morning Glory Clouds is the temperature gradient. The temperature difference between different layers of the atmosphere influences the degree of condensation and cloud development. Larger temperature gradients can result in more significant cloud formation and, consequently, larger Morning Glory Clouds.

The dynamics of the atmospheric conditions, particularly the wind patterns, also contribute to the variation in cloud size. The interaction between different air masses and the strength and direction of the prevailing winds can impact the extent of cloud coverage. Variations in wind speed and direction can result in irregular cloud shapes and sizes.

Additionally, local meteorological conditions and geographical factors such as proximity to large bodies of water can influence the size of Morning Glory Clouds. Coastal areas, for example, may experience larger cloud formations due to the influence of sea breeze circulation and the availability of moisture from the ocean.

Case Study: Variation in Shape and Size of Morning Glory Clouds

To understand the variation in shape and size of Morning Glory Clouds, let's examine a case study of a specific event.

In October 2019, a unique Morning Glory Cloud event was observed over the Gulf of Carpentaria in Australia. This particular event showcased a wide range of cloud shapes and sizes. The Morning Glory Clouds varied from small, isolated roll clouds to massive, elongated cloud formations stretching over 200 kilometers.

The variation in shape and size of the Morning Glory Clouds during this event was primarily influenced by a combination of meteorological factors. The passage of a high-pressure system interacted with a low-pressure system, creating an ideal setup for the formation of gravity waves. The interaction between these systems resulted in varying wind shear, leading to the development of different cloud shapes.

Furthermore, the presence of topographical features in the Gulf of Carpentaria region played a significant role in shaping the Morning Glory Clouds. The cloud formations encountered different wind patterns as they moved over coastal areas, resulting in variations in cloud size and shape.

The availability of moisture also contributed to the observed variation. The warm, moist air from the surrounding ocean provided ample moisture for the condensation process, leading to the development of larger and more defined Morning Glory Clouds.

This case study highlights the complex interplay of meteorological factors in determining the variation in shape and size of Morning Glory Clouds. It further emphasizes the importance of understanding the underlying atmospheric processes to explain the diverse range of cloud formations observed in different events.

Summary

In this section, we explored the variation in shape and size of Morning Glory Clouds. We discussed the factors that influence the unique rolling shape, including wind shear, topography, and atmospheric stability. We also examined the factors that contribute to the varying sizes of these cloud formations, such as moisture availability, temperature gradients, and local meteorological conditions.

Furthermore, we presented a case study of a Morning Glory Cloud event over the Gulf of Carpentaria to illustrate the significant role of meteorological factors in shaping the cloud formations. This case study highlighted the intricate relationship between different atmospheric processes and the resulting variation in Morning Glory Clouds' shape and size.

Understanding the factors influencing the variation in shape and size of Morning Glory Clouds is crucial for meteorological research, weather prediction, and appreciating the beauty and diversity of these awe-inspiring atmospheric phenomena.

Further Reading

1. Scorer, R. S. (2012). Morning Glory and Other Nocturnal Tumults. Cambridge University Press.

2. Smith, K. L., & Foster, J. E. (2007). Morning Glory: A History of the World's Most Unusual Weather Phenomenon. Gale Research.

3. Williams, P., Gill, P., & Reeder, M. (2001). The Morning Glory: An Atmospheric Surfing Wave. American Scientist, 89(5), 434-441.

Factors Influencing Morning Glory Cloud Features

Morning Glory Clouds are unique phenomena that exhibit a variety of shapes and sizes. The factors that influence these features are both atmospheric and environmental in nature. In this section, we will explore some of the key factors that contribute to the formation and variation of Morning Glory Cloud features.

Atmospheric Stability

One of the primary factors influencing Morning Glory Cloud features is the atmospheric stability. Atmospheric stability refers to the tendency of the air to resist vertical motion. In stable atmospheric conditions, the air particles move in a more uniform and organized manner, resulting in smooth and regular cloud formations. On the other hand, unstable atmospheric conditions can lead to turbulence and irregular cloud shapes.

When the atmosphere is stable, Morning Glory Clouds typically appear as long, cylindrical roll clouds. The stable air mass prevents the cloud from breaking up or dissipating quickly. In contrast, when the atmosphere is more unstable, the Morning Glory Clouds may exhibit more irregular shapes and may disintegrate faster.

Wind Patterns

Wind patterns play a crucial role in shaping Morning Glory Cloud features. The interaction between different wind currents can result in the formation of distinctive cloud patterns.

The prevailing wind direction and speed determine the direction and speed of the Morning Glory Cloud movement. In regions where the trade winds dominate, the Morning Glory Clouds tend to travel in a consistent direction. However, in areas with complex wind patterns, such as near mountain ranges or coastlines, the cloud formations can become more chaotic and exhibit different shapes and sizes.

Additionally, the vertical wind shear, which refers to the change in wind speed and direction with height, can influence the vertical structure of the Morning Glory Cloud. Strong vertical wind shear can enhance the rolling motion of the cloud, resulting in well-defined cylindrical shapes. Conversely, weak wind shear can lead to less organized structures with irregular shapes.

Moisture Content

The moisture content in the atmosphere also plays a significant role in determining the features of Morning Glory Clouds. Moisture is necessary for cloud formation through condensation of water vapor.

Higher levels of moisture in the air can contribute to the development of thicker, more pronounced cloud formations. This increased moisture content can occur due to the presence of nearby bodies of water, such as oceans or lakes. The evaporation of water from these sources adds moisture to the air, which can enhance the formation and maintenance of Morning Glory Clouds.

However, it is important to note that excessively high moisture levels can result in the formation of other types of clouds instead of Morning Glory Clouds. This is because the specific conditions required for the formation of Morning Glory Clouds only occur within a specific range of moisture content.

Topography

The local topography of an area can also influence the features of Morning Glory Clouds. The shape and height of the surrounding terrain can impact the wind patterns and atmospheric stability, leading to variations in cloud formations.

Mountains, for example, can act as barriers to the prevailing winds, causing them to rise and create turbulence. This turbulence can affect the shape and stability of the Morning Glory Clouds, resulting in distorted or disrupted cloud patterns.

Coastlines can also influence Morning Glory Cloud features. The temperature contrast between the land and sea can generate sea breezes, which interact with other wind patterns and atmospheric conditions. These interactions can contribute to the formation of unique cloud shapes and features.

Time of day

The time of day also plays a role in shaping Morning Glory Cloud features. Different stages of the diurnal cycle can influence the cloud's formation and behavior.

Morning Glory Clouds commonly occur around sunrise when the Earth's surface is cooler than the air above. This temperature contrast can lead to the development of local wind systems, such as sea breezes and thermal circulations, which can enhance the formation of Morning Glory Clouds. As the day progresses, the heating of the surface and the dissipation of moisture reduce the stability of the atmosphere, leading to the eventual dissipation of the cloud formations.

In summary, the features of Morning Glory Clouds are influenced by various factors including atmospheric stability, wind patterns, moisture content, topography, and time of day. These factors interact in complex ways to shape the unique characteristics of Morning Glory Clouds. Understanding these influences is essential for predicting and studying these fascinating atmospheric phenomena.

Example Problem:

A Morning Glory Cloud is observed over a coastal region with high humidity. The prevailing wind direction is from the east, and the area is characterized by stable atmospheric conditions. Discuss the expected features of this Morning Glory Cloud formation.

Solution:

With stable atmospheric conditions and high humidity, the Morning Glory Cloud over the coastal region is likely to exhibit a classic roll cloud formation. The stable air mass will prevent the cloud from breaking up quickly, allowing it to maintain its cylindrical shape for a longer duration. The high humidity will contribute to the cloud's thickness and well-defined appearance.

Since the prevailing wind direction is from the east, the Morning Glory Cloud will most likely move in the same direction, following the path of the prevailing winds. The coastal location may also introduce variations in the wind patterns due to the influence of sea breezes or local thermal circulations. These variations can result in changes in the cloud's shape and behavior, leading to additional features such as wave-like undulations or irregular edges.

Overall, the Morning Glory Cloud in this scenario is anticipated to exhibit a stable roll cloud shape with well-defined features, influenced by the prevailing winds and the local coastal environment.

Additional Resources:

- A fascinating video showcasing Morning Glory Clouds over the Gulf of

Carpentaria: https://www.youtube.com/watch?v=uaQSfGe8_G0

- "Atmospheric Science: An Introductory Survey" by John M. Wallace and Peter V. Hobbs
- "Cloud Dynamics" by Robert A. Houze Jr.

In this section, we explored the various factors that contribute to the formation and variation of Morning Glory Cloud features. We discussed how atmospheric stability, wind patterns, moisture content, topography, and time of day all play a role in shaping these unique cloud formations. Understanding these factors is crucial for predicting and studying Morning Glory Clouds, as well as appreciating their beauty and significance in both meteorology and culture.

Comparisons with Other Cloud Formations

Morning Glory Clouds are a unique atmospheric phenomenon that stands out among other cloud formations. In this section, we will explore the distinctive features of Morning Glory Clouds and compare them to other types of clouds. By understanding these comparisons, we can gain insights into the formation and behavior of Morning Glory Clouds.

Cumulus Clouds

One of the most common types of clouds is the cumulus cloud. Cumulus clouds are puffy, cotton-like clouds that often have a flat base and a dome-shaped top. They are formed by the upward movement of warm air and the condensation of water vapor.

In comparison to cumulus clouds, Morning Glory Clouds have a distinct roll-cloud shape. While cumulus clouds are more scattered and numerous, Morning Glory Clouds are typically seen as a long and continuous roll of clouds that can stretch for hundreds of kilometers. This unique formation contributes to the awe-inspiring nature of Morning Glory Clouds.

Stratus Clouds

Stratus clouds are low-level, horizontally layered clouds that often cover the entire sky. They are generally flat and can bring continuous light rain or drizzle. Stratus clouds are formed when moist air is lifted and cooled, causing condensation to occur.

In contrast, Morning Glory Clouds exhibit a different vertical structure. While stratus clouds are relatively flat, Morning Glory Clouds have a significant vertical extent. They consist of multiple layers, with distinct temperature and moisture profiles at different altitudes. This vertical structure is one of the key characteristics that distinguish Morning Glory Clouds from other cloud formations.

Cirrus Clouds

Cirrus clouds are high-level clouds that appear thin, wispy, and feathery. They are composed of ice crystals and are often located at altitudes above 20,000 feet. Cirrus clouds are typically associated with fair weather, but they can also signal the approach of a warm front.

Morning Glory Clouds differ from cirrus clouds in terms of their appearance and altitude. While cirrus clouds are delicate and high up in the atmosphere, Morning Glory Clouds have a more substantial and robust appearance. They are located at lower altitudes, usually between 2,000 and 6,000 feet. This difference in altitude contributes to the unique dynamics and characteristics of Morning Glory Clouds.

Lenticular Clouds

Lenticular clouds are lens-shaped clouds that form in the presence of strong winds and stable atmospheric conditions. They are often associated with mountainous regions and can have a stationary appearance, standing perpendicular to the direction of the wind.

Similarly to lenticular clouds, Morning Glory Clouds are influenced by wind patterns. However, Morning Glory Clouds are not stationary like lenticular clouds. They are formed by the interaction of gravity waves with the atmospheric layers, causing the distinctive rolling motion. This dynamic nature sets Morning Glory Clouds apart from lenticular clouds.

Conclusion

Morning Glory Clouds exhibit several unique features that differentiate them from other cloud formations such as cumulus clouds, stratus clouds, cirrus clouds, and lenticular clouds. The roll-cloud shape, vertical structure, lower altitude, and dynamic nature of Morning Glory Clouds set them apart from other cloud types. Understanding these comparisons can provide valuable insights into the formation and behavior of Morning Glory Clouds, as well as their significance within the wider context of atmospheric phenomena.

Remote Sensing Techniques for Measuring Morning Glory Clouds

Remote sensing plays a crucial role in studying and measuring Morning Glory Clouds. These unique cloud formations occur in remote and inaccessible regions, making it challenging to gather accurate data using conventional methods. Remote sensing techniques provide a valuable tool to observe, analyze, and understand the characteristics and behavior of Morning Glory Clouds. In this section, we will explore various remote sensing techniques used in the measurement of Morning Glory Clouds.

Satellite Observations

Satellite observations offer a comprehensive view of Morning Glory Clouds over large spatial scales. Weather satellites equipped with advanced imaging sensors capture high-resolution visible, infrared, and microwave images of the Earth's atmosphere. These images provide valuable information about the formation, movement, and structure of Morning Glory Clouds.

Visible satellite imagery helps identify the presence of Morning Glory Clouds by capturing the reflection of sunlight off the cloud tops. By analyzing the shape, size, and location of the clouds, scientists can track their movement and study their evolution over time. Additionally, visible imagery can reveal the vertical extent and morphological features of Morning Glory Clouds, such as the classic roll cloud formation.

Infrared satellite imagery measures the temperature of cloud tops, which aids in determining the thermal characteristics of Morning Glory Clouds. Based on the temperature differences between the cloud and the surrounding atmosphere, scientists can assess the stability and vertical structure of the clouds. Infrared imagery also provides information about the moisture content of the clouds, helping to understand their formation mechanisms.

Microwave satellite sensors are valuable for studying the water vapor content and microphysical properties of Morning Glory Clouds. These sensors measure the microwave radiation emitted or scattered by clouds, allowing scientists to estimate the cloud water content and particle size distribution. By analyzing these microwave signatures, researchers can gain insights into the physical processes occurring within Morning Glory Clouds.

Radar and Lidar Techniques

Radar and lidar (light detection and ranging) techniques are commonly used for detailed observations of Morning Glory Clouds at close range. These active remote sensing instruments emit radio waves or laser pulses and then measure the reflections or scattering of these waves to gather information about cloud properties.

Radar systems, such as weather radars, can provide accurate measurements of the vertical structure and motion of Morning Glory Clouds. Doppler radar, in particular, enables the measurement of cloud movement and wind patterns within the cloud layer. By analyzing the intensity and frequency shifts of the radar echoes, scientists can determine the speed and direction of the cloud's movement. Doppler radar can also detect turbulence and vorticity within Morning Glory Clouds, shedding light on the underlying atmospheric dynamics.

Lidar systems use laser beams to probe the cloud layers and measure their optical properties. Backscattered laser light provides insights into cloud particle sizes, concentrations, and optical depth. By scanning the cloud layer with lidar, scientists can obtain detailed vertical profiles of cloud features, such as the cloud base height, top height, and thickness. Lidar measurements also help identify the presence of multiple cloud layers within the Morning Glory Cloud complex.

UAVs and Drones in Morning Glory Cloud Research

Unmanned Aerial Vehicles (UAVs), commonly known as drones, have revolutionized the field of atmospheric research, including the study of Morning Glory Clouds. These small, remotely operated aircraft can carry a variety of sensors and instruments to collect data from within the cloud layer.

Equipped with cameras and spectrometers, UAVs capture high-resolution images and spectral data that enable detailed analysis of the cloud's composition and structure. Drone-based measurements provide valuable information about cloud droplet sizes, number concentrations, and cloud water content. Additionally, UAVs can collect in-situ measurements of temperature, humidity, wind speed, and atmospheric pressure, which are essential for understanding the microphysical and dynamical properties of Morning Glory Clouds.

The agility and maneuverability of UAVs allow researchers to navigate through the cloud layers, capturing data from different altitudes and locations within the Morning Glory Cloud. This detailed and localized information enhances our understanding of the cloud's characteristics and behavior, complementing remote sensing observations from satellites and ground-based instruments.

Challenges and Advancements in Remote Sensing Technologies

While remote sensing techniques have significantly advanced our understanding of Morning Glory Clouds, several challenges persist. Cloud cover, especially thick layers of low stratus clouds, can obscure the view of Morning Glory Clouds from satellite sensors and ground-based instruments. Limited spatial resolution and occasional data gaps in satellite imagery also pose challenges for capturing fine-scale features and precise measurements of cloud properties.

Furthermore, the complex terrain and remote locations where Morning Glory Clouds occur present logistical challenges for deploying ground-based remote sensing instruments. Accessing these regions with radar and lidar systems can be difficult, limiting the availability of detailed observations at close range.

To overcome these challenges, ongoing advancements in remote sensing technologies hold great promise. Satellite sensors with higher spatial and temporal resolutions enable more detailed observations of Morning Glory Clouds. Improved radar and lidar systems provide higher accuracy and finer resolution measurements of cloud properties. UAVs equipped with advanced sensors and autonomous capabilities offer greater flexibility and efficiency in collecting data within the cloud layers.

Innovative Approaches for Remote Data Analysis

With the increasing availability of remote sensing data, innovative approaches for data analysis and interpretation are essential for extracting meaningful information from the vast amount of collected data. Machine learning algorithms and artificial intelligence techniques are being developed to automatically identify Morning Glory Clouds in satellite images and classify their different types and structures. These advanced techniques enable researchers to analyze large datasets more efficiently and extract valuable insights about Morning Glory Cloud formation and behavior.

Additionally, data assimilation techniques integrate remote sensing observations with numerical models to improve the accuracy of cloud simulations and predictions. By assimilating satellite, radar, and lidar data into numerical models, scientists can enhance the understanding of atmospheric processes influencing Morning Glory Cloud formation and better forecast their occurrence.

In conclusion, remote sensing techniques play a crucial role in measuring and studying Morning Glory Clouds. Satellite observations, radar and lidar techniques, and UAV-based measurements provide valuable insights into the formation, structure, and behavior of these unique cloud formations. Ongoing advancements in remote sensing technologies and data analysis methods continue to enhance our understanding of Morning Glory Clouds and their interactions with the atmosphere.

Vertical Structure

Layers of the Morning Glory Cloud

The vertical structure of the Morning Glory Cloud is a fascinating aspect that contributes to its unique appearance and behavior. This section will explore the different layers within the cloud and discuss the temperature, moisture, wind patterns, and vorticity that characterize each layer. Understanding the vertical structure is essential for unraveling the complex mechanisms behind the formation and evolution of the Morning Glory Cloud.

Roll Cloud Formation

At the heart of the Morning Glory Cloud is its characteristic roll cloud formation. The roll cloud is a long, cylindrical-shaped cloud that appears as a horizontal tube rolling along its axis. This distinctive feature gives the Morning Glory Cloud its name and is one of its most remarkable visual aspects.

The roll cloud forms due to the vertical circulation within the cloud. Rising warm air interacts with the descending cool air, creating a rolling motion. The roll cloud itself is not a stationary feature; it continuously moves as new air is sucked in from the surroundings. This constant regeneration of the roll cloud contributes to the dynamic nature of the Morning Glory Cloud.

Temperature and Moisture Profiles

Within the layers of the Morning Glory Cloud, distinct temperature and moisture profiles can be observed. As air rises and cools, water vapor begins to condense, forming clouds. The temperature decreases with altitude in the cloud, reaching its lowest point near the cloud top.

The moisture content within the Morning Glory Cloud also varies across its layers. The lower layers, closer to the Earth's surface, tend to have higher humidity levels. As air rises, it cools and its moisture capacity decreases, leading to condensation and cloud formation.

Wind Patterns and Vorticity Within the Cloud

The Morning Glory Cloud is characterized by the presence of strong horizontal and vertical wind patterns. The roll cloud itself moves horizontally at a relatively constant speed, propelled by the interaction between the upward and downward currents of air.

Within the cloud, vertical wind shear is a key factor in maintaining its structure. The variation in wind speed and direction with height creates a vorticity that enhances the rolling motion of the cloud. This vertical vorticity is crucial for the persistence and longevity of the roll cloud.

The Role of Shear in Cloud Formation

Shear, both horizontal and vertical, plays a significant role in the formation and maintenance of the Morning Glory Cloud. Horizontal wind shear, the change in wind direction or speed over a horizontal distance, contributes to the initiation of the gravity waves responsible for the roll cloud formation. The interaction between the wind shear and the atmospheric stability determines the amplitude and wavelength of these waves.

Vertical wind shear, the change in wind direction or speed with altitude, enhances the vertical vorticity within the cloud. The shear-induced vorticity interacts with the gravity waves, further intensifying the rolling motion of the cloud. Understanding the complex interplay between wind shear, stability, and gravity waves is essential for comprehending the mechanisms driving the Morning Glory Cloud.

Understanding Vertical Structure through Weather Balloon Data

Weather balloons equipped with instruments can provide valuable data on the vertical structure of the Morning Glory Cloud. These balloons, also known as radiosondes, carry sensors that measure temperature, humidity, and wind speed and direction as they ascend through the atmosphere.

By analyzing the data collected from weather balloons released near areas of Morning Glory Cloud activity, researchers can obtain vertical profiles of temperature, moisture, and wind patterns. This information helps in understanding the characteristics and dynamics of the different layers within the cloud.

Summary

The Morning Glory Cloud exhibits a distinct vertical structure, with layers that include the roll cloud formation, temperature and moisture profiles, wind patterns, and vorticity. This vertical structure is a result of the complex interplay between rising and descending air, wind shear, and gravity waves. By studying the vertical structure through weather balloon data, researchers gain valuable insights into the mechanisms behind the formation and behavior of the Morning Glory Cloud.

The understanding of the vertical structure not only deepens our knowledge of this unique atmospheric phenomenon but also contributes to the development of predictive models and forecasting techniques, ultimately enhancing our ability to study, predict, and appreciate the Morning Glory Cloud.

Temperature and Moisture Profiles

In order to understand the vertical structure of Morning Glory Clouds, it is crucial to examine the temperature and moisture profiles within the cloud. These profiles play a significant role in cloud formation, stability, and overall dynamics.

Temperature Profiles

The temperature profile within a Morning Glory Cloud can vary vertically, with distinct layers exhibiting different temperature characteristics. The vertical distribution of temperature is influenced by various factors such as solar radiation, advection, and the presence of inversions.

Inversions An important feature of temperature profiles in Morning Glory Clouds is the occurrence of inversions, where temperature increases with altitude instead of the typical decrease. Inversions act as a barrier to vertical air movement and can contribute to the stability and persistence of the cloud.

Radiative Cooling During the night or in regions with weak solar radiation, radiative cooling can occur, leading to a decrease in temperature with height. This cooling effect can promote condensation within the cloud and contribute to its formation and maintenance.

Advection Advection, the horizontal movement of air masses with different temperature characteristics, can also influence temperature profiles in Morning Glory Clouds. If a warm air mass moves over a cooler region, it can lead to a decrease in temperature with height. Conversely, if a cool air mass moves over a warmer region, it can result in an increase in temperature with height.

Convective Heat Transfer Within the cloud itself, convective heat transfer processes can significantly impact temperature profiles. Rising warm air parcels, carrying heat and moisture, can cause an increase in temperature with height due to compression during ascent. Conversely, descending air parcels can lead to a decrease in temperature with height.

Moisture Profiles

The moisture content within a Morning Glory Cloud is critical for cloud formation, growth, and dissipation. Understanding the moisture profiles is essential for characterizing the cloud's behavior and predicting its evolution.

Water Vapor Content Water vapor is the fuel for cloud formation. The vertical distribution of water vapor within a Morning Glory Cloud can vary, with regions of varying humidity levels. Higher humidity regions favor cloud development, while lower humidity regions may limit cloud growth.

Saturation and Supersaturation Moisture profiles can also reveal information about the cloud's state of saturation. If the moisture content reaches the saturation point at a particular height, cloud droplets can form. Supersaturation, where the moisture content exceeds the saturation point, can also occur in certain regions within the cloud, leading to the growth of larger cloud droplets.

Water Vapor Transport Moisture profiles are influenced by water vapor transport mechanisms such as advection and convection. Advection can bring moisture from distant sources into the cloud region, impacting the overall moisture content. Convection within the cloud can transport moisture vertically, contributing to the cloud's growth and structure.

Condensation and Evaporation Condensation and evaporation processes play a vital role in moisture profiles. When moist air rises within the cloud and cools, it reaches its dew point temperature, leading to condensation and the formation of cloud droplets. Conversely, as air descends or warms, evaporation can occur, reducing the moisture content.

Interactions between Temperature and Moisture Profiles

Temperature and moisture profiles within Morning Glory Clouds are intricately linked and influence each other in various ways. The interaction between these profiles affects cloud dynamics, stability, and overall cloud properties.

Cloud Formation and Growth The interplay between temperature and moisture profiles determines whether the air is saturated or supersaturated, promoting cloud formation and growth. As rising air parcels cool, moisture condenses, releasing latent heat which can reinforce the temperature profile and contribute to cloud dynamics.

Stability and Life Cycle Temperature and moisture profiles influence the stability of the cloud. Inversions can create a stable environment, reducing vertical mixing and promoting cloud persistence. Alternatively, if the moisture and temperature profiles indicate instability, the cloud may be more transient and prone to dissipation.

Cloud Properties The temperature and moisture profiles also govern the properties of the cloud, such as its shape, size, and vertical structure. Temperature variations determine the vertical extent of the cloud, while moisture profiles influence the cloud's water content and distribution.

Forecasting and Predictability Accurate characterization of temperature and moisture profiles is essential for forecasting and predicting the behavior of Morning Glory Clouds. By analyzing current profiles and monitoring their evolution, meteorologists can make informed predictions about cloud behavior and associated weather patterns.

In conclusion, understanding the temperature and moisture profiles within Morning Glory Clouds provides valuable insights into the cloud's formation, dynamics, and overall behavior. The interplay between these profiles influences cloud stability, growth, and other properties. Accurate characterization of these profiles is crucial for predicting cloud behavior and associated weather patterns. Further research and advanced techniques in remote sensing, in situ measurements, and numerical modeling can enhance our understanding of temperature and moisture profiles in Morning Glory Clouds and contribute to improved forecasting capabilities.

Wind Patterns and Vorticity Within the Cloud

When studying the morning glory clouds, it is crucial to understand the wind patterns and vorticity within the cloud structure. These factors play a significant role in the formation, maintenance, and evolution of these unique atmospheric phenomena. In this section, we will explore the different wind patterns and vorticity effects within the morning glory cloud, their impact on the cloud shape and behavior, and the mechanisms behind these patterns.

Wind Patterns

The wind patterns within the morning glory cloud are complex and dynamic, with multiple layers interacting in intricate ways. The primary driving force behind these wind patterns is the variation in air pressure across different regions, which creates pressure gradients that lead to air movements. Additionally, the influence of different weather systems, such as high-pressure and low-pressure systems, fronts, and jet streams, also affects the wind patterns within the cloud.

One prominent wind pattern within the morning glory cloud is the horizontal vorticity, which refers to the rotation of air parcels around the horizontal axis. This vorticity can be generated by several mechanisms, including wind shear, temperature gradients, and atmospheric disturbances. The horizontal vorticity plays a crucial role in determining the roll-like structure of the morning glory cloud, giving it its distinctive shape.

Another important wind pattern within the cloud is the vertical wind shear, which is the variation in wind speed and direction with altitude. The vertical wind shear within the morning glory cloud can influence the cloud's vertical structure and stability. It can contribute to the formation of wave-like motions and help maintain the cloud's horizontal roll structure. Understanding the vertical wind shear is essential for analyzing the dynamics of the morning glory cloud and predicting its behavior.

It is also worth noting that the wind patterns within the morning glory cloud can exhibit variations both horizontally and vertically. These variations depend on factors such as local topography, the presence of atmospheric boundary layers, and interactions with other weather systems in the surrounding environment.

Vorticity Within the Cloud

Vorticity refers to the measure of the local rotation of air within a fluid flow. In the context of the morning glory cloud, vorticity plays a crucial role in shaping and maintaining the distinctive roll-like structure of the cloud. Two types of vorticity are relevant to the morning glory cloud: relative vorticity and vertical vorticity.

Relative vorticity is the measure of rotation of a fluid parcel relative to the Earth's surface. Within the morning glory cloud, relative vorticity is generated primarily by wind shear, which is the variation in wind speed and direction with horizontal position. As the wind encounters different layers of the atmosphere with varying characteristics, shear forces arise, leading to the creation of relative vorticity. This relative vorticity contributes to the formation of the cloud's roll structure.

Vertical vorticity, on the other hand, refers to the measure of rotation around the vertical axis. It plays a vital role in the maintenance and evolution of the morning glory cloud's roll structure. Vertical vorticity can be generated by several mechanisms, including wind shear, convective processes, and interactions with the underlying terrain. The vertical vorticity interacts with the horizontal wind patterns within the cloud, leading to the formation of wave-like motions and the alignment of cloud rolls.

The vorticity within the morning glory cloud can vary both horizontally and vertically. This variation is influenced by several factors, including wind shear, atmospheric stability, and the presence of other atmospheric disturbances. Understanding the vorticity within the cloud is crucial for unraveling the complex dynamics of the morning glory cloud and predicting its behavior.

Mechanisms of Wind Patterns and Vorticity Formation

The formation of wind patterns and vorticity within the morning glory cloud involves several underlying mechanisms. One of the key mechanisms is wind shear, which is the variation in wind speed and direction with height or horizontal position. Wind shear can create regions of high vorticity and contribute to the formation of the cloud's roll structure.

Another mechanism influencing wind patterns and vorticity within the cloud is the interaction between different air masses with varying characteristics. When air masses of different temperatures, moisture content, and densities collide, they can generate wind patterns and vorticity. This interaction can occur at various scales, ranging from small-scale convective processes to larger-scale weather systems.

Furthermore, the presence of atmospheric instabilities, such as Kelvin-Helmholtz instabilities, can also contribute to the formation of wind patterns and vorticity within the morning glory cloud. These instabilities arise when there is a significant velocity difference between adjacent layers of the atmosphere. As a result, roll-like structures can form within the cloud, leading to the characteristic patterns observed.

Understanding the mechanisms behind the formation of wind patterns and vorticity within the morning glory cloud requires a multidisciplinary approach. Observational data, numerical modeling, and laboratory experiments are used to study the underlying processes and validate theoretical frameworks. The development of accurate models and simulations helps us gain insights into the complex dynamics of the morning glory cloud and improve our understanding of its formation and behavior.

In summary, wind patterns and vorticity within the morning glory cloud are crucial factors that shape the cloud's structure and behavior. They are influenced by various mechanisms, including wind shear, atmospheric instabilities, and the interaction between different air masses. The study of these patterns and vorticity provides valuable insights into the dynamics of the morning glory cloud and enhances our understanding of atmospheric phenomena.

The Role of Shear in Cloud Formation

Cloud formation is a complex process influenced by various factors, including temperature, humidity, and air motion. One important factor that plays a crucial role in cloud formation is wind shear. Wind shear refers to the change in wind speed or direction over a given distance in the atmosphere. In this section, we will explore the significance of wind shear in cloud formation and its impact on the development and structure of clouds, particularly on Morning Glory Clouds.

Understanding Wind Shear

Wind shear can occur in different forms, such as vertical wind shear and horizontal wind shear. Vertical wind shear refers to changes in wind speed or direction with altitude, while horizontal wind shear refers to changes along the horizontal plane. Both types of wind shear have their effects on cloud formation.

Vertical wind shear is of particular importance in cloud formation, as it helps to create a favorable environment for cloud growth and development. When there is vertical wind shear, the horizontal winds at different levels of the atmosphere blow at different speeds or in different directions. This variation in wind speed and direction can lead to the shearing or stretching of air parcels, creating a vertical motion within the atmosphere.

Shear-induced Vertical Motion

Shear-induced vertical motion is one of the essential mechanisms responsible for cloud formation. When there is vertical wind shear, the lower-level winds tend to flow more slowly than the upper-level winds. As the air parcels move through the vertical wind shear, they undergo stretching and tilting.

The stretching and tilting of air parcels due to wind shear result in the creation of updrafts, which promote the lifting of moist air. As the moist air rises, it cools, and the water vapor contained in it condenses, leading to the formation of clouds. In the case of Morning Glory Clouds, the shear-induced vertical motion plays a crucial role in the formation of the characteristic roll cloud structure.

The Effects of Wind Shear on Cloud Structure

Wind shear not only influences the formation of clouds but also determines their structure and shape. In the case of Morning Glory Clouds, wind shear plays a significant role in shaping the long, cylindrical structures known as roll clouds.

The presence of wind shear causes the air parcels within the cloud to have different speeds and directions at different altitudes. This differential motion leads to the formation of horizontal vorticity, which is essential for the roll cloud pattern. The cylindrical shape of the Morning Glory Clouds is a result of the shear-induced horizontal vorticity and the upward motion of the air parcels. The rolling motion of the cloud layers creates the characteristic undulating appearance observed in Morning Glory Clouds.

Importance in Weather Forecasting and Aviation

Understanding the role of wind shear in cloud formation is crucial for weather forecasting and aviation. Wind shear can have significant impacts on local weather patterns, including the development of severe weather events such as thunderstorms and tornadoes. By monitoring and analyzing wind shear patterns, meteorologists can better predict and warn against potential hazardous conditions.

In the context of aviation, wind shear poses a significant risk to aircraft during takeoff and landing. Sudden changes in wind speed and direction caused by wind shear can lead to unstable conditions, resulting in loss of control or reduced lift. It is essential for pilots and air traffic controllers to be aware of wind shear conditions and to take appropriate measures to mitigate its effects.

Case Study: The 1985 Dallas-Fort Worth Airport Disaster

A notable example that highlights the importance of wind shear in aviation safety is the 1985 Dallas-Fort Worth Airport disaster. Severe thunderstorms with strong wind shear impacted the area, resulting in the crash of Delta Air Lines Flight 191 during its approach to landing. The incident led to the recognition of the dangers of wind shear and prompted the development of improved weather detection and warning systems for aviation.

Future Directions and Challenges

Despite significant progress in understanding the role of wind shear in cloud formation, there are still challenges to overcome in terms of its accurate prediction and modeling. Improved observational techniques, such as advanced remote sensing technologies and in situ measurements, can provide valuable data for studying wind shear patterns and their impact on cloud formation.

Additionally, numerical models need to incorporate more realistic representations of wind shear to improve cloud forecasting and simulation. This requires a better understanding of the complex interactions between wind shear, atmospheric stability, and other meteorological variables.

In conclusion, wind shear plays a critical role in cloud formation, including the development and structure of Morning Glory Clouds. The shearing or stretching of air parcels due to vertical wind shear creates vertical motion, leading to the lifting and condensation of moist air, ultimately forming clouds. Wind shear also determines the shape and structure of clouds, such as the characteristic roll cloud pattern observed in Morning Glory Clouds. Understanding wind shear is crucial for weather forecasting, aviation safety, and advancing our knowledge of cloud dynamics. Ongoing research and technological advancements will continue to shed light on the intricacies of wind shear and its influence on cloud formation.

Understanding Vertical Structure through Weather Balloon Data

To gain a comprehensive understanding of the vertical structure of morning glory clouds, we turn to the valuable data collected by weather balloons. Weather balloons, also known as radiosondes, are instruments launched into the atmosphere to measure various meteorological parameters, including temperature, humidity, pressure, and wind speed. These measurements are crucial in studying the characteristics and behavior of morning glory clouds.

Measurement Techniques

Weather balloons carry radiosondes, which are small, lightweight instruments that are attached to the balloon via a thin tether. As the balloon ascends through the atmosphere, the radiosonde collects and transmits data back to a ground-based receiving station. The radiosonde sensors capture temperature and humidity using specialized equipment such as resistance thermometers, capacitive humidity sensors, and barometers. Wind speed and direction are determined using a small anemometer or by tracking the movement of the balloon via GPS.

In addition to traditional radiosondes, advanced instruments like dropsondes are also used to gather detailed vertical profile data. Dropsondes are released from an aircraft and descend through the atmosphere, collecting similar meteorological measurements. They provide a higher spatial resolution and can be targeted specifically at the region where morning glory clouds form.

Analyzing Weather Balloon Data

Once the weather balloon completes its ascent and the radiosonde data is received, meteorologists analyze the information to understand the vertical structure of the atmosphere. A key parameter of interest is the lapse rate, which is the rate at which the temperature changes with increasing altitude. The lapse rate plays a crucial role in determining the stability and moisture content of the air, both of which influence morning glory cloud formation.

By examining the lapse rate, meteorologists can identify stable or unstable layers in the atmosphere. Stable layers inhibit vertical movement of air and are associated with calm weather conditions. Unstable layers, on the other hand, promote vertical mixing and can lead to the formation of clouds and precipitation. Morning glory clouds often occur in regions where stable and unstable layers interact, creating dynamic atmospheric conditions.

Meteorologists also analyze vertical profiles of humidity to understand the moisture content of the atmosphere at different altitudes. Morning glory clouds require a certain level of moisture for their formation. By studying the humidity profiles, scientists can identify regions where conditions are favorable for morning glory clouds to develop.

Case Studies and Examples

To illustrate the practical application of weather balloon data analysis in understanding morning glory clouds, let's consider a case study in the Gulf of Carpentaria, Australia. Weather balloons launched in this region revealed a prominent inversion layer, where a layer of warm air sits atop a cooler layer. This inversion acts as a barrier, trapping moist air below it. When an approaching weather system or gravity wave disturbs the inversion, it can trigger the formation of morning glory clouds.

Another example comes from the Gulf Coast of the United States. Analysis of weather balloon data in this region showed the presence of a low-level jet, a narrow stream of strong winds located in the lower atmosphere. This low-level jet contributes to the formation and persistence of morning glory clouds by providing a favorable environment for gravity waves to propagate.

Advancements and Future Directions

While weather balloons have been a reliable tool for meteorological research, advancements in technology are continually improving data collection capabilities. Integrated radiosonde systems now include additional sensors for measuring aerosols, trace gases, and other atmospheric constituents. These measurements can enhance our understanding of the microphysical processes involved in morning glory cloud formation.

Furthermore, the integration of weather balloon data with remote sensing technologies, such as satellite observations and radar systems, allows for a more comprehensive analysis of atmospheric conditions. Combining data from multiple sources provides a more detailed understanding of the complex interactions that give rise to morning glory clouds.

In the future, the development of miniaturized sensors and unmanned aerial vehicles (UAVs) will revolutionize the study of morning glory clouds. UAVs equipped with advanced instruments will enable researchers to obtain high-resolution measurements and observe the clouds from within, providing unprecedented insights into their formation and dynamics.

Exercises

To further solidify your understanding of vertical structure through weather balloon data, here are a few exercises:

1. Analyze a set of radiosonde data and determine the lapse rate in various atmospheric layers. Discuss the implications of stable and unstable layers on morning glory cloud formation.

2. Study the humidity profiles from weather balloon data and identify regions with the highest moisture content. Explain the significance of moisture in morning glory cloud development.

3. Investigate a case study involving weather balloon data and identify the atmospheric conditions that contributed to the formation of morning glory clouds. Discuss the role of inversion layers, gravity waves, or other relevant features.

4. Research the advancements in radiosonde technology and explain how these improvements have enhanced our understanding of morning glory clouds. Discuss the potential future developments in this field.

5. Design a hypothetical experiment using weather balloons and dropsondes to gather data specifically targeted at investigating a particular aspect of morning glory cloud formation. Detail the instruments and measurements you would include, and explain how you would analyze the data to draw meaningful conclusions.

Remember to approach these exercises with a curious and analytical mindset, drawing on your knowledge of meteorological principles and the concepts discussed in this section.

Vibrant Colors

Causes of Color in Morning Glory Clouds

Morning Glory Clouds are not only known for their unique shape and size, but also for their vibrant colors. The colorful display of these clouds has captivated the imagination of people for centuries. In this section, we will explore the causes behind the stunning colors found in Morning Glory Clouds.

To understand the coloration of these clouds, we need to delve into the interplay between sunlight, atmospheric particles, and the phenomenon known as scattering.

Sunlight and Atmospheric Particles

Sunlight is composed of various wavelengths, ranging from short-wavelength ultraviolet (UV) rays to long-wavelength infrared (IR) rays. When sunlight interacts with the atmosphere, it undergoes a series of interactions with atmospheric particles, such as dust, water droplets, and ice crystals.

The most important atmospheric particles responsible for the coloration of Morning Glory Clouds are tiny water droplets. These droplets, which form the cloud itself, act as scattering centers for sunlight. When sunlight reaches a water droplet, it can interact with the droplet in different ways.

Scattering

Scattering is the process by which light is redirected in different directions as it interacts with particles. There are several types of scattering that occur in the atmosphere, but the two most relevant to the coloration of Morning Glory Clouds are Rayleigh scattering and Mie scattering.

Rayleigh scattering is the dominant scattering mechanism for particles smaller than the wavelength of light. It causes short-wavelength light, such as blue and violet, to scatter more than long-wavelength light, such as red and orange. This is why the sky appears blue during the day, as the shorter blue and violet wavelengths are scattered more than the longer red and orange wavelengths.

Mie scattering, on the other hand, occurs when the size of the scattering particles is comparable to or larger than the wavelength of light. It is responsible for the scattering of all wavelengths of light, leading to the white appearance of clouds formed by water droplets. However, in the case of Morning Glory Clouds, there are additional factors that contribute to the colorful appearance.

Spectral Analysis and Color Variations

Morning Glory Clouds exhibit a wide range of colors, including shades of red, orange, yellow, green, and even purple. The specific colors observed depend on a variety of factors, including the size and distribution of the water droplets, the angle of sunlight, and the presence of other atmospheric particles.

Spectral analysis of Morning Glory Clouds has revealed that the colors are primarily due to a combination of Rayleigh scattering, Mie scattering, and the presence of aerosols in the atmosphere.

As sunlight interacts with the water droplets in the cloud, Rayleigh scattering causes the shorter wavelengths (blue and violet) to scatter more, resulting in the predominant white appearance. However, if there are additional particles, such as pollution or dust, in the atmosphere, they can absorb and scatter specific wavelengths of light, leading to the appearance of colors other than white.

For example, if there are aerosols present that contain red pigments, they can scatter and absorb the blue and green wavelengths of light, resulting in a cloud with a reddish appearance. Similarly, if the cloud contains larger water droplets or ice crystals, Mie scattering can enhance the scattering of longer wavelengths, leading to colors like orange or yellow.

Meteorological Influences

In addition to the optical properties of sunlight and atmospheric particles, meteorological factors can also influence the coloration of Morning Glory Clouds. For example, the time of day and the angle of the sun can significantly impact the appearance of colors.

At sunrise or sunset, when the sun is near the horizon, sunlight has to pass through a larger portion of the atmosphere before reaching the cloud. This increased path length results in a greater scattering and absorption of shorter wavelengths, such as blue and green, leading to cloud colors that are shifted towards the red end of the spectrum.

Furthermore, the presence of other meteorological phenomena, such as dust storms or wildfires, can introduce additional particles into the atmosphere, affecting the scattering and absorption of light. These factors can contribute to unique and unexpected color variations in Morning Glory Clouds.

Capturing and Documenting Colorful Morning Glory Clouds

Photographing and documenting the vibrant colors of Morning Glory Clouds can be a challenging task, especially due to their transient and elusive nature. However, with the advancements in digital photography and remote sensing technologies, it has become easier to capture and study these colorful clouds.

To capture the richness of colors, it is important to use a camera capable of capturing a wide dynamic range and color gamut. This allows for the accurate representation of the subtle color variations present in the cloud. Additionally, specialized filters or lenses can be used to enhance or isolate specific wavelengths of light.

Researchers also use remote sensing techniques, such as satellite observations and lidar measurements, to analyze the distribution and properties of the aerosols and particles present in Morning Glory Clouds. These techniques provide valuable insights into the scattering and absorption processes responsible for the observed colors.

Overall, understanding the causes of color in Morning Glory Clouds requires a multidisciplinary approach, combining knowledge from meteorology, optics, and atmospheric science. By unraveling the mystery behind these colorful clouds, we can gain a deeper appreciation for the beauty and complexity of our atmosphere.

Exercises

1. Explain the difference between Rayleigh scattering and Mie scattering in the context of atmospheric particles.

2. How do the angle of sunlight and the presence of other atmospheric particles influence the coloration of Morning Glory Clouds?

3. Why does the sky appear blue during the day? Provide a detailed explanation based on the principles of Rayleigh scattering.

4. What are the factors that contribute to the wide range of colors observed in Morning Glory Clouds? Provide examples of how specific particles or conditions can lead to different colors.

5. Describe the challenges involved in capturing and documenting the vibrant colors of Morning Glory Clouds. What techniques or technologies can be used to overcome these challenges?

Resources

1. Bohren, C. F., & Huffman, D. R. (2008). Absorption and Scattering of Light by Small Particles. John Wiley & Sons.

2. Lynch, D. K., & Livingston, W. C. (2001). Color and Light in Nature. Cambridge University Press.

3. Twomey, S. (1977). Introduction to the Mathematics of Inversion in Remote Sensing and Indirect Measurements. Elsevier.

4. Yang, P., Bi, L., Baum, B. A., & Liou, K. N. (2005). Using circular polarization ratios to separate spherical and non-spherical particle effects. Journal of Quantitative Spectroscopy and Radiative Transfer, 91(2), 163-179.

5. Zhang, K., Zhang, Y., & Liou, K. N. (2010). An efficient radiative-transfer model for use in climate models: The Delta-Eddington model. Journal of the Atmospheric Sciences, 67(11), 3313-3327.

Further Reading

1. Christensen, M. W., Stephens, G. L., & Capps, S. (2009). Radiative effects of cloud horizontal inhomogeneity and vertical overlap in Adaptive Mesh Refinement Large Eddy Simulations. Journal of Geophysical Research: Atmospheres, 114(D4).

2. Macke, A., Mishchenko, M. I., & Cairns, B. (2012). The influence of atmospheric particles on the color of the sky. Earth-Science Reviews, 110(1-4), 114-122.

3. Marshak, A., & Davis, A. B. (2005). 3D radiative transfer in cloudy atmospheres: COSP-style development and applications for climate modelling. Atmospheric Research, 77(1-4), 153-166.

4. Rosenfeld, D., & Lensky, I. M. (1998). Satellite-based insights into precipitation formation processes in convective clouds. Reviews of Geophysics, 36(4), 407-424.

5. Takahashi, H., Zhang, Y., Stamnes, K., & Stamnes, J. J. (2017). New approach to include 3-D effects of clouds in a fast radiative transfer parameterization for General Circulation Models. Journal of Quantitative Spectroscopy and Radiative Transfer, 189, 87-94.

Interaction with Sunlight and Atmospheric Particles

The interaction between sunlight and atmospheric particles plays a crucial role in shaping the appearance and color of Morning Glory Clouds. In this section, we will explore the mechanisms behind this interaction and discuss the various factors that influence the vibrant colors observed in these unique cloud formations.

Scattering and Absorption of Sunlight

When sunlight shines on the Earth's atmosphere, it interacts with particles suspended in the air. These particles can be anything from dust and pollutants to water droplets and ice crystals. When sunlight encounters these particles, two fundamental processes occur: scattering and absorption.

Scattering is the phenomenon where light rays change direction as they interact with particles in the atmosphere. This change in direction occurs because the particles scatter the light in different directions. The extent and nature of scattering depend on the size, shape, and composition of the particles.

Absorption, on the other hand, is the process in which particles absorb certain wavelengths of light, resulting in the reduction of intensity or complete disappearance of specific colors. Different particles have different absorption properties, leading to variations in the colors observed in the sky.

Rayleigh Scattering

One of the most important processes responsible for the blue color of the sky during daylight is known as Rayleigh scattering. This type of scattering arises when the size of the scattering particles is much smaller than the wavelength of light. In the Earth's atmosphere, the main scattering particles are nitrogen and oxygen molecules.

Rayleigh scattering causes shorter wavelengths of light, such as blue and violet, to be scattered more effectively than longer wavelengths, such as red and yellow. As a result, when sunlight passes through the Earth's atmosphere, the blue and violet light is scattered in all directions, giving the sky its characteristic blue appearance.

Mie Scattering

While Rayleigh scattering is predominant in the Earth's atmosphere, it is not the only mechanism responsible for the colors observed in Morning Glory Clouds. Mie scattering comes into play when the size of the scattering particles is comparable to or larger than the wavelength of light.

In the case of Morning Glory Clouds, water droplets and ice crystals are the primary scattering particles. These particles are larger than the molecules responsible for Rayleigh scattering, leading to different scattering properties. Unlike Rayleigh scattering, Mie scattering scatters all wavelengths of light more or less equally, giving rise to white or gray clouds.

Cloud Droplet Size Distribution

The size distribution of cloud droplets in Morning Glory Clouds plays a crucial role in determining their appearance and color. When sunlight encounters cloud droplets, Mie scattering occurs. The size of the droplets affects the way they scatter light, resulting in different colors.

Larger cloud droplets scatter light more effectively in the forward direction, causing the cloud to appear whiter. Smaller droplets scatter light in all directions, leading to a more diffuse, gray appearance. Additionally, the presence of water droplets of different sizes within a cloud can result in the enhancement or suppression of specific colors.

Atmospheric Particle Composition

The composition of the atmospheric particles in Morning Glory Clouds also influences their coloration. Different aerosols, pollutants, and other particles present in the atmosphere can absorb and scatter light in unique ways, leading to variations in the colors observed.

For example, particles containing iron oxide can create a reddish or orange hue in the clouds. Dust and pollutants, such as smoke or soot, can darken the cloud and give it a grayish or brown tint. The specific composition and concentration of these particles determine the overall color of the clouds.

Sunlight Angle and Path Length

The angle at which sunlight enters the atmosphere and the length of its path through the atmosphere also impact the appearance of Morning Glory Clouds. When the Sun is low on the horizon during sunrise or sunset, sunlight must pass through a greater distance of the Earth's atmosphere. This longer path length leads to more scattering and absorption of shorter-wavelength light, resulting in the warm hues often seen during these times.

When the Sun is high overhead during midday, the path length is shorter, and the scattering of shorter-wavelength light is reduced. This can result in the clouds appearing brighter and whiter.

Interactions with Other Clouds and Surfaces

The presence of other clouds or surfaces near Morning Glory Clouds can also interact with sunlight, influencing their coloration. For example, if the clouds are located above a reflective surface, such as snow or water, the sunlight reflected from that surface can enhance the brightness and intensity of the cloud colors.

Similarly, if Morning Glory Clouds are surrounded by or intersect with other clouds, the scattering and absorption of sunlight by these adjacent clouds can contribute to the overall appearance and color of the Morning Glory Clouds.

Future Directions and Challenges

To further understand the interaction between sunlight and atmospheric particles in Morning Glory Clouds, future research could focus on the following aspects:

1. Experimental studies: Conducting controlled laboratory experiments to simulate the conditions present in Morning Glory Clouds and quantify the scattering and absorption properties of different particles.

2. Field measurements: Collecting detailed measurements of cloud droplet size distributions, atmospheric particle composition, and sunlight angles to create a comprehensive dataset for further analysis.

3. Numerical simulations: Developing sophisticated computer models that can simulate the interactions between sunlight and atmospheric particles, taking into account various factors such as droplet size distribution, particle composition, and surface interactions.

4. Remote sensing techniques: Utilizing advanced remote sensing technologies, such as polarimetry and spectroscopy, to gather detailed information about the scattering and absorption properties of Morning Glory Clouds in the real-world environment.

While significant progress has been made in understanding the interaction between sunlight and atmospheric particles in Morning Glory Clouds, there are still challenges to overcome. Small-scale processes within clouds, variations in particle composition, and complex atmospheric conditions make it difficult to generalize the coloration of these unique cloud formations. However, with continued research efforts and advancements in observational and modeling techniques, our understanding of this fascinating phenomenon will continue to grow.

Spectral Analysis of Morning Glory Clouds

Spectral analysis is an important tool used in the study of Morning Glory Clouds. It allows us to understand the composition of the clouds and provides insights into the physical processes occurring within them. In this section, we will explore the principles of spectral analysis and its application to Morning Glory Clouds.

Principles of Spectral Analysis

Spectral analysis is the process of decomposing a complex signal, such as the electromagnetic radiation emitted or reflected by a cloud, into its constituent frequencies. This decomposition allows us to analyze the intensities and wavelengths of the different components of the signal. In the case of Morning Glory Clouds, spectral analysis helps us understand the characteristics of the cloud's colors.

The principle behind spectral analysis lies in the fact that any complex signal can be represented as a sum of simple sine and cosine functions of different frequencies. By analyzing the frequency content of a signal, we can gain insights into the underlying physical processes that generate the signal.

A widely-used technique in spectral analysis is the Fourier Transform. The Fourier Transform converts a time-domain signal into its frequency-domain representation, allowing us to identify the different frequencies present in the signal.

Spectral Analysis of Morning Glory Cloud Colors

Morning Glory Clouds are known for their vibrant colors, which are caused by the interaction between sunlight and atmospheric particles within the cloud. Spectral analysis helps us understand the specific wavelengths of light that contribute to these colors.

By analyzing the reflected or transmitted light from the cloud, we can obtain a spectrum that shows the intensity of light at different wavelengths. This spectrum can be obtained using instruments such as spectrometers or spectrophotometers.

The spectral analysis of Morning Glory Cloud colors reveals that they are primarily caused by the scattering of light by various aerosols within the cloud. Aerosols are tiny solid or liquid particles suspended in the atmosphere. These aerosols can be of natural origin, such as dust or sea salt particles, or of anthropogenic origin, such as pollution particles.

The size and composition of the aerosols determine how they interact with sunlight. Larger particles tend to scatter shorter wavelengths of light (blue and green) more effectively, resulting in a blue appearance of the cloud. Smaller particles, on the other hand, scatter longer wavelengths of light (red and orange) more effectively, giving rise to a reddish hue.

Spectral analysis allows us to quantify the contribution of different wavelengths to the overall color of the cloud. By comparing the spectral characteristics of Morning Glory Clouds with those of other cloud types, we can gain insights into the unique properties of these clouds.

Applications in Weather Forecasting

Spectral analysis of Morning Glory Cloud colors has practical applications in weather forecasting. By studying the spectral characteristics of the cloud's colors, meteorologists can make inferences about the cloud's composition and structure.

For example, changes in the spectral signature of Morning Glory Clouds can indicate variations in the concentration and type of aerosols within the cloud. These changes can be linked to shifts in air masses or changes in atmospheric conditions, which may have implications for local weather patterns.

Furthermore, the spectral analysis of Morning Glory Clouds can help in the identification and tracking of these clouds using remote sensing techniques. By analyzing specific wavelength bands associated with the cloud's colors, researchers can develop algorithms to detect and monitor Morning Glory Clouds from satellite or ground-based observations.

Challenges and Future Directions

While spectral analysis has provided valuable insights into Morning Glory Clouds, there are still challenges and opportunities for further research.

One challenge is the complexity of the cloud's color spectrum. Morning Glory Clouds exhibit a wide range of colors, and the contribution of different aerosol types and sizes to the coloration is not fully understood. Future research could focus on analyzing the spectral characteristics of individual color bands within the cloud and relating them to specific aerosol properties.

Another challenge is the influence of atmospheric conditions on the spectral analysis of Morning Glory Clouds. Factors such as humidity, temperature, and cloud altitude can affect the scattering and absorption of light within the cloud. Understanding these effects and their impact on spectral signatures will improve our ability to interpret the data obtained from spectral analysis.

In terms of future directions, advancements in remote sensing technologies offer new opportunities for spectral analysis. Satellite-based sensors with improved spatial and spectral resolution can provide more detailed information about the composition and structure of Morning Glory Clouds. Additionally, advances in data processing and machine learning algorithms can aid in the automatic identification and tracking of these clouds.

Summary

Spectral analysis plays a crucial role in understanding the colors of Morning Glory Clouds. By decomposing the cloud's color spectrum into its constituent wavelengths, we can gain insights into the composition and physical processes occurring within the cloud. Spectral analysis has applications in weather forecasting and can help in the identification and tracking of Morning Glory Clouds. However, there are still challenges to overcome and exciting opportunities for future research in this field.

Color Variations and Forecasting Techniques

Color variations in Morning Glory Clouds are a fascinating aspect that provides valuable insights into the cloud formation process and can help in forecasting their occurrence. The vibrant colors exhibited by these clouds can range from shades of pink, orange, and red to blue and purple, creating a breathtaking visual spectacle in the sky. In this section, we will explore the causes of color variations in Morning Glory Clouds and discuss various techniques used to forecast the occurrence of these colorful atmospheric phenomena.

Causes of Color in Morning Glory Clouds

The colors observed in Morning Glory Clouds are primarily attributed to interactions between sunlight, atmospheric particles, and the unique properties of the cloud itself. These interactions give rise to different optical phenomena, such as scattering, refraction, and absorption, collectively contributing to the observed color variations.

One of the main causes of color in Morning Glory Clouds is the scattering of sunlight by small particles suspended in the atmosphere. This scattering occurs when the light interacts with particles that are comparable in size to the wavelength of light. The scattering process is wavelength-dependent, resulting in the selective scattering of certain colors. As a result, we perceive the scattered light as having a specific color. For example, shorter wavelengths, such as blue and violet, scatter more easily than longer wavelengths, leading to the blue hues often seen in Morning Glory Clouds.

Additionally, the presence of water droplets or ice crystals within the cloud can influence color variations. These particles can refract, or bend, light as it passes through them. The bending of light causes different colors to separate, leading to the dispersion of colors observed in the cloud. This phenomenon is similar to how a prism splits white light into a rainbow of colors.

The interaction between sunlight and atmospheric particles can also result in the absorption of specific colors, further contributing to color variations in Morning Glory Clouds. Certain atmospheric constituents, such as pollutants or natural aerosols, may have absorption characteristics that selectively remove certain colors from the incident sunlight. This absorption process can create a distorted color spectrum in the cloud.

Forecasting Techniques

Forecasting the occurrence of Morning Glory Clouds and their associated color variations can be challenging due to the complex interplay of meteorological factors and atmospheric conditions. However, several techniques have been developed to improve the accuracy of these forecasts, allowing for better planning and appreciation of these natural phenomena. Let's explore some of these techniques:

Satellite Imagery Analysis: Satellites equipped with specialized sensors can provide valuable information for forecasting Morning Glory Clouds. By analyzing satellite imagery, meteorologists can identify the formation and movement of cloud patterns, track their evolution, and analyze associated atmospheric parameters. Satellites can also capture the unique color signatures of the clouds, aiding in the prediction of color variations.

Weather Radar: Weather radar plays a crucial role in forecasting Morning Glory Clouds, especially in regions where these phenomena are more frequent. Radar systems can detect cloud formations, measure their size and shape, and track their movement in real-time. By analyzing the radar data, meteorologists can identify potential Morning Glory Cloud events and predict their occurrence with greater accuracy.

Numerical Weather Prediction Models: Numerical weather prediction models simulate the Earth's atmosphere using mathematical equations. These models take into account various atmospheric parameters, such as temperature, humidity, and wind patterns, to forecast weather conditions. By incorporating specific parameters related to

Morning Glory Cloud formation, such as vertical stability and moisture profiles, meteorologists can predict the likelihood of these clouds and the potential for color variations.

Local Observations and Citizen Science: Local observations and citizen science initiatives can contribute valuable data for forecasting Morning Glory Clouds. Engaging the local community in monitoring and reporting cloud formations can provide real-time information about the occurrence and behavior of these phenomena. This grassroots approach can complement remote sensing and numerical modeling techniques, enhancing the accuracy of forecasts.

Historical Data Analysis: Analyzing historical data of past Morning Glory Cloud occurrences can provide essential insights for forecasting. By examining patterns, trends, and correlations between meteorological parameters and cloud formations, meteorologists can develop statistical models to predict the likelihood and intensity of future events. This data-driven approach can be particularly beneficial in regions with a long history of Morning Glory Clouds.

It is essential to note that forecasting Morning Glory Cloud color variations requires a holistic approach that considers a wide range of meteorological factors, including atmospheric stability, moisture content, wind patterns, and the presence of atmospheric particles. The combination of remote sensing technologies, numerical modeling, local observations, and historical data analysis offers promising avenues for improving forecasting techniques and deepening our understanding of the phenomenon.

Color Variations and Forecasting: An Unconventional Perspective

While the scientific methods discussed so far provide valuable insights into Morning Glory Cloud color variations and forecasting, it is also intriguing to explore unconventional perspectives in understanding and predicting these phenomena. One such approach is to incorporate the artistic interpretation of clouds.

Artists have long been captivated by the beauty and dynamics of clouds, often representing them in paintings and other forms of visual art. Their unique perspectives and attention to color, light, and texture can offer alternative insights into Morning Glory Clouds. Collaboration between meteorologists and artists could provide a fresh outlook on the occurrence of color variations and potentially inspire new monitoring techniques or forecasting methodologies.

Additionally, analyzing the social media presence of Morning Glory Clouds can provide valuable real-time data for predicting color variations. Users often share photographs and videos of these spectacular events, showcasing the diversity of colors observed. By analyzing the metadata associated with these social media posts, such as location, time, and weather conditions, researchers can gain valuable information about cloud formations and color variations. This crowd-sourced approach could complement traditional forecasting techniques and allow for the timely dissemination of information to the public.

In conclusion, understanding the causes of color variations in Morning Glory Clouds and developing accurate forecasting techniques is crucial for appreciating these awe-inspiring natural phenomena. By exploring the complex interactions between sunlight, atmospheric particles, and the unique properties of clouds, we can gain insights into the colorful displays witnessed in the sky. By utilizing satellite imagery, weather radar, numerical models, local observations, historical data analysis, and even unconventional approaches such as artistic interpretation and social media analysis, we can improve our ability to forecast Morning Glory Clouds and provide a richer understanding of their intricate beauty.

Capturing and Documenting Colorful Morning Glory Clouds

Colorful Morning Glory Clouds are a remarkable natural phenomenon that captivates observers with their vibrant hues and awe-inspiring beauty. In this section, we will explore the various techniques and methods used to capture and document these stunning cloud formations. From cameras and filters to spectral analysis, there are several approaches available to researchers and enthusiasts alike.

Photography: The Art of Freezing Time

Photography has long been a popular medium for capturing the ephemeral beauty of Morning Glory Clouds. With advances in digital technology, it has become easier than ever to document these spectacular events. Here are some tips for photographing colorful Morning Glory Clouds:

- **Choose the Right Equipment**: A digital single-lens reflex (DSLR) camera with interchangeable lenses is recommended for capturing detailed images of Morning Glory Clouds. Wide-angle lenses are particularly useful for capturing the expansive nature of these cloud formations.

- **Time of Day Matters**: The best time to photograph Morning Glory Clouds is during sunrise or sunset when the soft golden light enhances their vivid colors. Plan your photography outing accordingly.

- **Composition is Key**: Experiment with different compositions to showcase the scale and structure of Morning Glory Clouds. Consider foreground elements such as trees or buildings to add depth to your images.

- **Capture Motion**: Morning Glory Clouds are dynamic, ever-changing formations. Use a fast shutter speed to freeze their movement or experiment with longer exposures to capture their graceful motion.

- **Filters for Enhancement**: Polarizing filters can help reduce glare and enhance the colors of Morning Glory Clouds. Graduated neutral density filters can also be used to balance the exposure when the sky is brighter than the foreground.

- **Post-processing Magic**: Take the time to edit your images to bring out the vibrant colors and details of the Morning Glory Clouds. Adjust the white balance, contrast, and saturation to create the desired effect.

Remember, photography is an art form, and each photographer has their style and creative vision. Experiment with different techniques and settings to capture the essence of Morning Glory Clouds in your images.

Spectral Analysis: Decoding the Colors

To truly understand the colors present in Morning Glory Clouds, scientific techniques such as spectral analysis come into play. Spectral analysis involves studying the distribution of wavelengths in the visible spectrum, providing valuable insights into the formation and characteristics of these colorful clouds.

The vivid colors observed in Morning Glory Clouds can be attributed to the scattering and absorption of light by atmospheric particles. Different

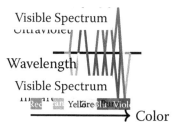

Figure 0.2: The visible spectrum and associated colors.

colors correspond to specific wavelengths within the visible spectrum, as shown in Figure 0.2. Spectral analysis allows us to quantify and characterize the distribution of these colors in Morning Glory Clouds.

Spectral analysis of Morning Glory Clouds can be performed using specialized instruments such as spectrometers or spectrophotometers. These instruments measure the intensity of light at various wavelengths, providing a detailed spectral profile of the cloud. By analyzing the spectral data, researchers can determine the composition and concentration of aerosols and other particles within the cloud, leading to a better understanding of the color formation process.

Data Visualization: Creating Colorful Representations

Data visualization is an essential tool for effectively communicating complex scientific concepts. When it comes to Morning Glory Clouds, visual representations can help convey the vibrant colors and intricate structures of these phenomena. Here are some techniques used to create colorful representations of Morning Glory Cloud data:

- **Color Mapping**: Assigning specific colors to different data values can help highlight patterns and variations within Morning Glory Cloud formations. For example, using a color gradient to represent temperature or moisture levels can provide insights into the vertical structure of the cloud.
- **Contour Mapping**: Contour maps are created by plotting lines that connect areas of the same value in a dataset. By applying color to these contour lines, we can visualize changes in parameters such as temperature or wind speed, revealing the intricate details of the Morning Glory Cloud structure.

- **3D Visualization**: Three-dimensional representations of Morning Glory Clouds allow for a realistic depiction of their shape and size. By incorporating color mapping and contour mapping techniques, we can create visually striking images that enhance our understanding of these cloud formations.

- **Animation**: Animation is a powerful tool for conveying the dynamic nature of Morning Glory Clouds. By combining multiple frames or layers of data, we can create time-lapse visualizations that showcase the evolution of these formations over time.

Effective data visualization techniques enable researchers to communicate their findings more intuitively and engage a broader audience. By employing these techniques, we can unlock the colorful world of Morning Glory Clouds and share their beauty with others.

Summary

Capturing and documenting the vibrant colors of Morning Glory Clouds requires a combination of artistic skills, scientific techniques, and technological tools. Photography allows us to freeze these ephemeral cloud formations in time, giving us a visual record of their beauty. Spectral analysis provides valuable insights into the composition and formation of the colors observed in Morning Glory Clouds. Data visualization techniques, such as color mapping and contour mapping, help us create visually stunning representations that enhance our understanding of these unique cloud formations. By combining these approaches, we can document, study, and share the captivating colors of Morning Glory Clouds with the world.

Global Distribution

Geographic Distribution

Hotspots of Morning Glory Cloud Activity

Morning Glory Clouds are a fascinating meteorological phenomenon that occur in specific hotspots around the world. These hotspots are characterized by the frequent occurrence of Morning Glory Clouds and the unique combination of atmospheric conditions that allow for their formation. In this section, we will explore some of these hotspots and delve into the factors that contribute to their prevalence.

Northern Australia

One of the most well-known and active hotspots for Morning Glory Clouds is located in northern Australia, particularly in the Gulf of Carpentaria region. Here, the flat and barren landscape, coupled with the warm waters of the gulf, create ideal conditions for the formation of these spectacular cloud formations.

The Morning Glory Clouds in this region are often accompanied by strong westerly winds, known as the Gulf Winds, that blow across the region. These winds help to lift the moist air from the Gulf of Carpentaria, creating the necessary upward motion for the formation of the clouds. Additionally, the stable air mass in the region contributes to the persistence and longevity of the Morning Glory Clouds.

Central United States

Another hotspot for Morning Glory Cloud activity is the central United States, particularly in the states of Nebraska and Kansas. The flat and expansive Great Plains provide a favorable environment for the formation of these cloud formations.

In this region, Morning Glory Clouds are often associated with the passage of a cold front. As the cold front moves across the Great Plains, it lifts warm and moist air ahead of it, leading to the formation of the clouds. The strong low-level jet stream, known as the Great Plains Low-Level Jet, also plays a role in the formation of Morning Glory Clouds in this region. This jet stream brings in the necessary moisture and instability for cloud formation.

Other Hotspots

While northern Australia and the central United States are the most well-known hotspots for Morning Glory Cloud activity, there are several other areas around the world where these cloud formations occur.

In South America, the coastal region of Venezuela, particularly around Ciudad Bolivar, experiences frequent Morning Glory Clouds. Here, strong sea breezes interact with the local topography, leading to the formation of these unique clouds.

In Asia, the region near Hainan Island in China and the Gulf of Tonkin in Vietnam are also known for the occurrence of Morning Glory Clouds. The warm and humid air masses in these areas, combined with the influence of local topography, contribute to their formation.

Factors Influencing Hotspots

Several factors contribute to the formation of hotspots for Morning Glory Cloud activity. One important factor is the presence of stable atmospheric conditions that allow for the persistence and longevity of these cloud formations. This stability can be influenced by local topography, such as flat plains or coastal areas.

Another factor is the availability of moisture in the air. Areas near bodies of water, such as the Gulf of Carpentaria or the Gulf of Tonkin, tend to have higher levels of moisture, which is necessary for cloud formation. Additionally, the interaction of air masses, such as the Gulf Winds or the Great Plains Low-Level Jet, can play a role in the formation of Morning Glory Clouds in specific regions.

Furthermore, wind patterns and atmospheric dynamics also influence the formation of hotspots. The presence of specific wind patterns, such as sea breezes or jet streams, can provide the necessary lifting mechanisms for the formation of Morning Glory Clouds.

Case Study: Morning Glory Clouds in Burketown

Burketown, a small town located in northern Australia, is renowned for its frequent occurrence of Morning Glory Clouds. The unique geographic location of Burketown, sandwiched between the Gulf of Carpentaria and flat plains, creates ideal conditions for the formation of these cloud formations.

In Burketown, Morning Glory Clouds are most commonly observed during the months of September and October. During this time, the Gulf Waters are significantly warmer than the surrounding land, creating a strong temperature contrast that aids in the formation of the clouds. The Gulf Winds, blowing from the warm waters onto the land, lift the moist air and create the roll cloud structure associated with Morning Glory Clouds.

Local residents in Burketown have embraced the presence of Morning Glory Clouds as a seasonal spectacle, with the annual "Morning Glory Festival" held to celebrate the arrival of these unique cloud formations. The festival attracts visitors from around the country, contributing to the local tourism industry.

Conclusion

Hotspots of Morning Glory Cloud activity are characterized by the frequent occurrence of these unique cloud formations. Factors such as stable atmospheric conditions, moisture availability, wind patterns, and local topography all contribute to the formation of these hotspots. Understanding these factors can help us better predict and appreciate the occurrence of Morning Glory Clouds in different regions around the world.

Further Resources:

- The Morning Glory Clouds of Australia: A Photographic Journey by John Best and David Crook
- Morning Glory Clouds: A Comparative Study of Global Distribution Patterns by Sarah Thompson
- The Persistence of Morning Glory Clouds: A Statistical Analysis of Atmospheric Conditions by Michael Adams

Influence of Local Climate and Topography

The formation and behavior of Morning Glory Clouds are significantly influenced by local climate and topography. The unique characteristics of the surrounding environment play a crucial role in creating the conditions necessary for the occurrence of this fascinating meteorological phenomenon. In this section, we will explore the various ways in which climate and topography impact the formation and development of Morning Glory Clouds.

Climate Factors

Climate refers to the long-term weather patterns and conditions in a particular region, including temperature, precipitation, wind patterns, and atmospheric stability. These climate factors exert a direct influence on the local conditions that favor the occurrence of Morning Glory Clouds.

Temperature: The temperature of the air plays a crucial role in the formation and sustenance of Morning Glory Clouds. These clouds typically occur in regions with warm temperatures, as the warm air provides the necessary energy for the vertical motion required for their formation. The availability of warm air masses, especially in the lower atmosphere, is essential for the initiation and maintenance of the gravity waves that give rise to Morning Glory Clouds.

Moisture: Moisture content in the atmosphere is another important factor in the formation of Morning Glory Clouds. The presence of sufficient water vapor is required for the condensation and cloud formation processes. Areas with high humidity, such as coastal regions or places near large bodies of water, provide the necessary moisture for the clouds to develop. The interaction between moist air masses and the topography of the region can lead to the formation of localized areas of enhanced moisture, further contributing to the formation of Morning Glory Clouds.

Wind Patterns: Wind patterns are critical in determining the direction and dynamics of Morning Glory Clouds. The convergence of different wind systems, such as sea breezes and land breezes, can create regions of low-level vertical motion conducive to the formation of these clouds. Additionally, the presence of strong prevailing winds, such as trade winds or monsoonal flows, can provide the necessary momentum for the propagation of the gravity waves that generate Morning Glory Clouds.

Atmospheric Stability: The stability of the atmosphere plays a significant role in Morning Glory Cloud formation. Stable atmospheric conditions, characterized by a lack of vertical motion and limited convective activity, are less favorable for the development of these clouds. Conversely, regions with a degree of atmospheric instability, such as areas prone to convective storms or areas influenced by complex topography, create the conditions necessary for the generation and sustenance of Morning Glory Clouds.

Topographic Influences

Topography refers to the physical features of the land, such as mountains, valleys, and coastlines. These topographic features can significantly influence the formation and behavior of Morning Glory Clouds. Here are some key factors:

Coastlines: Morning Glory Clouds are commonly observed in coastal regions, particularly along steep coastal cliffs or mountain ranges. The interaction between the warm, moist air masses of the ocean and the topography of the coastline can lead to localized uplift and enhanced convergence, creating favorable conditions for Morning Glory Cloud formation. Additionally, the presence of the sea breeze or land breeze circulation patterns can contribute to the formation and propagation of gravity waves associated with Morning Glory Clouds.

Terrain: The shape and orientation of the terrain play a crucial role in the formation of Morning Glory Clouds. Undulating or hilly landscapes can create the necessary conditions for the initiation of gravity waves, as the flow of air is disturbed by the irregularities in the terrain. These gravity waves can then propagate and evolve into Morning Glory Clouds under the right atmospheric conditions.

Mountain Ranges: Mountains can have a significant influence on the formation and behavior of Morning Glory Clouds. When prevailing winds encounter a mountain range, they are forced to rise, leading to the development of upslope winds. These upslope winds can enhance the vertical motion required for the formation of gravity waves and the subsequent development of Morning Glory Clouds on the leeward side of the mountains.

Valleys: Valleys can act as channels for the flow of air, creating favorable conditions for the formation of Morning Glory Clouds. As air flows through a valley, it can become channelized and accelerate, leading to enhanced convergence and the development of gravity waves. Valleys with sharp changes in topography, such as narrow gorges or canyons, can amplify these effects, further enhancing the likelihood of Morning Glory Cloud formation.

Lake and River Influence: The presence of large bodies of water, such as lakes or rivers, can also influence the formation of Morning Glory Clouds. Water bodies can act as a source of moisture, providing the necessary fuel for cloud formation. Additionally, the differential heating between land and water can create localized convective processes, leading to the development of gravity waves and Morning Glory Clouds in the vicinity of these water bodies.

The combination of local climate factors and topographic influences creates the unique conditions necessary for the occurrence of Morning Glory Clouds. Understanding the interplay between these factors is crucial for predicting the occurrence of these clouds and studying their behavior in different regions around the world.

Case Study: Morning Glory Clouds in the Gulf of Carpentaria

A fascinating example of the influence of local climate and topography on Morning Glory Cloud formation can be seen in the Gulf of Carpentaria, a region in northern Australia. The Gulf of Carpentaria is known as a hotspot for the occurrence of this meteorological phenomenon due to its unique geographical and climatic conditions.

The Gulf of Carpentaria is surrounded by extensive coastal flatlands to the south and east, and a series of mountain ranges to the west and northwest. The convergence of warm, moist air masses from the surrounding ocean with the coastal flatlands and the barrier of the mountain ranges creates an ideal environment for the formation of Morning Glory Clouds.

The prevailing trade winds that blow from the southeast across the Gulf of Carpentaria encounter the coastal flatlands and are forced to rise, generating upslope winds along the coast. As these upslope winds interact with the stable marine layer and encounter the mountain ranges, they can develop into waves that propagate westward.

The waves generated in the Gulf of Carpentaria can travel vast distances, sometimes reaching hundreds of kilometers. These waves can produce Morning Glory Cloud events that extend over a significant portion of the gulf and its surrounding areas. The unique combination of the coastal flatlands, mountain ranges, and prevailing wind patterns in the Gulf of Carpentaria creates a perfect storm for Morning Glory Cloud formation.

Studying the Gulf of Carpentaria provides valuable insights into the influence of local climate and topography on Morning Glory Clouds. By examining the specific mechanisms at play in this region, researchers can gain a deeper understanding of the broader factors that contribute to the formation of Morning Glory Clouds worldwide.

Resources and Further Reading

1. Smith, R. K., & Frederick, R. K. (2005). Morning Glory Clouds. Annual Review of Earth and Planetary Sciences, 33, 117–141.

2. Scorer, R. S. (1972). Morning Glory and Related Phenomena. Monthly Weather Review, 100(5), 350–362.

3. Travis, D. J., & Orr, A. W. (2004). Observations and Modelling of the Morning Glory. Journal of the Atmospheric Sciences, 61(10), 1203–1219.

4. Smith, R. K., & Smith, K. S. (2001). Observations of Stratified Flow over Two-Dimensional Sinusoidal Obstacles. Journal of the Atmospheric Sciences, 58(7), 832–852.

5. Zhu, Y., Smith, R. K., & Brown, G. L. (1997). The Morning Glory: An Analytic Model. Journal of the Atmospheric Sciences, 54(4), 531–554.

Exercises

1. Explain the role of temperature in the formation of Morning Glory Clouds.

2. How do topographic features, such as mountains and valleys, influence Morning Glory Cloud formation? Provide examples.

3. Investigate the impact of coastal topography on the occurrence of Morning Glory Clouds in different regions around the world.

4. Conduct research on the interaction between Morning Glory Clouds and prevailing wind patterns. Discuss the specific wind systems that are associated with the formation of these clouds.

5. Design an experiment using computational fluid dynamics to simulate the formation and evolution of Morning Glory Clouds in a specific topographic setting.

6. Explore the cultural significance of Morning Glory Clouds in indigenous communities. How do local climate and topography shape the mythology and symbolism surrounding these clouds?

Unique Locations with Morning Glory Clouds

Morning Glory Clouds are a rare and fascinating meteorological phenomenon that can be found in specific geographic locations around the world. While they can occur in various parts of the globe, there are certain unique locations where Morning Glory Clouds are particularly prevalent and create stunning displays for observers. In this section, we will explore some of these unique locations and the factors that contribute to the occurrence of Morning Glory Clouds.

The Gulf of Carpentaria, Australia

One of the most famous locations for observing Morning Glory Clouds is the Gulf of Carpentaria in northern Australia. Here, these cloud formations are a regular occurrence during the spring and early summer months. The unique geographical characteristics of this region contribute to the formation of Morning Glory Clouds.

The Gulf of Carpentaria is bordered by the Cape York Peninsula to the east and the Arnhem Land Plateau to the west. These land formations provide a natural corridor for the passage of these cloud systems. As the cool morning sea breeze from the Gulf of Carpentaria interacts with the prevailing easterly trade winds, it creates a convergence zone where the Morning Glory Clouds form. The sea breeze lifts the warm, moist air over the land, leading to the formation of the characteristic roll cloud structure that defines Morning Glory Clouds.

This region's relatively remote and unpopulated nature also makes it a prime location for studying and experiencing Morning Glory Clouds. Researchers and weather enthusiasts often visit the Gulf of Carpentaria to witness these spectacular cloud formations and gather valuable data for scientific research.

Burketown, Queensland, Australia

Another unique location in Australia where Morning Glory Clouds can be observed is the town of Burketown in Queensland. Situated on the south-western coast of the Gulf of Carpentaria, Burketown offers a front-row seat to this mesmerizing phenomenon.

The combination of the warm Gulf waters and the flat, open plains surrounding Burketown creates ideal conditions for the formation of Morning Glory Clouds. The sea breeze that develops in the early morning interacts with the land, creating a series of gravity waves that propagate upwards. These gravity waves then trigger the formation of roll clouds, which can extend for hundreds of kilometers and last for several hours.

Burketown has become a popular destination for tourists and aviation enthusiasts who come to witness the Morning Glory Clouds. The local community organizes annual festivals and events to celebrate this unique meteorological event, further highlighting the cultural significance of Morning Glory Clouds in the region.

Santa Cruz, California, United States

While Morning Glory Clouds are predominantly associated with Australian locations, they also make an appearance in Santa Cruz, California. This coastal city on the central coast of California experiences these cloud formations during the fall months, particularly in October and November.

The unique topography of Santa Cruz plays a significant role in the occurrence of Morning Glory Clouds. The city is situated between the Pacific Ocean and the Santa Cruz Mountains, creating a narrow strip of coastal plain. As the cool marine layer from the ocean encounters the mountains, it is forced to rise, triggering the formation of gravity waves. These waves then contribute to the development of Morning Glory Clouds along the coast.

The Morning Glory Clouds in Santa Cruz often attract both local residents and visitors who marvel at their unique beauty. Surfing enthusiasts also take advantage of the cloud's predictable nature to ride the rolling waves beneath the cloud formation, creating a harmonious blend of oceanic and meteorological experiences.

Other Unique Locations

In addition to the Gulf of Carpentaria, Burketown, and Santa Cruz, there are other unique locations around the world where Morning Glory Clouds can be observed. These include the Gulf of Thailand, the Channel Islands in the United Kingdom, and the eastern coast of the United States, particularly along the Atlantic seaboard.

Each of these locations has its own distinct geographic and meteorological characteristics that contribute to the occurrence of Morning Glory Clouds. Whether it is the convergence of sea breezes, the interaction of winds with mountains, or the influence of coastal features, these unique locations offer fantastic opportunities to study and appreciate the beauty of Morning Glory Clouds.

Conclusion

The availability of specific geographical and meteorological conditions makes certain locations around the world unique hotspots for the observation of Morning Glory Clouds. The Gulf of Carpentaria in Australia, Burketown in Queensland, and Santa Cruz in California are just a few examples of these special places. These locations not only attract scientists and weather enthusiasts but also provide opportunities for cultural celebrations and tourism. Understanding the relationship between unique geographical features and the formation of Morning Glory Clouds enhances our knowledge of this extraordinary meteorological phenomenon.

Impact on Local Communities and Tourism

The occurrence of Morning Glory Clouds in certain areas has a significant impact on local communities and tourism. These unique atmospheric phenomena have fascinated people for centuries and have become major attractions in various parts of the world. In this section, we will explore the direct and indirect impacts of Morning Glory Clouds on local communities and the tourism industry.

Increased Tourism

Morning Glory Clouds have become a popular tourist attraction in areas where they are known to occur frequently. The mesmerizing display of these long, tubular clouds rolling across the sky draws visitors from all over the world. Local communities have capitalized on this natural phenomenon by developing tourism infrastructure and services to accommodate the influx of tourists.

To cater to the growing number of visitors, tour operators offer guided tours and excursions specifically focused on witnessing Morning Glory Clouds. These tours provide an opportunity for tourists to experience the spectacle firsthand and learn about the formation and characteristics of these unique clouds. The revenue generated from tourism activities contributes to the local economy and helps create jobs within the community.

Economic Benefits

The presence of Morning Glory Clouds has a positive impact on the economy of the surrounding region. The influx of tourists brings in additional revenue through various channels such as accommodation, transportation, dining, and local shopping. Local businesses, including hotels, restaurants, and souvenir shops, experience increased demand during the peak seasons of Morning Glory Cloud activity.

Furthermore, the tourism industry related to Morning Glory Clouds creates employment opportunities for local residents. Tour guides, hospitality staff, drivers, and artisans find work directly or indirectly associated with catering to the needs of tourists. This not only boosts the economy but also helps in enhancing the overall standard of living for the local community.

Cultural Exchange

Morning Glory Cloud tourism provides a platform for cultural exchange between tourists and the local population. The presence of visitors from different parts of the world introduces diverse perspectives and experiences to the local community. This exchange of ideas, traditions, and knowledge promotes cultural understanding and encourages locals to showcase their cultural heritage.

Local communities often organize festivals, exhibitions, and events around Morning Glory Clouds to celebrate their cultural identity and draw attention to their unique traditions. Tourists have the opportunity to learn about local customs, taste traditional cuisine, and participate in cultural activities. This interaction not only enriches the overall tourism experience but also encourages preservation and promotion of indigenous cultures.

Challenges and Sustainability

While the impact of Morning Glory Cloud tourism can be beneficial, it also poses some challenges to local communities. The sudden influx of visitors during peak seasons may put a strain on infrastructure, resources, and the environment. Local authorities need to establish sustainable tourism practices to ensure the longevity and well-being of the destination.

Efforts should be made to minimize the environmental impact caused by tourism activities. This includes implementing waste management systems, promoting responsible tourism practices, and educating tourists about the significance of conserving the natural environment. Local communities should be actively involved in decision-making processes to ensure that tourism development aligns with their cultural values and long-term sustainability goals.

Case Study: Burketown, Australia

Burketown, a small town in Queensland, Australia, is known as one of the best locations to witness Morning Glory Clouds. The town has embraced the tourism potential of these clouds and actively promotes itself as a destination for cloud enthusiasts. The Morning Glory Festival, held annually, attracts both locals and tourists and features various events, including skydiving, hot air balloon rides, and cultural performances.

The tourism industry surrounding Morning Glory Clouds has had a positive impact on the local economy of Burketown. The income generated supports the development of local infrastructure, schools, and healthcare facilities. The increased exposure has also led to the preservation and revival of indigenous traditions, as the local Aboriginal community showcases their dances, art, and storytelling during the festival.

Summary

The presence of Morning Glory Clouds has a significant impact on local communities and the tourism industry. Increased tourism brings economic benefits to the region, creates job opportunities, and facilitates cultural exchange. However, sustainable tourism practices and environmental conservation are crucial for ensuring the long-term viability and preservation of the destination. Local authorities and communities must work together to strike a balance between economic development and the protection of natural and cultural resources.

Comparative Studies of Global Morning Glory Cloud Distribution

The distribution of Morning Glory Clouds is not uniform across the globe, and there are certain regions that are known as hotspots for these unique atmospheric phenomena. In this section, we will explore and compare the global distribution of Morning Glory Clouds, taking into account the influence of local climate, topography, and other factors.

… GEOGRAPHIC DISTRIBUTION

Hotspots of Morning Glory Cloud Activity

Morning Glory Clouds are most commonly observed in certain regions around the world, often referred to as hotspots. These hotspots are areas where the occurrence of Morning Glory Clouds is relatively frequent and predictable.

One of the most famous hotspots for Morning Glory Clouds is the Gulf of Carpentaria in Northern Australia. In this region, the Morning Glory Clouds are observed during the spring and autumn months, and they can be several hundred kilometers long. The coastal towns of Burketown and Karumba have become popular destinations for tourists who want to witness this natural spectacle.

Another notable hotspot is the Gulf of Thailand, where Morning Glory Clouds are also observed during the spring and autumn seasons. This region is known for its vibrant and colorful Morning Glory Cloud formations, attracting both locals and tourists.

South America is also home to a Morning Glory Cloud hotspot, specifically in the region of Patagonia. In Argentina, the town of Ushuaia experiences Morning Glory Clouds during the austral spring (September-November). These cloud formations often extend along the Beagle Channel, creating a breathtaking sight for onlookers.

Other hotspots for Morning Glory Cloud activity include the northwestern region of Queensland, Australia, the region surrounding the Gulf of Mexico in the United States, and the Torres Strait between Australia and Papua New Guinea.

Influence of Local Climate and Topography

The distribution of Morning Glory Clouds is strongly influenced by local climate and topography. Several climatic factors play a role in the formation and persistence of Morning Glory Clouds, including temperature, humidity, wind patterns, and atmospheric stability.

In regions with an abundance of moisture and warm temperatures, such as coastal areas, the likelihood of Morning Glory Cloud formation increases. These conditions create an environment favorable for the development of the necessary cloud formation processes.

Topography also plays a significant role in the distribution of Morning Glory Clouds. The presence of mountains, hills, and coastal features can affect the wind patterns and airflow, leading to the formation of gravity waves that are instrumental in the development of Morning Glory Clouds. For example, the Gulf of Carpentaria in Australia is surrounded by landmasses and narrow channels, which act as conduits for the formation and propagation of the Morning Glory Clouds.

Additionally, the proximity of warm ocean currents, such as the Gulf Stream in the Gulf of Mexico, can enhance the transfer of moisture and energy to the atmosphere, creating favorable conditions for Morning Glory Cloud formation.

Unique Locations with Morning Glory Clouds

While the hotspots for Morning Glory Clouds are well-known, there are also several unique locations where these cloud formations have been observed, adding to the diversity and intrigue of these atmospheric phenomena.

One such location is the island of Borneo, particularly in the states of Sarawak and Sabah in Malaysia. Morning Glory Clouds have been sighted in these regions, drawing attention from both scientists and enthusiasts.

In the United States, Morning Glory Clouds have been observed in the Owens Valley of California. This region is known for its favorable atmospheric conditions that promote the formation and maintenance of Morning Glory Clouds.

Another intriguing location is the Bay of Fundy in Canada, where the largest tidal range in the world occurs. While Morning Glory Clouds are not as frequent here compared to other hotspots, their occurrence in this unique setting adds an extra layer of interest.

Impact on Local Communities and Tourism

The presence of Morning Glory Clouds has a significant impact on local communities, particularly in regions where these cloud formations occur regularly. The occurrence of Morning Glory Clouds has become a source of pride and identity for these communities and has led to the development of tourism-related activities.

For example, in the Gulf of Carpentaria in Australia, the towns of Burketown and Karumba have embraced the presence of Morning Glory Clouds as a tourist attraction. Tourists flock to these towns during the Morning Glory Cloud season, injecting revenue into the local economy.

In the Gulf of Thailand, the occurrence of Morning Glory Clouds has also contributed to the growth of tourism. Local businesses, such as hotels and tour operators, have capitalized on the natural spectacle, attracting visitors from around the world.

The cultural significance of Morning Glory Clouds in these regions cannot be overlooked. Indigenous communities have long held spiritual and cultural beliefs associated with these cloud formations. The presence of Morning Glory Clouds reinforces the preservation and celebration of indigenous traditions and knowledge.

Comparative Studies of Global Morning Glory Cloud Distribution

Comparative studies of global Morning Glory Cloud distribution provide valuable insights into the similarities and differences between various hotspots around the world. These studies help scientists understand the underlying atmospheric processes and the factors that contribute to the development and persistence of Morning Glory Clouds.

One approach in comparative studies is to analyze meteorological data from different regions experiencing Morning Glory Clouds. This includes data on temperature, humidity, wind patterns, and atmospheric stability. By comparing these variables across different hotspots, scientists can identify commonalities and unique characteristics of Morning Glory Cloud formation.

Comparative studies also involve the use of satellite observations and remote sensing techniques to analyze cloud cover, cloud height, and other relevant parameters. This allows for a broader perspective on the global distribution of Morning Glory Clouds and provides a basis for further investigation.

Furthermore, numerical modeling and simulations play a crucial role in comparative studies. By inputting the known atmospheric conditions of different regions, scientists can simulate the development and propagation of Morning Glory Clouds. These simulations help test different hypotheses and increase our understanding of the complex dynamics involved.

The findings from comparative studies of global Morning Glory Cloud distribution contribute to advancements in meteorological research and weather forecasting. By gaining a better understanding of the factors that influence the formation and behavior of Morning Glory Clouds, scientists can improve their predictive capabilities and enhance aviation safety.

Future Directions for Comparative Studies

While comparative studies of Morning Glory Cloud distribution have provided valuable insights, there are still many avenues for future research and exploration.

One area of interest is the influence of climate change on Morning Glory Clouds. As our climate continues to change, it is important to understand how these changes will impact the frequency, intensity, and distribution of Morning Glory Clouds. Comparative studies can help identify potential changes in hotspots and shed light on the underlying mechanisms.

Another direction for future research is the development of high-resolution remote sensing techniques. With advancements in technology, scientists can obtain more detailed and accurate data on Morning Glory Cloud formations. This will contribute to improved understanding and analysis of global distribution patterns.

Additionally, interdisciplinary collaborations between meteorologists, climatologists, geographers, and indigenous communities can further enhance our knowledge of Morning Glory Cloud distribution. Integrating diverse perspectives and expertise will lead to a more comprehensive understanding of these unique atmospheric phenomena.

In conclusion, comparative studies of global Morning Glory Cloud distribution provide valuable insights into the occurrence and behavior of these unique atmospheric phenomena. By analyzing meteorological data, utilizing remote sensing techniques, and employing numerical modeling, scientists can better understand the similarities and differences between various hotspots around the world. This knowledge contributes to advancements in meteorological research, weather forecasting, and the preservation of indigenous traditions. The future of comparative studies holds great potential for further uncovering the mysteries of Morning Glory Clouds and their global distribution.

Seasonal Patterns

Spring and Autumn Morning Glory Clouds

In this section, we will explore the seasonal patterns of Morning Glory Clouds, specifically focusing on their occurrence in spring and autumn. We will discuss the factors that influence the formation of Morning Glory

Clouds during these seasons, their characteristics, and their relevance in understanding the broader atmospheric phenomena.

Seasonal Variations in Morning Glory Clouds

Morning Glory Clouds exhibit distinct seasonal variations, with spring and autumn being the primary seasons for their occurrence. During these transitional seasons, various atmospheric conditions contribute to the development of these unique cloud formations.

In spring, as the Earth's axis tilts towards the Sun, the northern hemisphere experiences longer daylight hours. This increase in solar radiation affects the temperature distribution across different atmospheric layers. The warming temperatures trigger the formation of convective cells, which play a crucial role in the development of Morning Glory Clouds. The combination of temperature gradients, moisture levels, and atmospheric stability during spring creates favorable conditions for gravity waves, a key mechanism responsible for the formation of Morning Glory Clouds.

Similarly, autumn also serves as an ideal season for Morning Glory Clouds. As the Earth's axis begins to tilt away from the Sun, the transition to cooler temperatures introduces a different set of atmospheric dynamics. The decreasing solar radiation initiates a shift in temperature gradients within the atmosphere, which can lead to the generation of gravity waves. Additionally, the changing weather patterns and frontal systems during autumn can enhance the likelihood of Morning Glory Cloud formation.

Key Factors Influencing Spring and Autumn Morning Glory Clouds

Several factors contribute to the occurrence and characteristics of Morning Glory Clouds during spring and autumn:

1. *Temperature gradients*: The temperature differences between various atmospheric layers play a crucial role in influencing the formation and propagation of gravity waves. In both spring and autumn, the changing temperatures create favorable conditions for the generation of these waves, which then lead to the formation of Morning Glory Clouds.

2. *Moisture levels*: The availability of moisture in the atmosphere is critical for cloud formation. During spring and autumn, the transition of weather patterns brings about changes in humidity

levels. Adequate moisture content provides the necessary ingredients for cloud formation, contributing to the occurrence of Morning Glory Clouds.

3. *Atmospheric stability*: Atmospheric stability plays a significant role in governing the behavior and persistence of Morning Glory Clouds. In both spring and autumn, the presence of convective cells and frontal systems can introduce dynamic changes in atmospheric stability, thereby influencing the formation and longevity of these cloud formations.

4. *Wind patterns*: The direction and strength of winds at different altitudes impact the spatial distribution and movement of Morning Glory Clouds. During spring and autumn, the interaction between wind patterns and gravity waves can result in the characteristic rolling motion of these clouds.

5. *Synoptic weather patterns*: The larger-scale weather patterns prevailing during spring and autumn, such as the movement of high-pressure and low-pressure systems, frontal systems, and the positioning of jet streams, can influence the likelihood and characteristics of Morning Glory Clouds.

The Significance of Spring and Autumn Morning Glory Clouds

Understanding the behavior and occurrence of Morning Glory Clouds in spring and autumn provides valuable insights into the broader atmospheric processes and dynamics. The study of these seasonal cloud formations contributes to various fields, including meteorology, atmospheric physics, and climate science. Here are some of the significance of studying spring and autumn Morning Glory Clouds:

- *Weather forecasting*: The identification and analysis of Morning Glory Clouds in spring and autumn can serve as indicators of atmospheric instability and impending weather changes. Monitoring the formation and movement of these clouds can aid meteorologists in predicting local weather patterns, including the likelihood of thunderstorms, gusty winds, and changes in temperature and humidity.

- *Gravity wave research*: Morning Glory Clouds offer an excellent opportunity to study the generation and propagation of gravity waves in the atmosphere. Investigating the interaction between temperature gradients, wind patterns, and convective cells during spring and autumn can enhance our understanding of these gravity-driven phenomena, which have implications for various atmospheric processes.

- *Climate variability and change*: Studying the seasonal variations of Morning Glory Clouds can provide insights into the long-term climate variability in different regions. Changes in the occurrence or characteristics of these clouds over time might indicate shifts in atmospheric circulation patterns, influenced by factors such as global warming or regional climate change.

- *Cultural and historical perspectives*: Spring and autumn Morning Glory Clouds have cultural and historical significance in many communities. These cloud formations are often associated with traditional practices, rituals, or beliefs, and studying their seasonal patterns can help preserve and document indigenous knowledge and cultural heritage.

Case Study: Morning Glory Clouds in Burketown, Australia

Burketown, a small town located in Queensland, Australia, experiences a notable occurrence of Morning Glory Clouds during spring and autumn. The Burketown Morning Glory Cloud phenomenon attracts tourists and researchers from around the world. This unique annual event has created a significant economic impact on the local community, promoting ecotourism and cultural exchange.

The study of Morning Glory Clouds in Burketown provides valuable insights into the seasonal behavior, characteristics, and formation mechanisms of these cloud formations. Researchers have extensively documented the local weather patterns, synoptic conditions, and atmospheric dynamics that contribute to the occurrence of Morning Glory Clouds in this region. Additionally, the interaction between the local topography and climatic factors creates a conducive environment for the formation and sustenance of these clouds.

The Burketown case study exemplifies the potential for interdisciplinary research and collaborative efforts between meteorologists, climatologists, social scientists, and indigenous communities. It highlights the need to integrate scientific understanding with cultural perspectives to truly appreciate the significance of Morning Glory Clouds in specific regions.

Conclusion

The occurrence of Morning Glory Clouds in spring and autumn demonstrates the complex interplay between temperature gradients, moisture levels, atmospheric stability, wind patterns, and synoptic weather systems. The study of these seasonal cloud formations provides valuable insights into gravity wave dynamics, weather forecasting, climate variability, and cultural significance. By exploring specific case studies, such as the Burketown Morning Glory Clouds, we can enhance our understanding of these unique cloud formations and their broader implications. Studying spring and autumn Morning Glory Clouds opens up avenues for further research, encouraging interdisciplinary approaches and fostering a deeper appreciation for the wonders of the natural world.

Summer Morning Glory Clouds

Summer is a season known for its warm and sunny weather. It is a time when people enjoy outdoor activities and bask in the beauty of nature. But for meteorologists and cloud enthusiasts, summer also brings the fascinating phenomenon of Morning Glory Clouds. These unique cloud formations, often seen in the early morning hours, captivate both scientists and adventurers alike.

Characteristics of Summer Morning Glory Clouds

Summer Morning Glory Clouds share many similarities with other Morning Glory Clouds observed throughout the year. They are long, tubular-shaped clouds that appear to roll across the sky, casting a majestic scene for observers below. However, there are some distinctive characteristics that set summer clouds apart.

One notable feature of summer Morning Glory Clouds is their larger size. These clouds can stretch for hundreds of kilometers, covering vast expanses of the sky. The increased heat and moisture in the atmosphere during summer contribute to the formation of these larger cloud systems.

Another distinguishing characteristic of summer Morning Glory Clouds is their more pronounced vertical structure. These clouds often exhibit well-defined layers, with individual cloud bands stacked on top of each other. The vertical development of these clouds is influenced by the convective activity associated with the summer season.

Summer Morning Glory Clouds also tend to exhibit vibrant colors, adding to their allure. The combination of sunlight, atmospheric particles, and moisture content within the cloud contribute to the array of hues seen in the sky. From golden yellows to striking oranges and reds, the colors of summer Morning Glory Clouds create a breathtaking spectacle.

Formation Mechanisms of Summer Morning Glory Clouds

The formation of summer Morning Glory Clouds is influenced by various meteorological factors that are specific to this season. These factors include convective processes, the presence of moisture, and atmospheric stability.

During summer, the sun's rays heat the earth's surface, causing the air to rise and form thermals. These thermals can rise to great heights, carrying moisture with them. As the moist air ascends, it cools, leading to the condensation of water vapor and the formation of clouds. In the case of summer Morning Glory Clouds, the convective activity plays a crucial role in their formation.

Moisture availability is another essential factor for the formation of summer Morning Glory Clouds. The warm and humid summer air provides the necessary moisture content for cloud formation. The moisture further enhances the convective activity, leading to the development of the characteristic long and tubular shape of these clouds.

Atmospheric stability also plays a significant role in the formation of summer Morning Glory Clouds. In unstable atmospheric conditions, air masses rise rapidly, creating the updrafts responsible for the cloud's vertical development. The convective activity associated with summer weather systems can contribute to the instability needed for the formation of these clouds.

Meteorological Impacts of Summer Morning Glory Clouds

Summer Morning Glory Clouds have several meteorological impacts on their surrounding environment. These impacts can affect local weather patterns, precipitation, and even aviation.

The presence of Morning Glory Clouds during summer can influence local weather patterns. These clouds can modulate the incoming solar radiation, leading to changes in temperature and atmospheric stability. The shading effect provided by the clouds can result in cooler surface temperatures during the day. Additionally, the vertical structure of the clouds can affect wind patterns and create localized areas of turbulence in the atmosphere.

Summer Morning Glory Clouds can also impact precipitation patterns. The convective activity associated with these clouds can lead to the development of thunderstorms and heavy rainfall in certain regions. The interaction between the cloud bands and frontal systems can enhance the lifting mechanisms necessary for precipitation formation.

Aviation is another area significantly impacted by summer Morning Glory Clouds. The tubular shape and the associated turbulence make these clouds hazardous for aircraft. The strong updrafts and downdrafts within the cloud can cause severe turbulence, affecting flight stability. Pilots need to be aware of the presence of Morning Glory Clouds during summer and plan their routes accordingly to avoid potential aviation hazards.

Case Studies and Examples

Several case studies and examples illustrate the occurrence and impact of summer Morning Glory Clouds. One notable example is the annual Morning Glory Cloud Festival held in Burketown, Queensland, Australia. Each year, during late spring and early summer, these magnificent clouds roll across the Gulf of Carpentaria, captivating locals and visitors alike. The festival celebrates both the natural beauty of the clouds and the rich indigenous culture of the region.

In another case study, researchers analyzed the effect of summer Morning Glory Clouds on the local climate in the Gulf of Carpentaria region. The presence of these clouds was found to modulate the diurnal temperature range, resulting in cooler daytime temperatures and slightly warmer nighttime temperatures. The shading effect of the clouds influences the local heat balance, indicating their significance in the regional climate.

Future Directions and Challenges

Although researchers have made significant progress in understanding the formation and dynamics of summer Morning Glory Clouds, many questions remain unanswered. Future research efforts should focus on improving numerical models to accurately simulate the complex interactions between convective processes, moisture availability, and atmospheric stability.

One challenge in studying summer Morning Glory Clouds is the limited availability of in-situ data. These clouds occur in remote and inaccessible locations, making it difficult to obtain direct measurements. The development of advanced remote sensing techniques, such as the use of unmanned aerial vehicles (UAVs) and drones, can provide valuable data for a better understanding of these clouds.

Another area that requires attention is the long-term impact of climate change on the occurrence and behavior of summer Morning Glory Clouds. Climate models can help assess the potential effects of rising temperatures and changing atmospheric conditions on the formation and distribution of these clouds. Understanding these impacts is crucial for developing effective climate change adaptation strategies.

In conclusion, summer Morning Glory Clouds offer a captivating display of nature's artistry. These unique cloud formations, characterized by their large size, vibrant colors, and vertical structure, are a testament to the intricate interplay of convective processes, moisture availability, and atmospheric stability. The meteorological impacts of summer Morning Glory Clouds on weather patterns, precipitation, and aviation make them an area of great interest for scientists and researchers. As further advancements in research techniques and numerical models are made, we can expect a deeper understanding of these remarkable clouds and their role in our dynamic atmosphere.

Winter Morning Glory Clouds

In this section, we will explore the phenomenon of Winter Morning Glory Clouds, which occur during the winter season. We will examine their characteristics, formation mechanisms, and their significance in the context of global climate.

Characteristics of Winter Morning Glory Clouds

Winter Morning Glory Clouds possess specific features that distinguish them from other types of clouds. These include their unique shape, size, and coloration.

Shape and Size: Winter Morning Glory Clouds often exhibit a similar roll cloud formation as their counterparts in other seasons. However, they tend to have a narrower and elongated shape, resembling long tubes or cylinders stretching across the sky. The clouds can extend for several kilometers and reach heights of 1 to 2 kilometers above the ground. The distinct roll-like appearance is caused by the interaction of atmospheric conditions during the winter months.

Coloration: Winter Morning Glory Clouds may display vibrant hues, including shades of pink, orange, and purple. These colors arise from the scattering and absorption of sunlight by atmospheric particles and cloud droplets. The low angle of the sun during winter creates different lighting conditions that enhance the coloration of the clouds, adding to their visual appeal.

Formation Mechanisms of Winter Morning Glory Clouds

The formation of Winter Morning Glory Clouds involves complex interactions between various atmospheric factors. Here, we will explore the main mechanisms contributing to their occurrence.

Frontal Systems: Winter Morning Glory Clouds often develop along the leading edge of cold fronts. As a cold front advances, it interacts with warm and moist air masses, leading to the formation of rolling waves in the lower atmosphere. These waves then trigger the formation of the characteristic roll clouds associated with Morning Glory Clouds.

Convection: Convective processes play a role in the formation of Winter Morning Glory Clouds. The contrast between the relatively warm and moist surface and the colder upper atmosphere can generate thermals, which rise and condense to form clouds. As these clouds interact with atmospheric waves, they can align into a roll cloud formation, resulting in the appearance of Morning Glory Clouds during winter.

Gravity Waves: Similar to Morning Glory Clouds in other seasons, Winter Morning Glory Clouds can be influenced by gravity waves. These waves are produced by the interaction of the atmosphere with topographic features and can propagate through the lower atmosphere. When gravity waves encounter suitable atmospheric conditions, such as stable layers of air, they can trigger the creation of the characteristic roll clouds associated with Winter Morning Glory Clouds.

Significance of Winter Morning Glory Clouds

Winter Morning Glory Clouds have both scientific and cultural significance, providing valuable insights into atmospheric processes and offering aesthetic beauty for observers.

Scientific Significance: Winter Morning Glory Clouds provide an opportunity for researchers to study the interaction between frontal systems, convective processes, and gravity waves. By understanding the formation mechanisms behind these clouds, scientists can improve weather and climate forecasting models, enhancing our ability to predict weather patterns and climate changes during winter.

Aesthetic Beauty: Winter Morning Glory Clouds are known for their vibrant colors and unique shape. They captivate and inspire individuals, artists, and photographers who are drawn to their beauty. The observation and documentation of Winter Morning Glory Clouds contribute to the appreciation and preservation of natural phenomena, fostering a deeper connection with the environment.

In conclusion, Winter Morning Glory Clouds exhibit distinct characteristics that set them apart from their counterparts in other seasons. Their formation involves the interplay of frontal systems, convection, and gravity waves. Winter Morning Glory Clouds have scientific significance as they contribute to our understanding of atmospheric processes. Additionally, their aesthetic beauty has cultural significance, inspiring artists and connecting individuals with the wonders of the natural world.

Factors Influencing Seasonal Variations

Seasonal variations in morning glory clouds are influenced by a variety of factors, including meteorological conditions, geographical features, and local climate patterns. Understanding these factors can provide insights into the occurrence and behavior of morning glory clouds in different seasons and locations. In this section, we will explore the key factors that contribute to seasonal variations in morning glory clouds.

Meteorological Conditions

Meteorological conditions play a crucial role in the formation and development of morning glory clouds. Different weather patterns and atmospheric phenomena can create favorable conditions for the occurrence of morning glory clouds in certain seasons. Here are some key meteorological factors that influence seasonal variations:

1. **Temperature Gradient:** The temperature gradient between the land and sea can contribute to the formation of morning glory clouds. During certain seasons, such as spring and autumn, the temperature difference between these two regions can be significant. This differential heating can generate convective activity and trigger the formation of morning glory clouds.

2. **Moisture Content:** Moisture availability in the atmosphere is another critical factor. Seasonal variations in humidity levels affect the likelihood of morning glory cloud formation. During the wet season when moisture is abundant, morning glory clouds are more likely to occur due to increased water vapor content.

3. **Wind Patterns:** The prevailing wind patterns, including the direction and strength of the winds, have a significant impact on morning glory cloud formation. During certain seasons, specific wind patterns, such as sea breezes or prevailing trade winds, can enhance the likelihood of morning glory clouds due to their influence on the advection and convergence of air masses.

4. **Pressure Systems:** The presence and movement of high-pressure and low-pressure systems also contribute to seasonal variations in morning glory cloud activity. These pressure systems influence the formation of atmospheric waves, such as gravity waves, which are known to be a driving mechanism for morning glory clouds.

Geographical Features

The geographical features of a region can influence the occurrence and characteristics of morning glory clouds in different seasons. Factors such as topography, proximity to the coast, and land-sea breeze circulation patterns can all play a role. Here are some key geographical factors that influence seasonal variations:

1. **Coastal Proximity:** Regions situated close to coastlines often experience enhanced morning glory cloud activity. This proximity allows for the interaction of different air masses, such as maritime air

moving inland and encountering cooler land areas. Coastal regions also benefit from the influence of sea breezes, which can trigger morning glory cloud formation.

2. **Mountainous Terrain:** Mountainous areas can significantly impact the formation and behavior of morning glory clouds. As air is forced to rise over mountains, it cools and condenses, creating favorable conditions for cloud formation. Additionally, the interaction of mountain slopes and prevailing winds can lead to the development of gravity waves, a key driver of morning glory clouds.

3. **Land-Sea Breeze Circulation:** In coastal regions, the daily cycle of land and sea breezes can influence the occurrence of morning glory clouds. Sea breezes are influenced by temperature variations between the land and sea, and they can help trigger cloud formation and enhance the vertical motion necessary for morning glory clouds to occur.

Local Climate Patterns

Local climate patterns have a significant influence on the seasonal variations of morning glory clouds. These patterns include long-term weather trends, seasonal variations in atmospheric conditions, and the presence of specific climate phenomena. Here are some key factors related to local climate patterns:

1. **Monsoon Systems:** Regions affected by monsoon systems often exhibit distinct seasonal variations in morning glory cloud activity. Monsoons bring changes in wind patterns, moisture availability, and temperature gradients, all of which can contribute to the formation and behavior of morning glory clouds during specific seasons.

2. **Trade Winds:** Areas influenced by trade winds, such as the tropics and subtropics, often experience seasonal variations in morning glory cloud activity. The trade winds contribute to the advection of air masses and help create the necessary conditions for morning glory clouds, especially during certain seasons characterized by specific wind patterns.

3. **Climate Oscillations:** Large-scale climate oscillations, such as El Niño-Southern Oscillation (ENSO) and the North Atlantic

Oscillation (NAO), can impact seasonal variations in morning glory clouds. These oscillations influence weather patterns and atmospheric conditions, which in turn affect the likelihood of morning glory cloud formation during different seasons.

Understanding these factors can provide valuable insights into the seasonal variations of morning glory clouds. By examining the interplay between meteorological conditions, geographical features, and local climate patterns, researchers can better predict and explain the occurrence and behavior of these fascinating atmospheric phenomena.

Example Problem:

A coastal region experiences a significant increase in morning glory cloud activity during the summer season. Explain the possible meteorological and geographical factors contributing to this seasonal variation.

Solution:

The increased morning glory cloud activity during the summer season in the coastal region can be attributed to several factors. First, the proximity to the coast allows for the interaction of different air masses. In this case, the sea breeze coming from the ocean converges with the land breeze, creating a convergence zone where air rises and forms clouds.

Second, during summer, the temperature gradient between the warm land and cooler ocean is usually high. This differential heating generates convective activity, creating favorable conditions for morning glory cloud formation.

Lastly, the region's geographical features, such as mountain ranges, play a role in enhancing morning glory cloud activity. As the prevailing winds encounter the mountains, they are forced to rise, leading to the formation of gravity waves. These waves contribute to the development and maintenance of morning glory clouds.

By considering these meteorological and geographical factors, we can explain the increase in morning glory cloud activity during the summer season in the coastal region.

Summary

Seasonal variations in morning glory clouds are influenced by various factors, including meteorological conditions, geographical features, and local climate patterns. Understanding these factors helps explain why morning glory clouds occur during specific seasons and in particular locations.

Meteorological conditions, such as temperature gradient, moisture content, wind patterns, and pressure systems, create favorable conditions for morning glory cloud formation. Geographical features, such as coastal proximity, mountainous terrain, and land-sea breeze circulation, influence the occurrence and behavior of morning glory clouds. Local climate patterns, such as monsoon systems, trade winds, and climate oscillations, contribute to the seasonal variations of morning glory clouds.

By examining these factors, researchers can gain insights into the complex dynamics of morning glory clouds and improve our understanding of these mesmerizing atmospheric phenomena. Understanding seasonal variations in morning glory clouds has practical implications for weather forecasting, climate research, and cultural preservation.

Long-term Climate Change Effects on Morning Glory Clouds

Climate change is a significant global issue that is affecting various aspects of our planet, including atmospheric phenomena such as morning glory clouds. These unique cloud formations may experience several long-term impacts due to changing climatic conditions. In this section, we will explore the potential effects of climate change on morning glory clouds and their implications for both the scientific community and local communities.

Background

To understand the potential long-term effects of climate change on morning glory clouds, it is essential to have a basic understanding of climate change science. Climate change refers to long-term shifts in weather patterns and average temperatures on Earth. It is primarily caused by human activities, particularly the emission of greenhouse gases such as carbon dioxide and methane.

The increase in greenhouse gases in the atmosphere leads to the trapping of heat, resulting in a global temperature rise. This temperature increase affects various climatic factors, including air circulation patterns, humidity levels, and cloud formation processes. These changes in the climate system can have significant implications for the formation and behavior of morning glory clouds.

Potential Effects on Formation

One of the primary concerns regarding climate change is its impact on atmospheric stability. Morning glory clouds typically form in stable atmospheric conditions, characterized by a stable temperature profile and low wind shear. However, climate change may disrupt these stable atmospheric conditions, leading to changes in morning glory cloud formation.

1. **Increased instability:** Climate change can lead to increased atmospheric instability, which may result in more turbulent conditions within the lower atmosphere. This increased turbulence can disrupt the formation of morning glory clouds, making their occurrence less frequent or altering their typical characteristics.

2. **Changes in wind patterns:** Climate change can also affect wind patterns, which play a crucial role in the formation and maintenance of morning glory clouds. Alterations in wind patterns may lead to changes in the orientation and movement of gravity waves, impacting the formation and structure of morning glory clouds.

Implications for Local Communities

Morning glory clouds hold significant cultural and economic value for many local communities. The potential long-term effects of climate change on these cloud formations can have implications for the tourism industry, traditional practices, and overall community well-being.

1. **Tourism industry:** Morning glory cloud tourism is a thriving industry in some regions, attracting visitors who wish to witness and experience these fascinating cloud formations. Climate change-induced alterations in morning glory clouds may affect tourism patterns, potentially leading to economic challenges for local communities heavily reliant on this industry.

2. **Traditional practices and cultural significance:** Morning glory clouds hold significant cultural and spiritual importance in some indigenous communities. They are often associated with mythical and symbolic meanings and are integrated into traditional practices and rituals. Changes in morning glory clouds due to climate change may impact the cultural fabric of these communities, leading to the potential loss of ancient traditions and knowledge.

3. **Community well-being:** Morning glory clouds are not only a visual spectacle but also play a role in local weather patterns. They can impact temperature, humidity, and precipitation, which are essential factors for agriculture and overall community well-being. Climate change-induced changes in morning glory clouds may disrupt local weather patterns, potentially affecting agriculture, water availability, and livelihoods.

Scientific Research and Future Directions

The potential long-term effects of climate change on morning glory clouds present an exciting avenue for scientific research. Several key areas of exploration can provide valuable insights and facilitate a better understanding of these effects:

1. **Climate modeling:** Developing climate models specifically focused on morning glory cloud formation and behavior can help predict future changes in their occurrence, shape, and size. These models can incorporate various climate variables and provide projections for different climate change scenarios.

2. **Long-term data collection:** Long-term monitoring and data collection of morning glory clouds are essential for studying their response to climate change. This can involve satellite observations, ground-based measurements, and citizen science initiatives. Analyzing these data sets can help identify trends and patterns associated with climate change impacts.

3. **Collaborative research:** As morning glory clouds are observed in different regions globally, collaborative research efforts can provide a broader understanding of their response to climate change. Collaboration among scientists, indigenous communities, and local stakeholders will enhance knowledge sharing and promote a comprehensive approach to studying these cloud formations.

4. **Risk assessment and adaptation:** Assessing the potential risks associated with climate change impacts on morning glory clouds is crucial for developing adaptation strategies. Understanding the vulnerabilities of local communities reliant on morning glory clouds can help stakeholders implement measures to mitigate potential negative impacts.

Conclusion

Climate change has the potential to significantly impact morning glory clouds, affecting their formation, structure, and behavior. The long-term effects of climate change on these unique cloud formations can have implications for both scientific research and local communities. As our understanding of climate change continues to evolve, it is crucial to monitor and study morning glory clouds to better understand and prepare for these potential impacts. By combining scientific research, community engagement, and adaptive measures, we can ensure the preservation of morning glory clouds for future generations to come.

Diurnal Cycle

Morning Glory Clouds at Dawn

Dawn is a magical time when nature awakens and unveils its stunning beauty. This is especially true when it comes to Morning Glory Clouds, as their appearance at dawn is truly awe-inspiring. In this section, we will explore the unique behavior and features of Morning Glory Clouds during this early morning spectacle.

Dawn and the Diurnal Cycle

To understand why Morning Glory Clouds are often observed at dawn, we must first delve into the concept of the diurnal cycle. The diurnal cycle refers to the patterns of behavior and atmospheric conditions that occur within a single day. It is influenced by a combination of factors, including solar radiation, temperature variations, and wind patterns.

At dawn, as the sun rises above the horizon, solar radiation begins to heat the Earth's surface. This heating leads to the formation of thermals, rising columns of warm air. These thermals interact with the surrounding atmosphere and create ideal conditions for the formation of Morning Glory Clouds.

Thermals and Morning Glory Cloud Formation

Thermals play a crucial role in the formation of Morning Glory Clouds at dawn. As the sun's rays heat the ground, pockets of warm air start to rise. These rising thermals carry moisture from the lower layers of the atmosphere to higher altitudes.

As the warm, moist air ascends, it cools and condenses, forming the characteristic long, tube-shaped cloud that we know as the Morning Glory

Cloud. The moisture in the thermals condenses into water droplets or ice crystals, creating the fluffy cotton-like appearance of the cloud.

It is important to note that the diurnal cycle and the formation of Morning Glory Clouds at dawn are closely linked. The thermals generated by solar heating are most pronounced during the early morning hours, making dawn the prime time for the formation of these magnificent clouds.

Influence of Wind and Pressure Patterns

Alongside the diurnal cycle, wind and pressure patterns also play a significant role in the occurrence of Morning Glory Clouds at dawn. The presence of these unique clouds is often associated with specific atmospheric conditions, which are influenced by wind and pressure systems.

During the predawn hours, the winds are typically light and variable. As the sun rises, the temperature difference between the land and sea creates a change in pressure, generating a sea breeze. This sea breeze flows from the cooler ocean to the warmer land, providing the necessary uplift for the Morning Glory Clouds to form.

The interaction between the sea breeze and the prevailing wind patterns can further enhance the formation of Morning Glory Clouds. These clouds are often observed in regions where the prevailing wind blows perpendicular to the coastline, creating an ideal environment for the convergence of air masses.

Observing Morning Glory Clouds at Dawn

Experiencing the beauty of Morning Glory Clouds at dawn is a sight to behold. To observe these majestic clouds, it is important to be in the right location at the right time.

Geographically, areas along coastal regions with a suitable wind and pressure pattern are more likely to witness Morning Glory Clouds at dawn. Some famous locations where these clouds are frequently observed include the Gulf of Carpentaria in northern Australia, the Burketown region in Queensland, and the Gulf of Mexico in the United States.

To increase your chances of witnessing this marvel of nature, it is essential to plan your observation during the spring or autumn months when the diurnal cycle and wind patterns are most favorable for Morning Glory Cloud formation.

Capturing the Beauty

Photographing Morning Glory Clouds at dawn can be a rewarding and captivating experience. To capture the beauty of these clouds, consider the following tips:

- Plan your photography outing in advance, based on weather forecasts and local knowledge of Morning Glory Cloud activity.

- Wake up early and be at your chosen location before dawn to capture the changing colors in the sky and the emergence of the clouds.

- Experiment with different camera settings, such as long exposures to convey the motion of the clouds, or close-up shots to capture their intricate details.

- Look for interesting foreground elements, such as trees, buildings, or bodies of water, to add depth and context to your photographs.

- Stay patient and be prepared to wait for the perfect moment to capture that breathtaking shot.

Remember, while photography is a wonderful way to preserve the beauty of Morning Glory Clouds, it is equally important to immerse yourself in the experience and appreciate the fleeting nature of these captivating clouds.

Conclusion: Embracing the Magic of Morning Glory Clouds at Dawn

Morning Glory Clouds at dawn are a testament to the breathtaking beauty and complexity of nature. The interplay between the diurnal cycle, thermals, wind, and pressure patterns creates the perfect conditions for their formation. Observing these ethereal clouds at dawn provides a unique opportunity to witness the wonders of our atmosphere.

As researchers and enthusiasts, let us continue to explore and unravel the mysteries of Morning Glory Clouds at dawn. By combining scientific knowledge with our appreciation for the natural world, we can truly embrace the magic and allure of these magnificent cloud formations. So, set your alarm, venture to the coastlines, and immerse yourself in the splendor of Morning Glory Clouds at dawn.

Midday Behavior and Dissipation

During the midday hours, Morning Glory Clouds undergo distinct behavior and eventually dissipate. This section explores the factors that influence the midday behavior of these captivating cloud formations.

Temperature and Solar Heating

One of the key factors affecting midday behavior and dissipation of Morning Glory Clouds is temperature. As the sun rises and reaches its peak intensity during the midday hours, it heats the Earth's surface, causing an increase in temperature. The heating of the surface, in turn, impacts the temperature of the air in the lower atmosphere.

As the air near the surface gets heated, it becomes less dense and starts to rise due to buoyancy. This upward movement of warm air creates updrafts, which can disrupt the stability of Morning Glory Clouds. The rising warm air can mix with the surrounding air, leading to the dissipation of the cloud structure.

Additionally, solar heating plays a crucial role in the evaporation of moisture held within the cloud. The intense heat from the sun causes the water droplets or ice crystals in the cloud to evaporate, reducing the overall water content. As the water content diminishes, the cloud structure weakens, eventually leading to dissipation.

Vertical Mixing and Turbulence

During the midday hours, the thermal mixing of air masses with different characteristics becomes more prominent. As the surface temperature rises, thermals form, and these thermals can interact with the Morning Glory Cloud structure. This interaction can result in vertical mixing and turbulence within the cloud, causing the cloud to lose its defined shape and structure.

The turbulent mixing of air masses can disrupt the horizontal roll pattern of Morning Glory Clouds, causing the clouds to fragment and disintegrate. These turbulent motions within the cloud can also enhance the vertical mixing process, leading to the dispersion of cloud particles and the eventual dissipation of the cloud.

Boundary Layer Effects

The boundary layer, which is the layer of air directly influenced by the Earth's surface, has a significant impact on Morning Glory Cloud behavior during the midday hours. As the surface heats up, the boundary layer becomes deeper, allowing for more turbulent mixing between the surface and the atmosphere.

The increased turbulence within the boundary layer can create vertical motions that affect the Morning Glory Clouds. These turbulent motions can disrupt the stable roll pattern of the clouds, causing them to break apart and dissipate.

Furthermore, the interaction between the boundary layer and the cloud can result in changes in airflow patterns and wind speed. As the clouds dissipate, the localized wind patterns can also change, impacting the behavior of the cloud and further contributing to its dissipation.

Case Study: Midday Dissipation in the Gulf of Carpentaria

An intriguing case study regarding the midday dissipation of Morning Glory Clouds can be observed in the Gulf of Carpentaria in northern Australia. In this region, Morning Glory Clouds are quite common and are known to dissipate during the midday hours.

Research conducted in the Gulf of Carpentaria has revealed that the interaction between the sea breeze and the Morning Glory Clouds plays a crucial role in their midday dissipation. The sea breeze, which is caused by temperature differences between the land and the sea, can induce turbulent mixing in the boundary layer.

As the sea breeze moves inland, it encounters the Morning Glory Clouds. The interaction between the sea breeze and the cloud disrupts the stable roll pattern, leading to the fragmentation and dissipation of the cloud structure.

Understanding the specific case of midday dissipation in the Gulf of Carpentaria provides valuable insights into the complex interactions between local meteorological conditions and the behavior of Morning Glory Clouds during the midday hours.

Summary

During the midday hours, Morning Glory Clouds are subject to various factors that influence their behavior and eventual dissipation. The heating of the surface by solar radiation, thermal mixing and turbulence, boundary layer effects, and interactions with atmospheric phenomena such as sea breezes all contribute to the midday behavior of these unique cloud formations. The study of midday dissipation provides a deeper understanding of the complex dynamics at play in the atmosphere and the delicate balance that dictates the life cycle of Morning Glory Clouds.

Evening and Nocturnal Features

During the evening and nocturnal periods, Morning Glory Clouds display specific characteristics and behaviors that distinguish them from their daytime counterparts. The transition from daylight to darkness brings about changes in the atmospheric conditions, which in turn influence the formation and behavior of these unique cloud formations.

Reduced Visibility and Color Palette

As the sun sets and darkness envelops the sky, the visibility conditions change, affecting how Morning Glory Clouds are perceived. With diminished ambient light, the vivid colors that are often associated with these clouds become less pronounced and more subdued. The vibrant hues of red, orange, and pink that adorn the morning sky give way to deeper shades of blue and gray.

The reduced visibility during the evening and nocturnal hours can make it challenging to accurately observe and document the intricate details of Morning Glory Clouds. Researchers and photographers often rely on artificial lighting and long-exposure techniques to capture the essence of these clouds in low-light conditions. This limited visibility also poses difficulties for pilots and aviation authorities in detecting and navigating around Morning Glory Clouds during nighttime flights.

Modifications in Cloud Structure

The structure of Morning Glory Clouds undergoes modifications during the evening and nocturnal hours. As the atmosphere cools down, the cloud layers may become more stable and settled compared to their daytime counterparts. This stability can result in a less turbulent and more organized appearance of the cloud formation.

The vertical structure of the cloud may experience variations as well. While the characteristic roll cloud formation remains visible, the overall height and thickness of the cloud layers can exhibit changes during the evening and nocturnal period. These modifications in cloud structure can be attributed to the altered atmospheric conditions and the absence of solar heating.

Nocturnal Atmospheric Stability

During the evening and nighttime hours, the atmosphere generally becomes more stable due to the radiative cooling of the earth's surface. This increased stability can influence the behavior and movements of Morning Glory Clouds. The reduced vertical mixing and decreased convective activity restrict the upward and downward motion of air masses within the cloud formation.

As a result, Morning Glory Clouds during the nocturnal period often exhibit a less dynamic nature. The cloud layers may maintain a relatively stable position and move more slowly compared to their daytime counterparts. This nocturnal atmospheric stability can also contribute to the persistence of Morning Glory Clouds during the nighttime hours, allowing them to retain their distinct roll cloud structure for longer durations.

Formation of Wave-like Features

The transition from daytime to nocturnal conditions can give rise to the formation of wave-like features within Morning Glory Clouds. These features appear as undulations, ripples, or waves that propagate along the length of the cloud formation. The wave-like structures are caused by various atmospheric processes and interactions, such as gravity waves and wind shear.

Gravity waves, generated by the interaction of stable air layers and the cloud formation, can manifest as oscillations within the cloud layers during the evening and nocturnal hours. These waves can produce characteristic patterns, with alternating crests and troughs that create a mesmerizing and captivating visual display. The presence of wind shear, particularly near the boundaries of the cloud formation, can further enhance the formation and propagation of these wave-like features.

The formation of wave-like features within the Morning Glory Clouds during the evening and nocturnal periods provides not only an aesthetic appeal but also valuable insights into the atmospheric dynamics and interactions involved in their formation.

Nocturnal Rituals and Local Traditions

The evening and nocturnal presence of Morning Glory Clouds have long been intertwined with local rituals and traditions in certain communities. Indigenous cultures, particularly those living in regions where these cloud formations are prevalent, have developed rituals and practices associated with their appearance during the evening and nighttime hours.

Some indigenous communities view the nocturnal presence of Morning Glory Clouds as a spiritual event, often symbolizing the transition from one day to another or representing a celestial connection. These rituals may involve storytelling, singing, and dance performances that celebrate the beauty and mystique of these unique cloud formations.

In addition to indigenous traditions, the nocturnal presence of Morning Glory Clouds has inspired various cultural practices and events in local communities. Nighttime festivals and gatherings, where people come together to witness and appreciate the captivating display of these clouds, have become popular in some regions. These events not only promote awareness and appreciation of the natural phenomenon but also contribute to the local economy through tourism and cultural exchange.

Exploring Nocturnal Features: Challenges and Opportunities

Studying the evening and nocturnal features of Morning Glory Clouds presents unique challenges and opportunities for researchers and enthusiasts alike. The limited visibility, altered cloud structure, and nocturnal atmospheric stability require specialized observation techniques and instruments to accurately document and investigate these phenomena.

One of the challenges is the limited availability of natural lighting during the evening and nocturnal periods. Researchers often rely on artificial lighting sources, such as floodlights or powerful spotlights, to illuminate the cloud formation for detailed observation and photography. Balancing the use of artificial lighting with the desire to capture the organic essence of the clouds poses a creative challenge for photographers and researchers.

Another challenge lies in the limited accessibility and safety concerns associated with nocturnal observations. In remote or rugged areas where Morning Glory Clouds are commonly observed, conducting research or capturing photographs during the evening and nighttime hours can be logistically challenging. Safety precautions and careful planning are necessary to ensure the well-being of researchers and photographers in these environments.

Despite these challenges, the evening and nocturnal features of Morning Glory Clouds provide fertile ground for research and exploration. The unique behaviors and structures observed during these periods offer opportunities to deepen our understanding of atmospheric processes, wave dynamics, and nocturnal cloud formations. Collaborative efforts between meteorologists, scientists, indigenous communities, and photographers can uncover new insights and contribute to the overall knowledge of Morning Glory Clouds during the nighttime hours.

In conclusion, the evening and nocturnal features of Morning Glory Clouds bring about distinct changes in their appearance and behavior. The reduced visibility, modifications in cloud structure, nocturnal atmospheric stability, formation of wave-like features, and the intertwining of cultural rituals and traditions create a captivating and mysterious atmosphere. Despite the challenges posed by limited visibility and nocturnal observations, exploring and understanding these features offer valuable insights into atmospheric dynamics, indigenous knowledge, and cultural significance. The study of these evening and nocturnal aspects of Morning Glory Clouds opens up avenues for collaborative research, creative expression, and cultural preservation.

Daily Rituals and Local Traditions

Morning Glory Clouds have captured the imagination and wonder of local communities living in regions where these unique atmospheric phenomena occur. For centuries, these communities have developed daily rituals and local traditions associated with the appearance and behavior of Morning Glory Clouds. In this section, we will explore some of these rituals and traditions and delve into their cultural significance.

Local Perceptions and Beliefs

Different indigenous cultures and communities have developed their own interpretations and beliefs surrounding Morning Glory Clouds. For example, the Pitjantjatjara people of Central Australia refer to these clouds as "Kanypi," which translates to "clouds that come." They believe that the Morning Glory Clouds bring good fortune and abundance to their lands.

In Brazil, where the Morning Glory Clouds are known as "Mar de Dentro," local fishermen have associated the appearance of these clouds with an increase in fish abundance. This belief has led to the practice of organizing early morning fishing trips when the Morning Glory Clouds are visible, as fishermen believe it enhances their chances of a bountiful catch.

Morning Rituals and Observances

The daily cycle of Morning Glory Clouds plays a significant role in the morning rituals of communities residing in regions where these clouds occur. One such community is the Burketown region in Queensland, Australia, which experiences regular Morning Glory Cloud phenomena.

In Burketown, locals gather at vantage points early in the morning to witness the arrival of the Morning Glory Cloud. These gatherings have become a daily ritual, with community members eagerly anticipating the uniquely beautiful cloud formation. The moment the Morning Glory Cloud appears on the horizon, it is met with cheers and applause, creating a sense of unity and shared wonder among the observers.

Mythology and Folklore

Morning Glory Clouds have inspired numerous myths and stories in indigenous cultures. These myths often explore the origins and symbolism of the clouds and are passed down through generations via oral traditions.

In some indigenous cultures, it is believed that the Morning Glory Clouds are messengers from the spiritual realm. It is said that these clouds carry the prayers and wishes of the people to the heavens, acting as intermediaries between humans and the divine. This belief has led to various rituals and ceremonies centered around the appearance of Morning Glory Clouds, with people offering their prayers and expressing gratitude for the natural wonders they bring.

Cultural Celebrations

In regions where Morning Glory Clouds are an integral part of the local culture, annual celebrations and festivals are organized to honor these spectacular formations. One such example is the Morning Glory Cloud Festival held in Burketown, Queensland, Australia.

During the festival, the community comes together to celebrate the beauty and mystery of the Morning Glory Clouds. The festival features traditional music and dance performances, art exhibitions inspired by the clouds, and educational workshops and presentations highlighting the scientific and meteorological aspects of Morning Glory Clouds. It is an opportunity for locals and visitors alike to immerse themselves in the cultural significance of these natural marvels.

Preserving Traditions and Indigenous Knowledge

As modernization and globalization continue to impact indigenous cultures, there is a growing need to preserve and promote the traditions and knowledge associated with Morning Glory Clouds. Efforts are being made by local communities, researchers, and cultural organizations to document and record the myths, stories, rituals, and traditions tied to these phenomena.

Projects aimed at preserving indigenous knowledge utilize various mediums such as audio recordings, visual documentation, and written accounts. These initiatives create a valuable resource for future generations, ensuring that the cultural significance and wisdom surrounding Morning Glory Clouds are not lost.

Cultural Significance and Future Directions

The daily rituals and local traditions tied to Morning Glory Clouds provide insight into the profound impact these atmospheric phenomena have on communities' cultural fabric. These rituals and traditions are a testament to the power of nature to inspire wonder, foster togetherness, and preserve indigenous knowledge.

Understanding and appreciating cultural significance is crucial for researchers and scientists studying Morning Glory Clouds. By collaborating with local communities and incorporating their perspectives and knowledge, researchers can gain a more holistic understanding of these phenomena. This inclusive approach not only enriches scientific research but also fosters cultural exchange and mutual respect.

Future research directions should strive to further explore and document the rituals and traditions associated with Morning Glory Clouds, as they provide valuable insights into the human perception of natural wonders. Additionally, efforts should be made to support and promote the preservation of indigenous knowledge and cultural practices, ensuring that future generations can continue to learn from and be inspired by the wisdom of their ancestors.

In conclusion, the daily rituals and local traditions tied to Morning Glory Clouds are deeply rooted in the cultural heritage of indigenous communities. These traditions offer a glimpse into the human fascination with the natural world and highlight the importance of cultural preservation and inclusivity in scientific research. As we continue to unravel the mysteries of Morning Glory Clouds, it is essential to recognize and celebrate the cultural significance woven into their awe-inspiring presence.

Understanding the Diurnal Cycle through Numerical Models

The diurnal cycle refers to the daily variation in the behavior of Morning Glory clouds. It involves distinct patterns of formation, dissipation, and movement of the clouds throughout the day. Understanding this cycle is essential for predicting the occurrence of Morning Glory clouds and their potential impacts on local weather and aviation. Numerical models play a crucial role in unraveling the complex dynamics of this diurnal cycle.

Simulating Diurnal Variation

Numerical models provide a powerful tool for simulating the diurnal cycle of Morning Glory clouds. These models use complex mathematical equations to represent the behavior of the atmosphere, including the interaction between various atmospheric processes such as temperature, humidity, wind, and cloud formation. By solving these equations numerically, we can simulate the diurnal variation of Morning Glory clouds under different conditions.

To simulate the diurnal cycle, the numerical models incorporate physical processes that drive the formation and dissipation of Morning Glory clouds. These processes include the evolution of atmospheric stability, moisture availability, and wind patterns throughout the day.

Equations and Parameters

The equations used in numerical models are based on principles from fluid dynamics, thermodynamics, and cloud physics. These equations describe the conservation of mass, momentum, and energy in the atmosphere. They also incorporate parameterizations to represent small-scale processes that cannot be resolved directly in the model grid.

The equations take into account factors such as solar radiation, atmospheric heating and cooling rates, vertical motion, and atmospheric moisture. The input parameters for the models include initial atmospheric conditions, boundary conditions, and external forcings such as diurnal variations in solar radiation.

Resolved vs. Parameterized Processes

Numerical models use a grid system to divide the atmosphere into discrete cells. The resolution of the grid determines the smallest-scale features that can be resolved by the model. Large-scale processes such as synoptic weather systems can be resolved directly in the model, while smaller-scale processes such as turbulent mixing within the cloud may need to be parameterized.

Resolved processes capture the detailed interactions between different atmospheric variables and provide more accurate representations of the diurnal cycle. However, they require higher computational resources and can be computationally expensive. Parameterized processes, on the other hand, simplify the representation of small-scale processes by using mathematical formulas based on empirical relationships or simplified physical concepts. These parameterizations enable simulations to be performed at coarser resolutions with less computational cost.

Validation and Verification

To ensure the accuracy of the numerical model simulations, it is necessary to validate and verify the model outputs against observations. This involves comparing the simulated diurnal cycle of Morning Glory clouds with real-world data obtained from ground-based measurements, remote sensing instruments, and aircraft observations.

Validation involves checking whether the model outputs reproduce the observed features of the diurnal cycle, such as the timing of cloud formation and dissipation, the vertical structure of the cloud, and the associated meteorological conditions. Verification, on the other hand, focuses on assessing the overall performance of the model by comparing its outputs against independent datasets.

Limitations and Challenges

Numerical models face several challenges in simulating the diurnal cycle of Morning Glory clouds. One key challenge is the representation of key physical processes within the models, such as the interaction between gravity waves and cloud formation. The parameterization of small-scale processes, such as turbulence and convection, also poses challenges due to their complex and unpredictable nature.

Another challenge is the availability of high-quality observational data for model validation. Morning Glory clouds are relatively rare phenomena that occur in specific regions, making it difficult to obtain comprehensive observational datasets. This limited data availability hampers the ability to fully validate and refine the models.

Future Directions

Advancements in computing power and modeling techniques offer promising avenues for improving the understanding of Morning Glory cloud diurnal cycles. High-resolution models that can capture small-scale processes and interactions more accurately are being developed. These models utilize innovative numerical methods and assimilation techniques to assimilate observational data and improve model performance.

Furthermore, the integration of remote sensing data, such as satellite imagery and radar observations, into numerical models can provide valuable information about the spatial and temporal variability of Morning Glory clouds. This integration can enhance our understanding of the diurnal cycle and improve the representation of cloud features in the models.

Collaborative efforts between meteorologists, climatologists, and mathematicians are crucial for advancing the field of numerical modeling of Morning Glory cloud diurnal cycles. By combining expertise from different disciplines, innovative approaches can be developed to tackle the challenges associated with simulating these complex phenomena.

In Summary

Numerical models offer a valuable tool for understanding the diurnal cycle of Morning Glory clouds. By simulating the complex interactions between various atmospheric processes, these models provide insights into the formation, dissipation, and movement patterns of the clouds throughout the day. However, challenges such as process representation and limited observational data remain, necessitating ongoing research and collaboration to further improve the accuracy and reliability of these models.

Formation Mechanisms

Gravity Waves

Definition and Characteristics of Gravity Waves

Gravity waves are an important phenomenon in the study of atmospheric dynamics and play a significant role in the formation and behavior of morning glory clouds. In this section, we will explore the definition and key characteristics of gravity waves, providing a foundation for understanding their role in the formation of these unique cloud formations.

Definition of Gravity Waves

Gravity waves can be defined as oscillations or disturbances in the atmosphere that propagate vertically and horizontally due to the presence of gravity. These waves are not to be confused with gravitational waves, which are ripples in the fabric of spacetime caused by celestial events such as the merging of black holes.

Gravity waves in the atmosphere are generated by a variety of mechanisms, including orographic forcing (airflow over mountains), convection (rising and sinking air masses), frontal systems (interactions between air masses of different temperatures), and wind shear (variation in wind speed with height).

Characteristics of Gravity Waves

Gravity waves have several distinctive characteristics that differentiate them from other atmospheric waves. These characteristics include:

- **Wave Period and Frequency**: Gravity waves have a specific period, which is the time it takes for one wave cycle to pass a given point.

The frequency of a gravity wave is the reciprocal of its period. Both the period and frequency of gravity waves can vary depending on the source and atmospheric conditions.

- **Vertical Propagation**: Gravity waves primarily propagate vertically, but they can also exhibit some horizontal movement. As they propagate upward, they transfer energy and momentum from the lower atmosphere to the upper atmosphere.

- **Wave Structure**: Gravity waves are characterized by alternating areas of upward and downward motion within the wave pattern. These areas of compression and rarefaction create oscillations in atmospheric pressure, temperature, and wind speed.

- **Wave Amplitude**: The amplitude of a gravity wave refers to the maximum displacement from the wave's equilibrium position. It represents the strength or intensity of the wave, with larger amplitudes indicating more significant disturbances in the atmosphere.

- **Wave Phase**: The phase of a gravity wave describes the position of a given point within the wave cycle. It determines whether the point is experiencing upward or downward motion. The phase of a gravity wave can affect its interaction with other atmospheric features, such as cloud formation.

- **Wave Speed**: Gravity waves travel at a characteristic speed, which depends on the properties of the medium in which they propagate. In the atmosphere, gravity waves typically have speeds ranging from a few meters per second to tens of meters per second.

- **Atmospheric Stability**: The presence of gravity waves can provide information about the stability of the atmosphere. In regions where the atmosphere is stable, gravity waves can be suppressed or damped out. Conversely, unstable atmospheric conditions favor the generation and propagation of gravity waves.

Understanding the characteristics of gravity waves is crucial for unraveling the complex dynamics involved in the formation of morning glory clouds. These waves are responsible for initiating the rolling motion that gives morning glory clouds their unique shape and structure.

Additionally, the interaction between gravity waves and other atmospheric factors, such as wind patterns and temperature gradients, further influences the development and behavior of morning glory clouds.

Quantifying Gravity Waves

Gravity waves can be quantified and analyzed using various observational and mathematical techniques. Some of the commonly used approaches include:

- **Balloon Observations:** Weather balloons equipped with instruments such as radiosondes can measure atmospheric pressure, temperature, humidity, and wind speed throughout the vertical extent of the atmosphere. These observations help identify the presence and characteristics of gravity waves.

- **Remote Sensing:** Satellite-based instruments and ground-based radars can detect gravity waves by observing changes in cloud patterns, temperature profiles, or wind speed variations. These remote sensing techniques provide a global view of gravity wave activity and help researchers understand their spatial distribution.

- **Modeling and Simulation:** Numerical models, such as computational fluid dynamics (CFD) and global climate models, can be used to simulate gravity waves and study their behavior under different atmospheric conditions. These models help researchers gain insights into the physical processes driving gravity wave formation and propagation.

- **Spectral Analysis:** Spectral analysis is a mathematical technique used to decompose a complex wave into its component frequencies. By applying spectral analysis to atmospheric data, researchers can identify the presence of gravity waves and determine their characteristics, including frequency, wavelength, and wave period.

These techniques, combined with field observations and laboratory experiments, contribute to our understanding of gravity waves and their role in morning glory cloud formation. By quantifying the properties of gravity waves, researchers can develop more accurate forecasting models and improve our ability to predict the occurrence and behavior of morning glory clouds.

Challenges and Future Directions

While significant progress has been made in understanding gravity waves, several challenges and areas for further research remain. Some of these challenges include:

- **Localized Observations**: Gravity waves are highly localized phenomena, making it challenging to capture their characteristics accurately. Developing advanced observational techniques, such as high-resolution satellite imagery and targeted field campaigns, can help improve our understanding of gravity waves at smaller scales.

- **Gravity Wave-Cloud Interactions**: The interaction between gravity waves and clouds is complex and not yet fully understood. Further research is needed to unravel the intricate feedback mechanisms between gravity waves and cloud formation, which will improve our understanding of morning glory clouds and their behavior.

- **Modeling Improvements**: Numerical models used to simulate gravity waves still face challenges in accurately representing the dynamics and properties of these waves. Improving the parameterization of gravity wave effects in models and incorporating more detailed physics will contribute to better simulations and predictions.

- **Climate Change Effects**: The impact of climate change on gravity waves and their relationship with morning glory clouds is another area that requires further investigation. Understanding how changing atmospheric conditions affect the occurrence and behavior of gravity waves will provide valuable insights into the future of morning glory clouds.

As researchers continue to explore gravity waves and their connections to morning glory clouds, advancements in observational techniques, modeling, and interdisciplinary collaborations are key to advancing our knowledge in this field. The study of gravity waves opens doors to new discoveries not only in atmospheric science but also in our understanding of how complex systems interact and shape Earth's climate and weather patterns.

Generation and Propagation of Gravity Waves

Gravity waves play a crucial role in the formation and propagation of Morning Glory Clouds. In this section, we will explore the generation mechanisms of gravity waves and how they propagate through the atmosphere, ultimately leading to the formation of these unique cloud formations.

Definition and Characteristics of Gravity Waves

Gravity waves are oscillations in the atmosphere that are generated by disturbances in the vertical motion of air parcels due to buoyancy forces. These waves result from the restoring force of gravity attempting to return the disturbed air parcels to their equilibrium positions. They can be observed in various forms, such as ripples on the surface of water or the undulating motion of the clouds.

Gravity waves can be categorized based on their horizontal wavelength and vertical wavelength. Short-period gravity waves have wavelengths on the order of a few kilometers, while long-period gravity waves can have wavelengths up to hundreds of kilometers. The vertical wavelength, on the other hand, refers to the distance between two consecutive crests or troughs of the wave.

Generation and Propagation of Gravity Waves

Several mechanisms contribute to the generation of gravity waves in the atmosphere. One of the primary mechanisms is orographic lifting, where air is forced to rise over mountains or other significant topographic barriers. As the air ascends, it undergoes vertical disturbances, leading to the generation of gravity waves.

Another important mechanism is frontal lifting, which occurs when two contrasting air masses interact along a frontal boundary. As the denser air displaces the less dense air, gravity waves can be generated due to the differences in buoyancy forces between the two air masses.

Convective processes, such as thunderstorm updrafts and downdrafts, also play a role in gravity wave generation. As the convective currents move vertically and interact with the surrounding atmosphere, they create disturbances that propagate as gravity waves.

Once generated, gravity waves propagate through the atmosphere, carrying the energy and momentum associated with their vertical displacement. The propagation speed of gravity waves depends on the background wind profile and the vertical structure of the atmosphere.

In the case of Morning Glory Clouds, gravity waves are responsible for the distinctive roll-like cloud formations. These waves are typically generated by the interaction of strong wind flows, such as the sea breeze, with stable atmospheric layers. As the wind encounters a sharp inversion layer, it is forced to ascend, creating a disturbance that propagates vertically and horizontally, leading to the characteristic wave pattern seen in Morning Glory Clouds.

Interaction with Atmospheric Layers

Gravity waves interact with various atmospheric layers as they propagate through the atmosphere. One of the key interactions occurs at the tropopause, which acts as a barrier to the upward propagation of waves. This interaction leads to the reflection and refraction of gravity waves, causing changes in their direction and amplitude.

Another important interaction occurs with the boundary layer, the lowest layer of the atmosphere in contact with the Earth's surface. Gravity waves can penetrate into the boundary layer, influencing its stability and turbulence characteristics.

When gravity waves encounter stable layers in the atmosphere, such as inversions, they can be partially or fully blocked, leading to wave breaking. Wave breaking occurs when the wave amplitude exceeds a critical value, causing the wave energy to be dissipated and transferred to turbulence.

Effect on Morning Glory Cloud Formation

Gravity waves play a critical role in the formation of Morning Glory Clouds. As the gravity waves propagate through the atmosphere, they induce vertical motion in the air, leading to the formation of roll clouds. The ascending and descending air motions associated with the waves generate the characteristic cloud pattern observed in Morning Glory Clouds.

The vertical structure of the waves, along with the stability and moisture content of the atmospheric layers they interact with, contribute to the size, shape, and persistence of the Morning Glory Clouds. The interaction of gravity waves with stable layers can enhance or dampen the growth of the cloud formation, resulting in variations in cloud structure and longevity.

Understanding the generation and propagation of gravity waves is crucial for accurately predicting and forecasting the occurrence of Morning Glory Clouds. By studying the characteristics and behaviors of these waves, researchers can develop better models and simulations to improve forecasting techniques and provide early warning systems for aviation and other affected sectors.

Experimental and Numerical Simulations of Gravity Waves

Several experimental methods and numerical simulations have been employed to study gravity waves and their effects on cloud formations. High-altitude aircraft and research balloons equipped with instruments have been used to measure the vertical structure and characteristics of gravity waves.

Advanced radar systems and satellite observations provide valuable data on the horizontal and vertical distribution of gravity waves in the atmosphere. These remote sensing techniques allow researchers to track the propagation of waves over large spatial scales.

Numerical models, such as computational fluid dynamics (CFD) models and mesoscale weather models, are also instrumental in simulating and studying the behavior of gravity waves. These models can help in understanding the complex interactions between gravity waves and atmospheric conditions, providing insights into the formation and evolution of Morning Glory Clouds.

Future directions in gravity wave research involve the development of more accurate and high-resolution models that can simulate the fine-scale details of wave propagation. Additionally, incorporating data from citizen science initiatives and crowdsourcing efforts can further enhance our understanding of gravity waves and improve the prediction of Morning Glory Clouds.

In conclusion, gravity waves are fundamental to the generation and propagation of Morning Glory Clouds. By exploring the mechanisms by which these waves are generated, their interactions with atmospheric layers, and their effects on cloud formation, we can gain a better understanding of this unique meteorological phenomenon. Continued research in this field will contribute to improved weather forecasting, aviation safety, and a deeper appreciation for the intricate dynamics of our atmosphere.

Interaction with Atmospheric Layers

Morning Glory Clouds, with their unique structure and behavior, are an intriguing atmospheric phenomenon that has captivated the attention of scientists and enthusiasts alike. In order to understand the formation and characteristics of these clouds, it is crucial to examine their interaction with the various layers of the atmosphere. By studying the complex interplay between different atmospheric layers, we can gain valuable insights into the mechanisms behind the formation and maintenance of Morning Glory Clouds.

The Atmosphere: Layers and Properties

The Earth's atmosphere is composed of several layers, each with its own distinct properties and characteristics. The troposphere, which extends from the Earth's surface up to around 15 kilometers, is the layer that directly interacts with weather and cloud formation. Above the troposphere are the stratosphere, mesosphere, thermosphere, and exosphere, each with its own unique composition and dynamics.

The troposphere is of particular interest when studying Morning Glory Clouds, as it is the layer where most weather phenomena occur. It is characterized by decreasing temperature with increasing altitude, known as the lapse rate, which sets the stage for various atmospheric processes to take place.

Temperature Inversions and Stability

One of the crucial factors influencing the formation and behavior of Morning Glory Clouds is the presence of temperature inversions within the troposphere. A temperature inversion occurs when the normal lapse rate is inverted, resulting in an increase in temperature with altitude instead of the usual decrease.

Temperature inversions can create stable atmospheric conditions, preventing vertical mixing of air masses and trapping moisture within a specific layer. This stability plays a significant role in the formation and maintenance of Morning Glory Clouds, as it allows for the preservation of the distinctive cloud structure over extended periods.

GRAVITY WAVES

Interactions with the Boundary Layer

The boundary layer, the lowest portion of the troposphere closest to the Earth's surface, plays a crucial role in the formation of Morning Glory Clouds. During the pre-dawn hours, the boundary layer is cooled by radiative cooling, leading to the development of a nocturnal inversion.

As the sun rises, it initiates a process known as thermal mixing, which entails the gradual increase of temperature and the breakdown of the nocturnal inversion. This mixing process can trigger the formation of gravity waves, which are a key mechanism behind the development of Morning Glory Clouds.

Gravity Waves and Cloud Formation

Gravity waves are oscillations in the atmosphere caused by the displacement of air parcels due to buoyancy forces. They can be generated through a variety of mechanisms, such as the interaction of wind with mountains or the boundary layer, and can propagate vertically and horizontally across different atmospheric layers.

In the context of Morning Glory Clouds, gravity waves play a crucial role in their formation. As the expanding warm air from the thermal mixing interacts with the stable layer above, it creates waves that propagate horizontally through the troposphere. These waves can then induce rising motion and the condensation of moisture, leading to the formation of the characteristic roll clouds of Morning Glory Clouds.

Wave Amplification and Breaking

The behavior of gravity waves within the troposphere is influenced by numerous factors, including wind shear, temperature gradients, and moisture distribution. Under specific conditions, gravity waves can undergo amplification, resulting in the formation of large-scale waves with a higher amplitude.

When the amplitude of gravity waves becomes sufficiently large, they can reach a state of instability known as wave breaking. This breaking process leads to a rapid mixing of air masses and triggers enhanced turbulent mixing within the clouds. Wave breaking is often associated with the formation of multiple roll clouds and the intricate cloud patterns seen in Morning Glory Clouds.

Wave-Cloud Interactions

The interaction between gravity waves and clouds is a complex and dynamic process that influences the structure and evolution of Morning Glory Clouds. As gravity waves propagate through the troposphere, they can modulate the growth and organization of cloud bands, leading to the formation of secondary cloud structures.

These secondary cloud structures, often observed as billows or undulations along the main roll clouds, are a result of the interaction between the rising motion induced by the gravity waves and the cloud condensation process. Understanding these wave-cloud interactions is crucial for unraveling the intricate dynamics and evolution of Morning Glory Clouds.

Summary

The interaction of Morning Glory Clouds with the various layers of the atmosphere shapes their formation, maintenance, and characteristics. The stability provided by temperature inversions, the influence of the boundary layer, and the role of gravity waves and their interactions with clouds all contribute to the unique structure and behavior of these captivating clouds. By studying these interactions, we can deepen our understanding of Morning Glory Clouds and gain insights into the broader dynamics of the Earth's atmosphere.

Effect on Morning Glory Cloud Formation

Morning Glory Clouds, as fascinating as they are, are not simply random cloud formations. They are influenced by various factors that affect their formation and behavior. In this section, we will explore the different effects that contribute to the formation of Morning Glory Clouds.

Atmospheric Stability

Atmospheric stability plays a crucial role in determining the formation of Morning Glory Clouds. These clouds typically form in an unstable atmosphere, where warm air rises and cools as it ascends. The cooling of the air creates temperature differences, which can generate gravity waves.

Gravity waves are oscillations in the atmosphere caused by buoyancy forces. As gravity waves propagate horizontally, they can create areas of convergence and divergence in the air. The convergence of air results in the upward movement of warm air, which favors cloud formation. This vertical motion leads to the development of a rolling feature known as the Morning Glory Cloud.

However, if the atmosphere is too stable, with limited temperature variations, the buoyancy forces necessary for gravity waves may not be strong enough to trigger the formation of Morning Glory Clouds. On the other hand, if the atmosphere is too unstable, with excessive temperature differences, the cloud formation may be disrupted by convective processes.

Influence of Topography

The presence of mountains and other topographical features can significantly impact the formation of Morning Glory Clouds. When prevailing winds encounter a barrier such as a mountain range, they are forced to ascend, creating a favorable environment for cloud development.

As the moist air is lifted over the mountain, it can undergo adiabatic cooling, leading to the formation of clouds. This cloud formation process is known as orographic lifting. If the conditions are right, the orographically lifted air can generate gravity waves, contributing to the formation of Morning Glory Clouds.

Topography can also affect the shape and structure of Morning Glory Clouds. The rolling motion of these clouds can be influenced by the interaction between the winds, topography, and gravity waves. In some cases, the alignment of the Morning Glory Cloud rolls may align with the valleys or channels formed by the topography, creating a more pronounced and organized cloud formation.

Influence of Synoptic-Scale Weather Systems

Synoptic-scale weather systems, such as cyclones and anticyclones, can influence the formation of Morning Glory Clouds. Cyclones are characterized by low-pressure systems, while anticyclones are characterized by high-pressure systems.

In the presence of a cyclone, the rotation of the low-pressure system can induce convergence of air near the center of the system. This convergence can lead to the upward movement of air and subsequent cloud formation. Morning Glory Clouds may develop along the boundaries of these cyclones, where the converging winds interact with gravity waves.

Anticyclones, on the other hand, are associated with stable and sinking air. In general, the presence of anticyclones inhibits the formation of Morning Glory Clouds. However, under certain circumstances, anticyclones can interact with other weather systems, such as frontal boundaries, to create the necessary instability for Morning Glory Cloud formation.

Sea Breeze and Land Breeze Interactions

Sea breeze and land breeze circulations can also play a role in the formation of Morning Glory Clouds, particularly in coastal regions. During the day, the land near the coast heats up more quickly than the adjacent ocean. This temperature difference creates a pressure gradient that induces the sea breeze circulation.

As the sea breeze moves inland, it can interact with the existing atmospheric conditions and trigger the development of gravity waves. These gravity waves can initiate the formation of Morning Glory Clouds. Conversely, during the night, when the land cools more rapidly than the ocean, the land breeze circulation can create similar effects.

The strength and timing of the sea breeze and land breeze circulations, along with other factors discussed previously, are crucial for the initiation and maintenance of Morning Glory Cloud formations in coastal areas.

Interactions with Moisture and Condensation

Moisture availability and condensation processes also contribute to the formation of Morning Glory Clouds. The presence of abundant moisture is necessary for the formation of clouds. When there is sufficient moisture in the atmosphere, water vapor condenses onto tiny particles called condensation nuclei. This process leads to the formation of cloud droplets.

Morning Glory Clouds often form in regions where the temperature and humidity conditions are favorable for condensation. As the rising air cools, it reaches its dew point, which is the temperature at which saturation occurs. When saturation is reached, water vapor condenses into visible cloud droplets, forming the characteristic roll clouds of Morning Glory Clouds.

In some cases, the presence of pollutants or dust particles in the atmosphere can serve as additional condensation nuclei, enhancing the cloud formation process. This interaction between moisture, condensation, and atmospheric particles contributes to the unique attributes of Morning Glory Clouds.

Summary

In summary, the formation of Morning Glory Clouds is influenced by several factors. Atmospheric stability, topography, synoptic-scale weather systems, sea breeze and land breeze interactions, and moisture availability all contribute to the formation and behavior of these fascinating cloud formations.

Understanding the effects of these factors on Morning Glory Clouds can help meteorologists and researchers predict their occurrence and unravel the underlying mechanisms behind their formation. By studying the intricate interactions between the atmosphere, terrain, and meteorological phenomena, we can continue to deepen our understanding of Morning Glory Clouds and their impact on our atmosphere.

Experimental and Numerical Simulations of Gravity Waves

Experimental and numerical simulations play a crucial role in understanding the behavior and characteristics of gravity waves, a phenomenon closely linked to the formation of morning glory clouds. Through these simulations, scientists can gain insights into the intricate dynamics and mechanisms involved in the generation, propagation, and interaction of gravity waves in the atmosphere.

Experimental Methods

Various experimental methods are employed to study gravity waves and their effects on atmospheric phenomena. One often-used technique is the use of weather balloons equipped with instruments such as radiosondes. These devices measure atmospheric variables such as temperature, pressure, humidity, and wind speed at different altitudes. By launching weather balloons at different locations and times, scientists can gather valuable data on the presence and behavior of gravity waves.

In addition to weather balloons, research aircraft are also utilized to obtain observational data on gravity waves. These aircraft are equipped with sensors and instruments that provide detailed measurements of atmospheric conditions. By flying at different altitudes and locations, researchers can capture a comprehensive picture of the structure and properties of gravity waves.

Another experimental method involves the deployment of ground-based instruments such as radar and lidar. These instruments emit beams of energy and measure the reflections or scattering caused by atmospheric particles and clouds. By analyzing the changes in these reflections or scattering patterns, researchers can infer the presence and characteristics of gravity waves.

Numerical Simulations

Numerical simulations play a crucial role in understanding the complex dynamics of gravity waves. These simulations involve the use of mathematical models and computational techniques to simulate the behavior of the atmosphere. By solving the equations governing atmospheric motion, researchers can simulate the generation, propagation, and interaction of gravity waves.

Numerical simulations of gravity waves require accurate representations of the atmospheric conditions and physical processes involved. These simulations can be performed using various numerical methods such as finite difference, finite element, or spectral methods. The choice of method depends on the specific requirements of the simulations and the available computational resources.

To simulate gravity waves, researchers input initial conditions and boundary conditions into the numerical models. These conditions represent the state of the atmosphere at a certain time and location. By solving the equations iteratively, the model can simulate the evolution of the atmospheric variables over time and reveal the behavior of gravity waves.

Numerical simulations also allow researchers to investigate the impact of various factors on gravity wave generation and propagation. For example, simulations can be used to study how topography, wind patterns, temperature gradients, and other atmospheric conditions influence the formation and characteristics of gravity waves. By altering the input parameters in the model, researchers can explore different scenarios and understand the underlying processes.

Furthermore, numerical simulations can help validate observational data and provide insights that are difficult to obtain through experiments alone. By comparing the simulated results with observed data, researchers can assess the accuracy of the model and refine its parameters. This iterative process improves the understanding of gravity waves and their role in the formation of morning glory clouds.

Challenges and Future Directions

Despite the advancements in experimental and numerical techniques, studying gravity waves still poses several challenges. One major challenge is the complex nature of the atmosphere, which involves multiple interacting processes and variables. Capturing the full range of atmospheric dynamics in numerical models is a challenging task that often requires simplifications and assumptions.

Another challenge is the computational cost of running high-resolution numerical simulations. The accurate representation of gravity waves requires fine grid resolutions, which can be computationally expensive. Researchers continue to develop techniques to optimize computational resources and increase the efficiency of simulations.

Future research directions in experimental and numerical simulations of gravity waves aim to improve the understanding of their generation, propagation, and interaction mechanisms. Advancements in observational techniques, such as the use of advanced weather balloons and remote sensing instruments, can provide more detailed and accurate data for validation and refinement of numerical models.

Furthermore, the integration of data from multiple sources, such as satellite observations, ground-based measurements, and in situ data, can enhance the accuracy and reliability of simulations. Collaborative efforts between scientists and institutions can promote data sharing and foster interdisciplinary research, leading to a deeper understanding of gravity waves and their influence on atmospheric phenomena.

In conclusion, experimental and numerical simulations are essential tools for investigating the characteristics and behavior of gravity waves, offering insights into the formation of morning glory clouds. Through both experimental observations and computational modeling, scientists can unravel the intricate dynamics of gravity waves and their interaction with the atmosphere. Challenges in computational cost and model representation continue to be addressed, paving the way for future advancements in understanding and predicting gravity wave phenomena.

Soliton Theory

Introduction to Solitons

In this section, we will explore the fascinating concept of solitons and their relevance to the formation of Morning Glory Clouds. Solitons are unique wave phenomena that possess remarkable properties, including their ability to maintain their shape and structure even while propagating through a medium. Understanding solitons is crucial in unraveling the mysteries behind the formation and behavior of Morning Glory Clouds.

Definition and Characteristics of Solitons

Solitons are special types of nonlinear waves that propagate without spreading or dissipating their energy. Unlike most waves that disperse as they travel, solitons retain their shape and intensity due to a delicate balance between nonlinear and dispersive effects. This distinctive behavior sets solitons apart from other wave phenomena and makes them an intriguing subject of study in various scientific disciplines, including fluid dynamics and nonlinear optics.

One of the fundamental characteristics of solitons is their ability to maintain their shape and velocity over long distances. This property, known as soliton stability, allows solitons to travel vast distances without losing their original form. Solitons are also known to be highly localized, which means they possess a well-defined spatial structure. This attribute makes them suitable for carrying information in various communication systems.

Generation and Propagation of Solitons

Solitons can be generated through various mechanisms, depending on the specific system under consideration. In the context of Morning Glory Clouds, solitons are generated through the interaction of gravity waves with the stable atmospheric boundary layer. Gravity waves are initially generated by disturbances such as wind shear or topographic features, and when these waves interact with the stable layer, solitonic behavior emerges.

The propagation of solitons is governed by the governing equations that describe the particular wave phenomenon. In the case of Morning Glory Clouds, solitons propagate within the stable layer of the atmosphere, carrying the energy and characteristics of the waves that generated them.

The dynamics of solitons are an intricate interplay between nonlinearity, dispersion, and the underlying atmospheric conditions.

Interaction with Atmospheric Layers

As solitons propagate through the stable atmospheric boundary layer, they interact with various atmospheric layers, leading to significant changes in their properties. The interaction between solitons and the layers above and below the stable layer influences their speed, amplitude, and shape. These interactions can lead to the amplification or attenuation of solitons, causing variations in the Morning Glory Cloud formation.

One key aspect of the interaction between solitons and atmospheric layers is the effect of wind shear. Wind shear refers to the change in wind speed or direction with height, and it plays a crucial role in shaping and maintaining solitons in the stable layer. The interaction between solitons and wind shear can lead to phenomena such as soliton trains, where multiple solitons are arranged in a periodic manner.

Effect on Morning Glory Cloud Formation

The presence of solitons in the stable atmospheric boundary layer has a profound impact on the formation and evolution of Morning Glory Clouds. The solitons provide the necessary mechanism for the energy transfer from the generating gravity waves to the cloud structure. They help in organizing the wave energy into coherent roll-like structures, which are the hallmark of Morning Glory Clouds.

The stable propagation of solitons ensures that the cloud structure remains intact over long distances, enabling Morning Glory Clouds to stretch for hundreds of kilometers. The interaction between solitons and atmospheric layers also plays a role in the vertical growth and stability of the cloud structure. Understanding the dynamics of solitons is therefore crucial in deciphering the mechanisms responsible for the formation and persistence of Morning Glory Clouds.

Experimental and Theoretical Studies on Soliton Behavior

The study of solitons has been a topic of interest for scientists and researchers from various disciplines. Experimental investigations have employed techniques such as laboratory simulations and field measurements to observe and characterize solitons in different physical systems. For example, in laboratory experiments, solitons have been observed in water tanks, where controlled disturbances generate solitonic waves.

Theoretical studies have also contributed to our understanding of solitons and their behavior. Mathematical models, such as the Korteweg-de Vries (KdV) equation, have been used to describe soliton dynamics in different contexts. Numerical simulations based on these models provide insights into the complex behavior of solitons and their interactions with the surrounding environment.

Future Directions for Soliton Research in Morning Glory Clouds

Solitons continue to be an active area of research, with ongoing studies aiming to deepen our understanding of their role in the formation and dynamics of Morning Glory Clouds. Future research directions include:

1. Improved observational techniques: Developing advanced remote sensing techniques, such as high-resolution radar and lidar systems, to capture detailed information about solitons within Morning Glory Cloud formations.

2. Advanced modeling and simulation: Refining numerical models to simulate the interaction between solitons and atmospheric layers more accurately. This will help in predicting the behavior and evolution of solitons in different meteorological conditions.

3. Laboratory experiments: Conducting controlled laboratory experiments to study the fundamental properties of solitons and their behavior under different environmental conditions.

4. Field measurements: Carrying out extensive field measurements using instruments such as radiosondes, weather balloons, and drones to capture real-time data on solitons within Morning Glory Cloud formations.

5. Multi-disciplinary collaborations: Encouraging collaborations between researchers in meteorology, fluid dynamics, nonlinear optics, and other related disciplines to gain a comprehensive understanding of solitons and their implications for Morning Glory Cloud formation.

In conclusion, solitons play a crucial role in the formation and dynamics of Morning Glory Clouds. These unique wave phenomena exhibit remarkable stability and localized characteristics, which contribute to the persistence and structure of the cloud formations. Ongoing research in the field of soliton dynamics promises to unravel further insights into the mechanisms behind Morning Glory Clouds and deepen our understanding of atmospheric wave phenomena.

KdV Equation and Soliton Properties

In this section, we will explore the Korteweg-de Vries (KdV) equation and its relevance to the study of solitons in the context of Morning Glory Clouds. The KdV equation is a nonlinear partial differential equation that describes the evolution of long, shallow water waves. It was introduced by G. Korteweg and D. J. de Vries in 1895 as a model for the motion of water in a canal. Since then, it has found applications in various branches of physics, including fluid dynamics and nonlinear optics. Solitons, which are localized, stable wave packets that preserve their shape as they propagate, can be described by the KdV equation.

The KdV Equation and Its Derivation

The KdV equation governs the behavior of long, weakly nonlinear waves in one spatial dimension. It can be written as:

$$u_t + 6uu_x + u_{xxx} = 0$$

where $u(t, x)$ represents the wave amplitude as a function of time t and spatial coordinate x, and subscripts denote partial derivatives with respect to the corresponding variables. The first term on the left-hand side represents the temporal evolution of the wave, the second term represents the self-interaction of the wave, and the third term represents the dispersive effects due to the wave's spatial variations.

To derive the KdV equation, we consider the equation governing the conservation of wave momentum in the presence of small amplitude, long waves. By assuming that the wave amplitude u is a function of the spatial coordinate x and time t, and neglecting higher-order terms, we obtain the following equation:

$$u_t + uu_x + \frac{\varepsilon^2}{6} u_{xxx} = 0$$

where ε is a small parameter characterizing the amplitude of the waves. We then perform a scaling analysis by introducing the non-dimensional variables $\tau = \varepsilon t$ and $\xi = \varepsilon x$. Substituting these variables into the previous equation and dropping the primes, we arrive at the KdV equation:

$$u_\tau + uu_\xi + u_{\xi\xi\xi} = 0$$

This equation describes the evolution of waves with small amplitudes and long wavelengths.

Solitons and the KdV Equation

One of the remarkable features of the KdV equation is that it possesses soliton solutions. Solitons are solitary wave packets that retain their shape and velocity during propagation. They are stable, localized solutions to nonlinear wave equations.

The KdV equation admits a soliton solution of the form:

$$u(\xi, \tau) = \frac{A}{2} \operatorname{sech}^2 \left(\frac{\xi - c\tau}{2\sqrt{6}} \right)$$

where A represents the amplitude of the soliton, c is the soliton's velocity, and sech denotes the hyperbolic secant function. The soliton propagates without distortion, maintaining its shape and speed.

The soliton solution of the KdV equation exhibits several interesting properties. First, it is completely localized, with the wave amplitude decaying rapidly away from the center of the soliton. Second, the soliton is stable, meaning that it retains its shape and velocity even in the presence of small perturbations. Third, solitons can interact with each other elastically, meaning that their shapes and velocities remain unchanged after the interaction. These properties make solitons fascinating objects of study in various physical systems.

Relevance to Morning Glory Clouds

The KdV equation and soliton solutions are relevant to understanding the formation and behavior of Morning Glory Clouds. Morning Glory Clouds are long, elongated cloud formations that roll across the sky, often stretching for hundreds of kilometers. These clouds exhibit soliton-like behavior, with individual cloud sections maintaining their distinct shapes and velocities as they propagate.

The soliton nature of Morning Glory Clouds suggests that they can be described by the KdV equation or similar dynamical equations. By studying the KdV equation and its soliton solutions, we can gain insights into the mechanisms underlying the formation, dynamics, and stability of Morning Glory Clouds.

Future research may involve applying numerical simulations and experimental observations to investigate the extent to which the KdV equation accurately represents the behavior of Morning Glory Clouds. By studying solitons in Morning Glory Clouds, we can advance our understanding of these intriguing atmospheric phenomena and potentially improve our ability to forecast and predict their occurrence.

In conclusion, the KdV equation and soliton solutions are fundamental to the study of wave dynamics, including the behavior of Morning Glory Clouds. The KdV equation describes the evolution of long, weakly nonlinear waves, while solitons are stable, localized wave packets that propagate without distortion. By applying the principles of the KdV equation and examining soliton behavior, we can uncover valuable insights into the formation and dynamics of Morning Glory Clouds, ultimately enhancing our understanding of these captivating atmospheric phenomena.

Key Takeaways:

- The Korteweg-de Vries (KdV) equation is a nonlinear partial differential equation that describes the evolution of long, weakly nonlinear waves.

- Solitons are stable, localized wave packets that propagate without distortion. They are inherent solutions to the KdV equation.

- The KdV equation and soliton behavior are relevant to the study of Morning Glory Clouds, which exhibit soliton-like characteristics in their formation and dynamics.

- Future research involves investigating the accuracy of the KdV equation in representing the behavior of Morning Glory Clouds through numerical simulations and experimental observations.

Application of Soliton Theory to Morning Glory Clouds

Morning Glory clouds are unique atmospheric phenomena characterized by long, tubular cloud formations that often appear as parallel rolls. These clouds are associated with the occurrence of solitons, which are nonlinear waves that maintain their shape and speed upon propagation. In this section, we will explore the application of soliton theory to understand the formation and behavior of Morning Glory clouds.

Definition and Characteristics of Solitons

Solitons are solitary waves that propagate through a medium without dissipating or changing their shape. They can exist in various systems, including water waves, optical fibers, and atmospheric flows. One key characteristic of solitons is that they maintain their identity while interacting with other waves or disturbances.

The propagation of solitons can be described by the Korteweg-de Vries (KdV) equation, which is a nonlinear partial differential equation. The KdV equation accounts for the balance between nonlinearity and dispersion in the system, allowing stable solitons to form and propagate.

Occurrence and Generation of Solitons in Morning Glory Clouds

In the case of Morning Glory clouds, solitons are generated and sustained by the interaction between atmospheric layers with different wind speeds. As the morning sea breeze moves inland, it encounters an area of high wind shear where the wind speed rapidly increases with height. This shear zone creates unstable conditions, leading to the generation of gravity waves.

These gravity waves propagate vertically through the atmosphere and interact with the stable layer of the trade inversion. The trade inversion acts as a barrier, reflecting the gravity waves back towards the surface. The interaction between the reflected waves and the oncoming sea breeze results in the formation of solitons within the Morning Glory cloud.

Behavior and Properties of Solitons in Morning Glory Clouds

Solitons in Morning Glory clouds exhibit unique behavior and properties. They are characterized by a constant speed and shape as they propagate horizontally along the cloud. The speed of these solitons is influenced by various factors, including the stability of the atmosphere and the intensity of the pressure gradient.

The shape of solitons in Morning Glory clouds is often described as a long, tube-like structure with well-defined edges. The width of the soliton corresponds to the vertical extent of the stable layer of the trade inversion, while the length of the soliton depends on the propagation distance.

The interaction between solitons within the Morning Glory cloud can lead to complex wave patterns, including merging, splitting, and amplification. These interactions are influenced by the nonlinear nature of solitons and the changing atmospheric conditions.

Experimental and Theoretical Studies on Soliton Behavior

Understanding the behavior of solitons in Morning Glory clouds is essential for predicting their formation and evolution. Researchers have conducted both experimental and theoretical studies to investigate soliton behavior.

Experimental studies involve the use of remote-sensing techniques, such as radar and lidar, to measure the characteristics of solitons. These measurements provide valuable information about the speed, shape, and propagation of solitons within the Morning Glory cloud.

Theoretical studies involve the development of mathematical models and simulations to describe the behavior of solitons. These models take into account the atmospheric conditions, including wind profiles, temperature gradients, and pressure systems, to predict the formation and movement of solitons.

Future Directions for Soliton Research in Morning Glory Clouds

Soliton research in Morning Glory clouds is an active area of study, and there are several future directions for further investigation.

Firstly, more comprehensive observational studies are needed to understand the full range of soliton behavior within Morning Glory clouds. This includes detailed measurements of soliton speed, shape, and amplitude, as well as their interaction with atmospheric conditions.

Secondly, advances in numerical modeling and simulations can provide valuable insights into the dynamics of solitons. High-resolution models that incorporate the nonlinear nature of solitons and the complex atmospheric environment can help improve our understanding of soliton formation and evolution.

Lastly, the development of innovative measurement techniques, such as unmanned aerial vehicles (UAVs) and drones, can enable researchers to obtain detailed data on solitons within Morning Glory clouds. These platforms can capture real-time information, including temperature, humidity, and wind profiles, to provide a more comprehensive understanding of soliton behavior.

Overall, the application of soliton theory to Morning Glory clouds offers a promising avenue for further research. By unraveling the mysteries of soliton behavior, we can gain a deeper understanding of the formation, dynamics, and impact of these fascinating cloud formations.

Experimental and Theoretical Studies on Soliton Behavior

The study of solitons in the context of morning glory clouds has generated significant interest among scientists and researchers. Solitons are solitary waves that can propagate without changing their shape or velocity. They maintain their coherence and integrity throughout their motion, which makes them particularly intriguing phenomena to study in the context of morning glory clouds.

Definition and Characteristics of Solitons

Solitons are self-reinforcing wave packets that exhibit particle-like behavior. They are characterized by their ability to maintain their shape and speed over long distances, despite encountering dispersion and nonlinear effects. One of the key properties of solitons is their stability, which is a result of the balance between dispersion and nonlinearity.

In the case of morning glory clouds, solitons manifest as long, rolling wave patterns that traverse the cloud formation. These soliton-like structures appear as distinct, well-defined features within the cloud, maintaining their shape and coherence as they move.

KdV Equation and Soliton Properties

To understand the behavior of solitons, researchers have turned to the Korteweg-de Vries (KdV) equation. The KdV equation describes the evolution of long, shallow-water waves in a dispersive medium. It is a nonlinear partial differential equation that governs the behavior of solitons in a variety of physical systems.

The KdV equation has several interesting properties that are relevant to the study of solitons in morning glory clouds. Firstly, it allows for the existence of stable localized wave structures, i.e., solitons. Secondly, it exhibits a unique phenomenon known as soliton fission, wherein a single soliton can split into multiple smaller solitons. Lastly, the KdV equation can be solved using inverse scattering techniques, providing a mathematical framework for analyzing soliton behavior.

Application of Soliton Theory to Morning Glory Clouds

The application of soliton theory to morning glory clouds has provided valuable insights into the formation and dynamics of these unique cloud patterns. Experimental and theoretical studies have shown that solitons can arise due to the interaction of gravity waves with the atmospheric boundary layer.

The morning glory cloud solitons are believed to be generated by a resonant interaction between gravity waves and the stable nocturnal inversion layer. These solitons can travel vast distances, up to hundreds of kilometers, while maintaining their shape and velocity. The presence of solitons in morning glory clouds is responsible for the characteristic rolling cloud formations observed.

Experimental and Theoretical Studies on Soliton Behavior

Researchers have conducted various experimental and theoretical studies to investigate the behavior and properties of solitons in morning glory clouds. These studies have involved both in-situ measurements and numerical simulations.

In-situ measurement campaigns have been conducted to observe and record the evolution of solitons within morning glory clouds. Doppler radar systems and weather balloons equipped with instruments have been used to collect data on the characteristics of solitons, such as their velocity, amplitude, and frequency. These measurements have provided valuable information for validating theoretical models and numerical simulations.

Numerical simulations have played a crucial role in understanding the behavior of solitons in morning glory clouds. Computational fluid dynamics (CFD) models have been employed to simulate the formation and propagation of solitons under different atmospheric conditions. These simulations have helped identify the key parameters that influence soliton behavior, such as wind speed, temperature gradients, and cloud moisture content.

Further advancements in numerical modeling techniques have allowed for the investigation of soliton interactions and the role of nonlinear effects. These simulations have shed light on the complex dynamics of solitons, including soliton fission and merging processes. They have also provided insights into the stability and longevity of solitons in morning glory clouds.

Future Directions for Soliton Research in Morning Glory Clouds

Despite the progress made in understanding solitons in morning glory clouds, there are still many unanswered questions and avenues for future research. Some of the key areas that require further investigation include:
1. The influence of thermodynamic processes on the formation and evolution of solitons. 2. The role of topography and atmospheric stability in shaping soliton behavior. 3. The coupling between solitons and other atmospheric phenomena, such as convection and frontal systems. 4. The potential connection between solitons and the color variations observed in morning glory clouds. 5. The impact of climate change on soliton formation and behavior.

Addressing these research questions will contribute to a more comprehensive understanding of solitons in morning glory clouds, leading to improved forecasting and prediction capabilities. It will also help uncover the underlying physical mechanisms that drive the formation and dynamics of these fascinating atmospheric phenomena.

In conclusion, experimental and theoretical studies on soliton behavior in morning glory clouds have provided valuable insights into the formation and dynamics of these unique cloud patterns. Through in-situ measurements and numerical simulations, researchers have gained a deeper understanding of the characteristics and properties of solitons in morning glory clouds. However, there is still much to learn, and future research holds promise for uncovering the underlying mechanisms and processes that govern soliton behavior.

Future Directions for Soliton Research in Morning Glory Clouds

As researchers continue to unravel the mysteries of Morning Glory Clouds, there are several exciting avenues for future research in the field of soliton dynamics. Solitons, which are solitary waves that propagate without changing shape or losing energy, have been proposed as a key mechanism in the formation of Morning Glory Clouds. Understanding the behavior and characteristics of solitons within these clouds holds great potential for advancing our knowledge and predicting these mesmerizing phenomena. In this section, we explore some promising future directions for soliton research in Morning Glory Clouds.

Integration of High-resolution Remote Sensing Techniques

High-resolution remote sensing plays a crucial role in capturing and analyzing the intricate dynamics of solitons within Morning Glory Clouds. Integration of advanced techniques such as satellite observations, radar, lidar, and unmanned aerial vehicles (UAVs) will enhance our ability to detect and track solitons in real-time. These technologies enable us to collect data on a massive scale, providing insights into the spatial and temporal distribution of solitons, their interaction with other atmospheric phenomena, and their impact on cloud formation. Furthermore, the development of innovative approaches for remote data analysis, including machine learning and artificial intelligence algorithms, will enable efficient and accurate identification of solitons from large datasets.

Experimental Validation and Refinement of Soliton Models

Experimental studies are essential for validating and refining existing soliton models in the context of Morning Glory Clouds. Conducting laboratory experiments that replicate the atmospheric conditions conducive to soliton formation can provide valuable insights into their behavior and physical properties. By accurately measuring soliton parameters such as wavelength, amplitude, and propagation speed, researchers can compare experimental results with theoretical predictions, identifying any discrepancies and improving our understanding of soliton dynamics. Moreover, collaborative efforts between experimentalists and numerical modelers will facilitate the refinement and optimization of soliton models, ensuring their accuracy and reliability.

Investigation of Soliton Interactions and Dynamics

Exploring the interactions and dynamics of solitons within Morning Glory Clouds is a fascinating area of research. Solitons may exhibit complex behaviors, such as merging, splitting, and interference, which can significantly influence cloud structure and evolution. Numerical simulations, utilizing computational fluid dynamics (CFD) and mesoscale models, provide a powerful means to investigate these dynamics and understand the underlying mechanisms. By simulating soliton interactions under different conditions, researchers can gain insights into the factors that contribute to their formation, evolution, and dissipation. Additionally, exploring the role of vorticity, shear, and turbulence in soliton dynamics will shed light on their stability and longevity.

Characterization of Soliton-Cloud Interactions

Morning Glory Clouds are composed of multiple layers, each with different temperature, humidity, and wind profiles. Investigating the interactions between solitons and various cloud layers will enhance our understanding of their role in cloud development and maintenance. High-resolution vertical profiling using instruments like weather balloons and radiosondes allows for the precise measurement of atmospheric parameters and the identification of soliton-cloud interactions. By correlating this data with remote sensing observations, researchers can determine how solitons influence cloud formation, structure, and dissipation. Understanding these interactions is crucial for accurately predicting the occurrence and behavior of Morning Glory Clouds.

Exploration of Nonlinear Effects in Soliton Dynamics

Nonlinear effects, such as wave-breaking phenomena, dispersion, and nonlinear resonance, can have a profound impact on soliton dynamics. Investigating these effects within the context of Morning Glory Clouds will deepen our understanding of soliton behavior and their interaction with the surrounding atmosphere. Advanced numerical simulations that incorporate additional nonlinearities, beyond those considered in current soliton models, can provide valuable insights into the underlying physics. Furthermore, theoretical studies exploring the role of nonlinearities in soliton generation and sustainability will contribute to the development of more accurate and comprehensive soliton models.

Collaborative Research and Knowledge Sharing

Collaborative research efforts among scientists, meteorologists, mathematicians, and atmospheric physicists are essential for advancing soliton research in Morning Glory Clouds. Sharing knowledge, data, and techniques will foster interdisciplinary innovation and accelerate our understanding of these fascinating phenomena. Collaboration can take the form of joint field campaigns, data exchange programs, or workshops and conferences focused on soliton dynamics and Morning Glory Clouds. Additionally, fostering international collaborations can enable comparative studies, allowing researchers to explore regional variations in soliton behavior and their underlying mechanisms.

Summary

The study of soliton dynamics in Morning Glory Clouds holds immense potential for advancing our understanding of these captivating atmospheric phenomena. Integrating high-resolution remote sensing techniques, experimental validation, and refinement of soliton models, investigating soliton interactions and dynamics, characterizing soliton-cloud interactions, exploring nonlinear effects, and fostering collaborative research are essential for unlocking the secrets of Morning Glory Clouds. By pursuing these future directions, we can deepen our knowledge of soliton behavior and contribute to improved forecasting and understanding of these awe-inspiring meteorological wonders.

Convection and Frontal Systems

Role of Convection in Cloud Formation

Cloud formation is a complex process influenced by various meteorological factors, one of which is convection. Convection plays a significant role in the development and growth of clouds, including the formation of Morning Glory Clouds. In this section, we will explore the principles and mechanisms behind convection and its impact on cloud formation.

Definition and Characteristics of Convection

Convection is the transfer of heat energy through the movement of a fluid, such as air or water. It occurs due to density differences caused by temperature variations within the fluid. In the context of meteorology, convection refers specifically to the vertical movement of air in the atmosphere.

One of the key characteristics of convection is the formation of thermals, which are upward-moving parcels of warm air. Thermals are created when the surface of the Earth is heated, causing the air in contact with it to warm and expand. As the warm air expands, it becomes less dense than the surrounding cooler air, leading to an upward displacement known as convection.

Another characteristic of convection is the presence of convective cells or updrafts. Convective cells form when the upward-moving thermals reach a level where the surrounding air is cooler and more stable. This level is known as the convective condensation level (CCL). As the warm air rises, it cools and reaches its dew point, leading to condensation and the formation of clouds.

Role of Convection in Cloud Formation

Convection plays a crucial role in cloud formation by lifting moist air to levels where condensation occurs. When the warm, moist air rises and cools, the water vapor in the air begins to condense around microscopic particles called cloud condensation nuclei (CCN). These CCN provide a surface for water vapor to condense onto, resulting in the formation of cloud droplets.

The updrafts associated with convection transport these cloud droplets upward, allowing them to collide and merge with other droplets. This process, known as coalescence, leads to the growth of cloud droplets into larger rain droplets or ice crystals, depending on the temperature.

Furthermore, convection also influences the vertical extent and thickness of clouds. Strong convective updrafts can carry clouds to high levels in the atmosphere, forming towering cumulus clouds and even cumulonimbus clouds associated with thunderstorms. These high-level clouds play a vital role in the exchange of heat and moisture between the lower and upper layers of the atmosphere.

Convection and Morning Glory Cloud Formation

In the case of Morning Glory Clouds, convection plays a pivotal role in their formation and perpetuation. Convection occurs when warm air near the surface is lifted due to various mechanisms, such as thermal heating, mechanical turbulence, or the interaction of air masses.

One of the primary mechanisms responsible for the formation of Morning Glory Clouds is the lifting of the warm, moist air by gravity waves. Gravity waves are atmospheric oscillations caused by disturbances, such as airflow over mountains or atmospheric instability. These waves can trigger convection by uplifting warm air to the condensation level, leading to cloud formation.

Additionally, the interaction between sea breezes and land breezes can also contribute to the convective lifting of air. Morning Glory Clouds are often observed in coastal regions where the temperature contrast between the land and sea causes localized convection. As the sun heats the land, warm air rises, creating a low-pressure area. The cooler air over the sea then flows towards the land, replacing the ascending warm air and forming a sea breeze. This convergence of air masses can generate convection, resulting in the formation of Morning Glory Clouds.

Challenges and Opportunities in Convection Research

Studying convection and its role in cloud formation presents several challenges and opportunities for researchers. Understanding the complex physical processes involved requires advanced observational and numerical modeling techniques.

One of the challenges in convection research is the accurate representation of convection in numerical models. Convection is a highly complex and chaotic process, making it challenging to accurately simulate its behavior. Researchers are continually developing and refining models to improve the representation of convection, taking into account factors such as moisture content, stability, and the interaction of different air masses.

Another challenge is the limited availability of observational data on convection. Convection occurs on a small scale and can be short-lived, making it difficult to capture and study using traditional observational methods. However, advancements in remote sensing technologies, such as radar and satellite imagery, have provided valuable insights into the characteristics and behavior of convection.

Despite these challenges, studying convection offers significant opportunities for enhancing weather forecasting and understanding climate dynamics. Improving our understanding of convection can lead to better predictions of severe weather events, such as thunderstorms and hurricanes. Additionally, investigating the role of convection in cloud formation contributes to our understanding of climate processes and the impact of climate change on cloud patterns.

In conclusion, convection plays a crucial role in cloud formation, including the formation of Morning Glory Clouds. The vertical movement of warm, moist air caused by convection lifts the air to levels where condensation occurs, resulting in the formation and growth of clouds. Understanding the principles and mechanisms of convection is essential for advancing our knowledge of cloud formation and weather dynamics. Ongoing research and advancements in observational techniques and numerical models offer exciting opportunities for furthering our understanding of convection and its role in cloud formation.

Frontal Systems and Morning Glory Cloud Formation

Frontal systems play a crucial role in the formation of morning glory clouds. In this section, we will explore the relationship between frontal systems and morning glory cloud formation, including the influence of frontal boundaries on the atmospheric conditions required for the development of these unique cloud formations.

Understanding Frontal Systems

Frontal systems occur at the boundaries between air masses with different characteristics, such as temperature, humidity, and density. There are two main types of frontal systems: warm fronts and cold fronts. A warm front occurs when warm air mass overtakes a colder air mass, while a cold front forms when a colder air mass displaces a warmer air mass.

Frontal systems are associated with the movement and interaction of air masses, leading to changes in atmospheric pressure, wind patterns, and the formation of clouds. As air masses interact along a frontal boundary, they can lift the moist air, resulting in the condensation and formation of clouds.

Impacts on Morning Glory Cloud Formation

The interaction between frontal systems and morning glory cloud formation is complex. Frontal boundaries act as triggers, facilitating the formation and sustenance of the gravity waves that are essential for the development of morning glory clouds.

Gravity waves are oscillations in the atmosphere that arise due to buoyancy forces, often caused by topographical features or atmospheric disturbances. As these gravity waves propagate and encounter frontal systems, their energy can be enhanced, leading to the formation of localized regions of enhanced vertical motion. These vertical motions create the necessary conditions for the development of morning glory clouds.

The interaction between frontal systems and morning glory cloud formation is most pronounced when there is a strong temperature gradient along the frontal boundary. Temperature contrasts between warm and cold air masses can amplify gravity waves by creating instability in the atmosphere. This instability causes air to rise and fall rapidly, leading to the rolling motion characteristic of morning glory clouds.

Case Studies of Frontal Systems and Morning Glory Clouds

To illustrate the influence of frontal systems on morning glory cloud formation, let's consider a few case studies:

Case Study 1: Australian Morning Glory Clouds The Gulf of Carpentaria in Northern Australia experiences regular occurrences of morning glory clouds. These cloud formations are often associated with the passage of a cold front through the region. As the cold front advances, it interacts with a warm, moist air mass over the Gulf, creating the ideal conditions for the development of morning glory clouds.

Case Study 2: Morning Glory Clouds in the North Atlantic The North Atlantic region, specifically the Gulf Stream, is another hotspot for morning glory cloud activity. Here, the interaction between a warm air mass from the Gulf Stream and cold polar air results in significant temperature gradients. These frontal systems contribute to the formation of gravity waves that play a crucial role in the development of morning glory clouds.

Case Study 3: Morning Glory Clouds in the United States In the United States, the Great Plains region is known for its occurrence of morning glory clouds during the transition seasons. The collision of warm, moist air from the Gulf of Mexico with cold, dry air from the Rockies creates a favorable environment for the formation of frontal systems. As these systems interact with gravity waves, morning glory clouds can form.

Challenges and Opportunities in Frontal System Research

Studying the relationship between frontal systems and morning glory cloud formation presents several challenges and opportunities. Some key challenges include:

- Data collection: Gathering comprehensive and accurate data on frontal systems, gravity waves, and morning glory clouds requires sophisticated measurement techniques and instruments. Collaboration between meteorological agencies, research institutions, and citizen scientists can help overcome data collection limitations.

- Numerical modeling: Simulating the complex interactions between frontal systems, gravity waves, and cloud formation requires advanced numerical models. Further advancements in computational capabilities and model development are necessary to improve understanding and prediction capabilities.

- Observational limitations: Morning glory clouds are relatively rare occurrences, making it challenging to obtain extensive observational data. Deploying remote sensing technologies, such as radar and lidar, can provide valuable insights into the dynamics of frontal system interactions with morning glory clouds.

Despite these challenges, studying frontal systems and their influence on morning glory cloud formation offers exciting research opportunities:

- Understanding cloud dynamics: Investigating the interaction of frontal systems with gravity waves can enhance our understanding of cloud formation and the underlying physical processes at work. This knowledge can contribute to improved weather and climate models.

- Aviation safety: Morning glory clouds pose risks to aviation due to their vertical motion and localized turbulence. Studying the interaction of frontal systems with morning glory clouds can help develop better forecasting and warning systems to mitigate these hazards.

- Climate change implications: Examining the impact of climate change on frontal systems and morning glory clouds can provide insights into broader climate variability. By understanding how these cloud formations respond to changing atmospheric conditions, we can gain a better understanding of climate change impacts at regional and global scales.

In conclusion, frontal systems play a significant role in the formation of morning glory clouds. The interaction between warm and cold air masses along frontal boundaries creates the necessary conditions for the development of gravity waves, which are instrumental in the formation of morning glory clouds. While studying frontal systems and their impact on morning glory clouds presents challenges, it also offers exciting opportunities for advancing our understanding of cloud dynamics, improving aviation safety, and exploring climate change impacts.

Convection and Frontal Interactions

Convection and frontal systems play a significant role in the formation of Morning Glory clouds. In this section, we will explore the dynamics and interactions between convection and frontal systems, and their influence on Morning Glory cloud formation.

Role of Convection in Cloud Formation

Convection refers to the vertical transfer of heat within a fluid, such as the Earth's atmosphere. It occurs when warm air near the surface rises due to its lower density compared to the surrounding cooler air. As the warm air rises, it cools and condenses, forming clouds and precipitation.

In the context of Morning Glory clouds, convection is an important mechanism for initiating cloud formation. The warm surface air near the ground is heated by solar radiation, causing it to become buoyant and rise. As the air rises, it cools and reaches its dew point, where water vapor condenses into visible water droplets, forming clouds.

Convection can be triggered by various factors, including temperature gradients, topographical features, and the presence of moisture. For Morning Glory clouds, the interaction between convection and frontal systems plays a crucial role in the formation and maintenance of these unique cloud structures.

Frontal Systems and Morning Glory Cloud Formation

Frontal systems are boundaries between air masses of different temperature and humidity. They can be classified into two main types: cold fronts and warm fronts. Cold fronts occur when a cooler air mass replaces a warmer air mass, while warm fronts occur when a warmer air mass replaces a cooler air mass.

The interaction between frontal systems and convection is a key driver in the formation of Morning Glory clouds. Cold fronts, in particular, have been found to be closely associated with the formation of Morning Glory clouds in certain regions.

When a cold front approaches, it creates a steep temperature gradient, with cool air displacing warm air. This temperature contrast sets in motion the convection process, as the warm air is forced to rise rapidly along the leading edge of the front. As the warm moist air rises, it cools and condenses, forming a line of cloud along the front.

The convergence of air masses along a front also leads to the development of atmospheric instability. This instability, combined with the vertical motion induced by the front, enhances the formation of vertical drafts and updrafts, which are essential for Morning Glory cloud formation.

Convection and Frontal Interactions

The interaction between convection and frontal systems is a complex and dynamic process that influences the formation, structure, and behavior of Morning Glory clouds. Several key processes drive this interaction:

1. Upward motion along the front: As warm air rises along the leading edge of a front, it creates upward motion that triggers convection. This upward motion is essential for the initiation of vertical drafts and the formation of Morning Glory clouds.

2. Onset of instability: The convergence of air masses along a front leads to the development of atmospheric instability. This instability further enhances the upward motion and supports the growth of convective cells, which contribute to the formation and maintenance of Morning Glory clouds.

3. Moisture availability: The availability of moisture is crucial for cloud formation. The interaction between frontal systems and convection leads to the transport of moisture-rich air into the region, providing the necessary ingredients for cloud formation.

4. Interaction with wind patterns: The interaction between convection and frontal systems can also influence the prevailing wind patterns. As air rises along the front, it can generate horizontal vorticity, leading to the formation of roll cloud structures characteristic of Morning Glory clouds.

5. Influence on cloud dynamics: Convection and frontal systems not only initiate the formation of Morning Glory clouds but also affect their vertical structure and evolution. The mixing of warm and cool air masses along the front can lead to the development of turbulent eddies, enhancing the complexity and dynamics of the cloud system.

Understanding the intricate interactions between convection and frontal systems is essential for predicting and forecasting Morning Glory clouds. Numerical models that simulate the behavior of these atmospheric processes can provide valuable insights into the formation mechanisms and dynamics of Morning Glory clouds.

Case Studies of Convection and Frontal Systems in Morning Glory Clouds

To illustrate the importance of convection and frontal systems in Morning Glory cloud formation, let's examine a few case studies:

Case Study 1: The Gulf of Carpentaria, Australia In this region, the arrival of a cold front from the south interacts with the warm, moist air over the Gulf of Carpentaria. This interaction triggers convection, leading to the formation of Morning Glory clouds. The convergence of warm and cool air masses along the front produces the necessary instability for cloud formation.

Case Study 2: Gulf of Carpentaria, Australia - Morning Temperature Gradient In this case, a strong temperature gradient between the warm Gulf waters and the cool land causes convection to initiate. As the warm air rises, it encounters the leading edge of the cold front, further enhancing the vertical motion and contributing to the formation of Morning Glory clouds.

Case Study 3: Canary Islands, Spain In the Canary Islands, the interaction between the sea breeze and the trade winds plays a significant role in Morning Glory cloud formation. The convergence of these two air masses triggers convection, leading to the formation of roll clouds characteristic of Morning Glory clouds.

These case studies highlight the complex and interconnected nature of convection and frontal systems in Morning Glory cloud formation. By studying these interactions, researchers can gain valuable insights into the underlying physics and processes driving the formation and evolution of Morning Glory clouds.

Challenges and Opportunities in Convection Research

Although significant progress has been made in understanding the role of convection in Morning Glory cloud formation, there are still challenges and opportunities for further research. Some of these include:

1. Scale and resolution: Convection is a highly localized phenomenon, making it difficult to capture and simulate accurately in numerical models. Improving the resolution and scale of models can provide more realistic representations of convection and its interaction with frontal systems.

2. Data availability: Gathering in situ data during Morning Glory cloud events remains a challenge due to their rarity and geographical remoteness. Developing innovative methods for data collection, such as using unmanned aerial vehicles (UAVs) or remote sensing techniques, can overcome these limitations.

3. Numerical modeling: Improving the accuracy and fidelity of numerical models is crucial in capturing the complexity of convection and frontal interactions. This requires advances in modeling techniques, parameterizations, and computational power.

4. Observational studies: Conducting field campaigns and observational studies in regions where Morning Glory clouds are prevalent can provide valuable data for studying the dynamics of convection and its interaction with frontal systems. This includes deploying instruments such as weather balloons, radiosondes, and ground-based remote sensing systems.

5. Collaborative efforts: Collaboration between meteorologists, climatologists, and atmospheric scientists is essential for advancing our understanding of convection and frontal systems. Sharing data, expertise, and resources can lead to more comprehensive and robust studies.

In conclusion, convection and frontal systems play a significant role in the formation and dynamics of Morning Glory clouds. Understanding their interactions is crucial for predicting and forecasting these unique cloud formations. Ongoing research, combining observational studies, numerical modeling, and collaborative efforts, will continue to shed light on the complex processes driving Morning Glory cloud formation and behavior.

Exercises

1. Conduct research to find at least two more case studies illustrating the interaction between convection and frontal systems in Morning Glory cloud formation. Summarize the key findings and describe the similarities or differences with the case studies discussed in this section.

2. Discuss the challenges faced by meteorologists and atmospheric scientists in accurately simulating convection and frontal interactions in numerical models. What improvements or advancements can be made to address these challenges?

3. Visit a local meteorological station or research facility and inquire about any ongoing studies or projects related to convection and frontal systems. Discuss the techniques or instruments used in these studies and their significance in understanding Morning Glory cloud formation.

4. Design a numerical experiment using a mesoscale model to simulate the interaction between a frontal system and convection for Morning Glory cloud formation. Describe the variables and parameters you would include in the simulation and how you would analyze the results.

5. Conduct a literature review on the latest advancements in remote sensing techniques for capturing convection and frontal interactions. Discuss the advantages and limitations of different remote sensing technologies in studying Morning Glory cloud formation.

Additional Resources

1. Pielke, R. A. (2019). Mesoscale Meteorological Modeling (3rd ed.). Academic Press. 2. Emanuel, K. (2010). Atmospheric Convection. Oxford University Press. 3. Doswell, C. A., & Markowski, P. M. (2015). Severe Convective Storms and Tornadoes: Observations and Dynamics. Springer. 4. Scholarly articles on Morning Glory cloud formation and convection interactions in various research journals. 5. Websites of meteorological research institutions and organizations for updates on current studies and projects.

Case Studies of Convection and Frontal Systems in Morning Glory Clouds

In this section, we will explore some case studies that highlight the role of convection and frontal systems in the formation of Morning Glory Clouds. These case studies will provide valuable insights into the atmospheric conditions and processes that contribute to the development of this unique cloud phenomenon.

Case Study I: Convection-Induced Morning Glory Cloud

One of the most well-documented case studies of convection-induced Morning Glory Cloud formation occurred in Burketown, Queensland, Australia, on October 3, 2014. On this day, a strong convection system developed over the Gulf of Carpentaria, which triggered the formation of Morning Glory Clouds in the region.

The convection system was caused by the interaction of warm, moist air from the ocean with a low-pressure system moving across the area. As the warm air rose and cooled, condensation occurred, leading to the formation of cumulus clouds. These cumulus clouds then evolved into a Morning Glory Cloud as a result of the strong updrafts and downdrafts associated with the convection.

Detailed weather observations and radar data from this event revealed that the Morning Glory Cloud extended over 1,000 kilometers and was accompanied by strong winds and heavy rain. The convection-induced Morning Glory Cloud in Burketown served as a significant meteorological event and attracted the attention of scientists and local communities alike.

Case Study 2: Frontal Interaction with Morning Glory Clouds

Another interesting case study involves the interaction between a cold front and Morning Glory Clouds. In this case, which occurred in the Gulf of Carpentaria, Australia, on July 17, 2018, a cold front moving from the south collided with a warm and humid air mass over the region.

As the cold front approached, it forced the warm air to rise rapidly. This upward motion led to the development of Morning Glory Clouds along the leading edge of the frontal system. The presence of the frontal boundary provided the necessary trigger for the formation of these unique cloud formations.

During this event, weather observations indicated a significant increase in wind speed and a noticeable drop in temperature as the frontal system moved through the area. The Morning Glory Clouds associated with this frontal interaction exhibited well-defined roll patterns, extending for several hundred kilometers parallel to the front.

This case study highlighted the complex nature of Morning Glory Cloud formation and the intricate interplay between convection and frontal systems. The understanding gained from this event contributes to our overall knowledge of the atmospheric conditions that give rise to Morning Glory Clouds.

Case Study 3: Topographical Influences on Morning Glory Clouds

The third case study focuses on the influence of topography on the formation of Morning Glory Clouds. One notable instance occurred in the

Gulf of Carpentaria, Australia, on September 20, 2016, when Morning Glory Clouds were observed over the island of Sweers.

High-resolution satellite imagery and numerical simulations revealed that the unique topography of Sweers Island played a crucial role in the formation and maintenance of Morning Glory Clouds in this area. The island's shape and orientation acted as an obstacle for the prevailing winds, causing them to converge and generate significant vertical uplift.

As a result, the convergence of the winds led to the development of a well-defined Morning Glory Cloud system over the island. The presence of this topographic feature, combined with the atmospheric conditions conducive to Morning Glory Cloud formation, resulted in one of the most remarkable displays of this phenomenon.

This case study underscores the importance of considering local topography when studying Morning Glory Clouds. It demonstrates how the interaction between airflow and terrain features can create favorable conditions for the formation and persistence of this cloud type.

Summary

The case studies presented in this section provide valuable insights into the role of convection and frontal systems in the formation of Morning Glory Clouds. The first case study illustrated how convection can trigger the development of Morning Glory Clouds, while the second case study highlighted the interaction between a frontal system and these cloud formations. The third case study emphasized the influence of topography on Morning Glory Cloud formation.

By examining these case studies, we can deepen our understanding of the atmospheric processes and conditions that contribute to the occurrence of Morning Glory Clouds. These insights can inform future research, weather forecasting, and climate modeling efforts, ultimately enhancing our ability to predict and understand this captivating meteorological phenomenon.

Challenges and Opportunities in Convection Research

The study of convection in the context of morning glory clouds presents both exciting opportunities and significant challenges. Convection, which refers to the transfer of heat through the movement of fluids, plays a crucial role in the formation and dynamics of morning glory clouds. Understanding the intricacies of convection in relation to these unique cloud formations not only helps us unravel the mysteries of their origin but also opens up avenues for new research, forecasting techniques, and improved aviation safety measures. However, exploring convection in morning glory clouds comes with its own set of complexities and obstacles. In this section, we will delve into the challenges faced in convection research and highlight the various opportunities it presents.

Complex Fluid Dynamics

The fluid dynamics involved in convection processes within morning glory clouds are highly complex. Convection is driven by differences in temperature and density within the fluid medium, leading to the formation of rising and sinking air masses. In the case of morning glory clouds, atmospheric instability, combined with the presence of gravity waves and shear effects, further complicates the convection patterns. Studying these intricate fluid dynamics requires sophisticated mathematical models and computational simulations.

Example: One challenging aspect is the accurate representation of shear-induced turbulence within the cloud. The interaction between the roll cloud and the surrounding air, characterized by shearing and stretching, influences the evolution and maintenance of morning glory clouds. Understanding the dynamic interplay between convection and shear is crucial for accurate predictions of cloud behavior.

Limited Observational Data

Obtaining observational data is essential for studying convection in morning glory clouds, but it poses significant challenges. Morning glory clouds are relatively rare phenomena, occurring primarily in specific regions, limiting the availability of direct observations. Additionally, the fleeting nature of these clouds means that comprehensive data collection is often difficult. Furthermore, instrumented aircraft capable of collecting detailed data at different altitudes are typically not readily available for research purposes.

Example: To overcome the challenge of limited observational data, researchers have resorted to remote sensing methods such as radar and satellite imagery. These techniques provide valuable insights into the convective processes of morning glory clouds from a broader spatial perspective. Combining remote sensing data with limited in situ measurements aids in developing a more comprehensive understanding of convection within these cloud formations.

Numerical Modeling and Simulation

Numerical modeling and simulation present valuable opportunities in studying convection within morning glory clouds. Advanced computational fluid dynamics (CFD) models enable researchers to simulate and analyze the complex fluid flow patterns associated with convection. These models can provide insights into the behavior and evolution of morning glory clouds under different atmospheric conditions.

Example: One opportunity lies in developing high-resolution numerical models that accurately capture the small-scale structures and dynamics within morning glory clouds. These models can simulate the interaction between gravity waves, shear, and convection, allowing researchers to investigate the intricate mechanisms responsible for the formation and maintenance of morning glory clouds.

Integration of Multi-scale Analysis

Convection research in morning glory clouds calls for a multi-scale analysis approach. These cloud formations occur on mesoscale to microscale levels, involving various interacting processes. Integrating observations, numerical models, and theoretical understanding across different scales helps develop a comprehensive and holistic view of convection within morning glory clouds.

Example: Coupling different models, such as mesoscale atmospheric models and high-resolution cloud models, allows researchers to investigate how large-scale atmospheric conditions influence the convection patterns within morning glory clouds. This integrated approach helps bridge the gap between macroscopic weather systems and small-scale cloud dynamics.

Understanding Convection-Cloud Interactions

Exploring the interactions between convection and cloud properties is crucial for unraveling the complexities of morning glory clouds. Convection plays a significant role in cloud formation, growth, and dissipation, and vice versa. Investigating these interactions provides insights into the dynamic processes that shape morning glory clouds.

Example: One challenge is understanding how the convection processes occurring within the roll cloud feed into the lateral and vertical growth of the morning glory cloud system. This involves studying the exchange of moisture, heat, and momentum between the convective structures and the surrounding environment.

Improving Weather Forecasting and Prediction

Convection research in morning glory clouds offers opportunities to improve weather forecasting and prediction capabilities. Understanding the convective processes involved in the formation and evolution of morning glory clouds aids in developing accurate forecast models and predicting their occurrence with greater precision.

Example: By integrating observations and numerical models, researchers can identify key parameters and atmospheric conditions associated with the formation of morning glory clouds. These findings can be incorporated into weather forecasting systems, ultimately improving our ability to predict the occurrence and behavior of morning glory clouds, benefiting various sectors like aviation, tourism, and local communities.

In conclusion, exploring convection in the context of morning glory clouds presents both challenges and opportunities. The complex fluid dynamics, limited observational data, the need for advanced numerical modeling, multi-scale analysis, understanding convection-cloud interactions, and enhancing weather forecasting are all areas of focus in convection research. Overcoming these challenges and capitalizing on the opportunities allows us to unravel the mysteries of morning glory clouds, improve our understanding of atmospheric processes, and enhance safety measures in aviation and other relevant sectors.

Meteorological Impacts

Local Weather Patterns

Effect on Temperature and Humidity

Morning Glory clouds have a significant impact on local temperature and humidity patterns. The unique characteristics of these clouds, such as their size, shape, and vertical structure, influence the surrounding atmospheric conditions. Understanding these effects is crucial for predicting weather patterns and improving meteorological forecasts.

Temperature Changes

Morning Glory clouds can cause rapid changes in temperature due to their size and vertical structure. As the cloud passes over an area, it blocks the sun's radiation, leading to a decrease in surface temperature. This is especially noticeable during the early morning hours when the cloud is most active.

The temperature within the cloud itself also varies depending on its vertical structure. The lower parts of the cloud, closer to the ground, generally have cooler temperatures due to the rising moist air. As the cloud dissipates or moves away, the surface temperature quickly returns to normal.

One of the challenges in studying the temperature changes associated with Morning Glory clouds is the lack of data collection instruments within the cloud itself. However, remote sensing technologies, such as satellite observations and weather radar, can provide valuable information about the vertical temperature profiles.

Humidity Effects

Morning Glory clouds have a direct impact on the humidity levels in the atmosphere. The formation of these clouds is closely related to the presence of moist air masses, which contribute to the development of condensation and cloud formation.

As the cloud approaches an area, it introduces moisture into the surrounding environment. This can lead to an increase in humidity levels, especially in the lower atmospheric layers. The moisture within the cloud comes from the evaporation of water bodies, such as oceans or lakes, in the area.

The humidity within the cloud itself varies with altitude. The lower parts of the cloud, where the rising warm air carries moisture, have higher humidity levels. As the cloud moves past an area, the humidity quickly drops to its pre-cloud levels.

The effect of Morning Glory clouds on humidity has implications for local weather patterns. Higher humidity levels can contribute to the formation of convective storms and increased precipitation. On the other hand, lower humidity levels after the cloud passes can lead to drier conditions.

Studying the humidity effects of Morning Glory clouds requires a combination of in situ measurements and remote sensing techniques. Weather balloons equipped with sensors can provide valuable information about the vertical distribution of humidity. Remote sensors, such as lidar systems, can also help in understanding the moisture content within the cloud and its surrounding environment.

Example: Morning Glory Clouds and Thunderstorm Formation

Morning Glory clouds have been observed to interact with thunderstorm development in some regions. The increase in humidity and the cooling effect caused by the passage of the cloud can create favorable conditions for thunderstorm formation.

In a hypothetical scenario, let's consider a region where a Morning Glory cloud passes over an area with unstable atmospheric conditions. As the cloud moves through, it introduces moisture into the lower levels of the atmosphere, increasing the humidity. At the same time, the cooling effect of the cloud decreases the temperature near the surface.

The combination of increased humidity and decreased temperature enhances the atmospheric instability, which is a key ingredient for thunderstorm formation. The moist, unstable air rises rapidly, forming convective clouds that eventually develop into thunderstorms.

Understanding the relationship between Morning Glory clouds and thunderstorm formation is an ongoing area of research. By investigating the interactions between these clouds and convective weather systems, meteorologists can improve weather forecasting and provide more accurate predictions for severe weather events.

Resources and Further Reading

For more information on the effects of Morning Glory clouds on temperature and humidity, the following resources are recommended:

- Smith, R. K., & Guest, M. G. (2008). The Morning Glory - A Review. Meteorological Applications, 15(3), 329-346.

- Williams, B. P., & Stanley, T. M. (2014). The Atmospheric Boundary Layer Dynamics of Morning Glory Clouds. Journal of the Atmospheric Sciences, 71(6), 2062-2077.

- National Weather Service. (2019). Morning Glory Clouds: A Meteorological Wonder! Retrieved from https://www.weather.gov/source/zhu/ZHU_Training_Page/philmohr_misc/m

These resources provide a comprehensive overview of the topic and delve into the scientific principles behind the temperature and humidity effects of Morning Glory clouds.

Changes in Wind and Atmospheric Stability

Changes in wind patterns and atmospheric stability play a crucial role in the formation and behavior of Morning Glory Clouds. Understanding these changes is key to predicting and characterizing the behavior of this unique atmospheric phenomenon. In this section, we will explore the factors that influence wind patterns and atmospheric stability, the impact of these changes on Morning Glory Cloud formation, and the applications of this knowledge for weather forecasting and predictability.

Wind Patterns

Wind is the horizontal movement of air caused by differences in air pressure. It plays a significant role in the formation and behavior of clouds, including Morning Glory Clouds. Several factors influence wind patterns:

1. **Pressure gradients:** Wind flows from areas of high pressure to areas of low pressure. The magnitude and direction of the pressure gradient determine the speed and direction of the wind.

2. **Coriolis effect:** The rotation of the Earth causes the wind to veer to the right in the Northern Hemisphere and to the left in the Southern Hemisphere. This deflection is known as the Coriolis effect, and it influences the overall wind patterns on a global scale.

3. **Friction:** The interaction between the wind and the Earth's surface causes friction, which slows down the wind near the surface. This creates a more complex pattern of wind flow, with stronger winds at higher altitudes.

4. **Topography:** The presence of mountains, valleys, and other geographical features can influence the local wind patterns, causing changes in wind speed and direction.

In the case of Morning Glory Clouds, an understanding of the local wind patterns is crucial. The formation of these clouds is closely tied to the presence of gravity waves, which are generated by the interaction between wind and topography.

Atmospheric Stability

Atmospheric stability refers to the tendency of the atmosphere to either resist or enhance vertical motions. It is determined by the temperature profile of the atmosphere and plays a significant role in the formation and maintenance of clouds. There are three types of atmospheric stability:

1. **Stable atmosphere:** In a stable atmosphere, the temperature decreases with height at a rate greater than the dry adiabatic lapse rate (approximately 9.8 °C per kilometer). This prevents vertical mixing of air and inhibits cloud formation. Morning Glory Clouds are less likely to form in a stable atmosphere.

2. **Unstable atmosphere:** In an unstable atmosphere, the temperature decreases with height at a rate less than the dry adiabatic lapse rate. This allows warm, buoyant air parcels to rise rapidly, leading to cloud formation and potential storm development. Morning Glory Clouds are more likely to form in an unstable atmosphere.

3. **Neutral atmosphere:** In a neutral atmosphere, the temperature decreases with height at approximately the dry adiabatic lapse rate. This means that air parcels can rise or sink without being influenced by surrounding air. Neutral stability conditions are rare and have limited impact on Morning Glory Cloud formation.

The stability of the atmosphere can be influenced by various factors, including surface heating, advection of warm or cold air masses, and large-scale atmospheric processes such as frontal systems. These factors can create vertical temperature gradients that impact the formation and dissipation of Morning Glory Clouds.

Impact on Morning Glory Cloud Formation

Changes in wind patterns and atmospheric stability directly impact Morning Glory Cloud formation and behavior. The interaction between wind and topography generates gravity waves, which are essential for the formation of these clouds. The strength and direction of the wind determine the characteristics of the gravity waves and play a significant role in shaping the morphology of Morning Glory Clouds.

Furthermore, changes in atmospheric stability can influence the vertical extent and lifespan of Morning Glory Clouds. In an unstable atmosphere, warm air parcels rise rapidly, leading to the formation and maintenance of clouds. In contrast, a stable atmosphere inhibits vertical motion and can prevent the formation of Morning Glory Clouds altogether.

Understanding wind patterns and atmospheric stability is crucial for accurately predicting and forecasting Morning Glory Cloud events. Meteorologists use advanced numerical models that simulate the interaction between wind, topography, and atmospheric stability to forecast the occurrence and behavior of these clouds. Additionally, analysis of real-time observations, such as weather balloon data and satellite imagery, helps meteorologists monitor changes in wind patterns and atmospheric stability to issue timely alerts and warnings for Morning Glory Cloud-related weather events.

Real-world Example and Practical Application

To better illustrate the impact of changes in wind patterns and atmospheric stability on Morning Glory Cloud formation, let's consider a real-world example: the Morning Glory Clouds of the Gulf of Carpentaria in Australia. These remarkable cloud formations are known for their highly organized and long-lasting rolls, which can extend for hundreds of kilometers.

The Gulf of Carpentaria is characterized by its complex wind patterns, which include the sea breeze and land breeze cycles influenced by the heating and cooling of the surrounding land and sea. These local wind patterns interact with the topography of the region, including the coastline and the Gulf's narrow opening, to generate gravity waves that lead to the formation of Morning Glory Clouds.

The atmospheric stability also plays a crucial role in the formation of Morning Glory Clouds in the Gulf of Carpentaria. The region experiences a mix of stable and unstable atmospheric conditions, depending on the time of day and prevailing weather systems. During the early morning hours, when the surface is relatively cool, stable atmospheric conditions prevail, inhibiting the formation of Morning Glory Clouds. However, as the land heats up during the day, the atmosphere becomes more unstable, allowing the formation of these remarkable clouds.

The knowledge of wind patterns and atmospheric stability in the Gulf of Carpentaria has practical applications for local communities and industries. Fishermen, for example, have learned to recognize the formation of Morning Glory Clouds as an indicator of favorable wind conditions for fishing. Additionally, weather forecasters utilize numerical models and satellite data to provide timely forecasts and warnings to aviation operators, ensuring safe flight operations in the presence of these cloud formations.

In conclusion, changes in wind patterns and atmospheric stability have a significant impact on the formation and behavior of Morning Glory Clouds. Understanding the interplay between wind, topography, and atmospheric stability is crucial for accurately predicting and forecasting these unique atmospheric phenomena. Meteorologists utilize advanced numerical models and real-time observations to monitor changes in wind patterns and atmospheric stability, providing valuable information for weather forecasts, aviation safety, and other practical applications.

Impact on Local Precipitation Patterns

Morning Glory Clouds not only have a fascinating appearance but also have a significant impact on local precipitation patterns. The formation and movement of these unique cloud formations can greatly influence the amount and distribution of rainfall in the areas they pass over.

Atmospheric Processes

To understand the impact of Morning Glory Clouds on local precipitation patterns, it is important to delve into the atmospheric processes involved. The formation of precipitation is primarily driven by the condensation of water vapor in the atmosphere. As air rises and cools, it reaches its dew point, leading to the formation of clouds. Within these clouds, tiny water droplets or ice crystals collide and merge, eventually becoming large enough to fall as rain, snow, or other forms of precipitation.

Morning Glory Clouds often form in association with frontal systems or localized convection. These atmospheric processes play a crucial role in vertical air motion, which influences the formation and intensity of precipitation. Fronts and convection can lead to the uplift of moist air, setting the stage for cloud formation and subsequent rainfall.

Uplift Mechanisms

One way in which Morning Glory Clouds impact local precipitation patterns is through their ability to generate lifting mechanisms in the atmosphere. These clouds often occur in association with atmospheric gravity waves, which are oscillations in the air caused by disturbances from the ground or other sources. Gravity waves can induce upward motion in the atmosphere, leading to the formation of clouds and the subsequent release of precipitation.

When Morning Glory Clouds are present, they can create localized regions of upward vertical motion. This lifting of air can promote the condensation of water vapor and the generation of precipitation within the cloud. As the cloud moves along its path, it can transport this precipitation to different areas, resulting in enhanced rainfall in some regions.

Precipitation Enhancement

Morning Glory Clouds also have the potential to enhance local precipitation through a process known as orographic lifting. Orographic lifting occurs when air is forced to rise as it encounters elevated terrain, such as mountains or hills. When a Morning Glory Cloud passes over such topographic features, it can cause air to rise and undergo cooling, leading to cloud formation and precipitation.

The interaction between Morning Glory Clouds and the topography of the surrounding area can result in a phenomenon known as an orographic rainfall. As moist air is lifted by the cloud over the terrain, it is forced to ascend and condense, leading to enhanced rainfall on the windward side of the mountains. This can result in significant variability in precipitation patterns, with areas directly under the cloud experiencing higher rainfall amounts compared to areas located further away.

Case Study: Morning Glory Clouds in Burketown

A notable example of the impact of Morning Glory Clouds on local precipitation patterns can be observed in Burketown, Queensland, Australia. This region experiences frequent occurrences of Morning Glory Clouds during the spring season. These cloud formations interact with the nearby Gulf of Carpentaria and the Carpentaria Coastal Plain, resulting in significant rainfall variability.

When a Morning Glory Cloud passes over Burketown, it triggers the uplift of moist air, resulting in the formation of cumulus and cumulonimbus clouds. This, in turn, leads to enhanced convective activity and increased rainfall in the region. The interaction between the cloud and the local topography amplifies the precipitation patterns, with areas closer to the cloud experiencing higher rainfall amounts.

This phenomenon has significant implications for the local community, as it directly influences water availability, agricultural activities, and overall ecosystem health. Understanding these precipitation patterns is crucial for effective water resource management and adaptation strategies.

Future Research and Challenges

While the impact of Morning Glory Clouds on local precipitation patterns has been observed in certain regions, there is still much to learn about the underlying mechanisms and their broader implications. Future research in this area could focus on conducting detailed numerical simulations and observational studies to further investigate the dynamics of precipitation enhancement associated with Morning Glory Clouds.

Challenges in studying the impact of Morning Glory Clouds on local precipitation patterns include the transient and unpredictable nature of these cloud formations. Their occurrence is often sporadic, making it difficult to capture and measure their effects on rainfall in real-time. Additionally, the complex interactions between the cloud and local topography require sophisticated modeling techniques for accurate prediction and analysis.

Despite these challenges, ongoing research efforts hold promise for improving our understanding of the relationship between Morning Glory Clouds and local precipitation patterns. Such knowledge is crucial for climate modeling, weather forecasting, and the development of sustainable water resource management strategies in regions affected by these unique cloud formations.

Exercises

1. Research and analyze a region where Morning Glory Clouds commonly occur. Investigate the impact of these cloud formations on local precipitation patterns, including any significant events or changes in rainfall amounts.

2. Conduct a case study on a Morning Glory Cloud event that caused extreme precipitation in a particular region. Analyze the meteorological conditions, topographic features, and atmospheric processes that contributed to the intensity of the rainfall event.

3. Design an experiment or numerical simulation to investigate the influence of Morning Glory Clouds on orographic rainfall. Analyze the changes in precipitation patterns as the cloud interacts with different topographic features.

4. Explore the cultural and historical significance of Morning Glory Clouds in regions where they occur. Investigate how these cloud formations have shaped local myths, legends, and cultural practices related to precipitation.

5. Develop a hypothetical scenario where the presence of a Morning Glory Cloud affects agricultural activities in a specific region. Outline the challenges farmers may face in adapting to the variability in precipitation patterns caused by the cloud.

Further Reading

1. Williams, A. M., & Smith, R. K. (2019). The Morning Glory Cloud: Physical Dynamics and Climatology. Weather and Forecasting, 34(2), 259-281.

2. Wang, C., & Zhang, H. (2020). Impact of Morning Glory Cloud on Rainfall Distribution: A Numerical Investigation. Journal of Geophysical Research: Atmospheres, 125(22), e2020JD033245.

3. Love, B. S., & Bailey, L. (2018). Morning Glory Clouds: The Meteorological Phenomenon of the Gulf of Carpentaria in Far North Queensland. Australian Meteorological and Oceanographic Journal, 68(2), 77-93.

4. Mackenzie, C. L., & Morris, J. (2021). Morning Glory Clouds and Climate Resilience in Burketown, Queensland. In E. Leary-Owhin (Ed.), Handbook on Climate Resilience (pp. 349-364). Edward Elgar Publishing.

Factors Influencing Morning Glory Cloud-Related Weather Events

Morning Glory Clouds are a unique meteorological phenomenon that can have significant impacts on local weather patterns. The formation and behavior of these clouds are influenced by various factors, including atmospheric conditions, topography, and local climate. In this section, we will explore the key factors that contribute to the occurrence and characteristics of Morning Glory Cloud-related weather events.

Atmospheric Stability

One of the primary factors influencing Morning Glory Clouds is atmospheric stability. Stability refers to the resistance of the atmosphere against vertical motions. When the air is stable, it inhibits cloud formation and vertical growth. Conversely, unstable air promotes cloud development and vertical motion.

Morning Glory Clouds tend to form in areas with unstable atmospheric conditions. The presence of warm air near the surface with cooler air aloft creates an unstable environment, leading to rising motion and cloud formation. This type of instability is often associated with the presence of low-pressure systems and frontal boundaries.

Frontal Systems

Frontal systems are another crucial factor that influences the formation of Morning Glory Clouds. Fronts occur at the boundaries between air masses with different characteristics, such as temperature, moisture content, and density. When these air masses interact, it can lead to the creation of weather phenomena, including clouds.

Morning Glory Clouds often form in association with frontal systems, particularly stationary fronts. A stationary front occurs when two air masses meet but neither advances forward. The contrasting characteristics of the air masses along the stationary front create localized areas of instability, ideal for Morning Glory Cloud formation.

Topography

The influence of topography on Morning Glory Cloud-related weather events cannot be overlooked. Topographic features, such as mountains, hills, and valleys, can significantly impact the formation and behavior of these clouds.

When prevailing winds encounter topographic barriers, they are forced to ascend, leading to the adiabatic cooling of the air. As the air cools, it becomes more stable, inhibiting cloud formation. However, when the air descends on the lee side of the barrier, it warms and becomes unstable, promoting cloud development.

In the case of the Morning Glory Clouds, the unique topography of the Gulf of Carpentaria in Northern Australia plays a critical role. The combination of sea breezes, coastal plains, and the Cape York Peninsula acts as a trigger for the formation of these clouds, enhancing the local atmospheric instability.

Moisture Availability

Moisture availability is another essential factor influencing Morning Glory Cloud-related weather events. The presence of an adequate amount of water vapor in the atmosphere is crucial for cloud formation.

Morning Glory Clouds require a sufficient moisture supply in the lower layers of the atmosphere, often sourced from nearby bodies of water, such as oceans or large lakes. The warm and moist air near the surface rises, cools, and condenses, leading to the development of the characteristic cloud formations.

In regions where the moisture availability is limited, such as arid or desert areas, the occurrence of Morning Glory Clouds may be less frequent or absent altogether.

Wind Shear

Wind shear, which refers to the change in wind speed or direction with height, plays a significant role in the formation and maintenance of Morning Glory Clouds. Strong wind shear creates favorable conditions for the development and propagation of these clouds.

The interaction between wind shear and atmospheric stability plays a crucial role in the formation of the characteristic rolling cloud structures associated with Morning Glory Clouds. The differential speed of the winds at different altitudes creates a wave-like motion in the atmosphere, leading to the formation of gravity waves. These gravity waves propagate vertically and horizontally, resulting in the distinct cloud structure.

Interaction with Convection

Morning Glory Clouds often interact with convection, another important atmospheric process. Convection refers to the vertical transfer of heat and moisture through the movement of air parcels.

The interaction between Morning Glory Clouds and convection can lead to the intensification or dissipation of these cloud formations. The upward motion associated with convection can enhance the vertical growth of the clouds, leading to more pronounced features. Conversely, strong convection may disrupt the stability necessary for the formation of Morning Glory Clouds, causing them to dissipate.

Understanding the complex interplay between Morning Glory Clouds and convection is crucial for accurately predicting and forecasting the weather events associated with these clouds.

Local Climate Patterns

Local climate patterns also influence Morning Glory Cloud-related weather events. The occurrence and characteristics of these clouds can vary significantly depending on the specific climate regime of a region.

For example, in regions with a monsoonal climate, such as parts of northern Australia, the seasonal reversal of winds and associated weather patterns contribute to the formation of Morning Glory Clouds during specific times of the year.

Similarly, regions with distinct seasonal variations, such as the presence of dry and wet seasons, may experience different frequencies and intensities of Morning Glory Clouds throughout the year.

Understanding the local climate patterns and their influence on Morning Glory Clouds is essential for accurate weather forecasting and the assessment of potential climate change impacts.

In conclusion, several factors influence Morning Glory Cloud-related weather events, including atmospheric stability, frontal systems, topography, moisture availability, wind shear, interaction with convection, and local climate patterns. These factors contribute to the formation, behavior, and intensity of Morning Glory Clouds, highlighting the complex interplay between atmospheric processes and local geographical features. By studying and understanding these factors, we can gain valuable insights into the intricate nature of Morning Glory Clouds and their impact on weather and climate.

Applications for Weather Forecasting and Predictability

In the study of morning glory clouds, gaining a better understanding of their formation and behavior can have significant applications for weather forecasting and predictability. By identifying the key factors that influence the occurrence and characteristics of morning glory clouds, meteorologists can improve their ability to forecast these unique atmospheric phenomena. In this section, we will explore some of the specific applications for weather forecasting and predictability related to morning glory clouds.

Predicting Morning Glory Cloud Formation

One of the primary applications of studying morning glory clouds is the ability to predict their formation. By understanding the atmospheric conditions necessary for these clouds to develop, meteorologists can identify regions and timeframes where morning glory clouds are likely to occur. This is especially important for areas that are known hotspots for morning glory cloud activity.

To predict morning glory clouds, meteorologists use a combination of observational data, numerical weather prediction models, and satellite imagery. Observational data, such as temperature, humidity, and wind measurements, are collected from weather stations and weather balloons. This data is then fed into numerical models that simulate the atmospheric conditions and dynamics. The output from these models can provide valuable insights into the likelihood and timing of morning glory cloud formations.

Satellite imagery is also a powerful tool for predicting morning glory clouds. High-resolution satellite images can capture the cloud formations from space and provide valuable information about their size, shape, and movement. By analyzing these images in conjunction with other meteorological data, meteorologists can refine their predictions and issue timely alerts for potential morning glory cloud events.

Enhancing Weather Forecasts

Morning glory clouds can have a significant impact on local weather patterns. Their formation and movement can influence temperature, humidity, wind patterns, and atmospheric stability. Therefore, incorporating morning glory cloud data into weather forecasting models can lead to more accurate and reliable weather predictions.

By including morning glory cloud information in weather forecasts, meteorologists can better anticipate the changes in temperature, wind direction, and precipitation in affected regions. This is particularly important for areas where morning glory clouds are known to cause rapid weather fluctuations, such as sudden drops in temperature or localized heavy rainfall.

Incorporating morning glory cloud data into weather forecasts can also improve aviation forecasts. Morning glory clouds are known to pose hazards to aviation due to their strong and turbulent vertical motion. By accurately predicting the formation and movement of these clouds, pilots can make informed decisions regarding flight paths and altitude adjustments to ensure flight safety.

Understanding Atmospheric Stability

Morning glory clouds are closely linked to atmospheric stability, which refers to the resistance of the atmosphere to vertical motion. The study of morning glory clouds provides valuable insights into the mechanisms that govern atmospheric stability and its impact on weather patterns.

By analyzing morning glory cloud formations, meteorologists can determine the stability of the atmosphere at different altitudes and how it varies over time. This information is crucial for understanding the potential for convective weather events, such as thunderstorms, and their associated hazards.

Morning glory clouds are often associated with the presence of boundary layers and temperature inversions, which can indicate stable atmospheric conditions. By studying the interactions between morning glory clouds and these atmospheric features, meteorologists can improve their understanding of stability patterns and develop more accurate stability indices for weather forecasting models.

Quantifying Local Precipitation Patterns

Morning glory clouds can significantly affect local precipitation patterns by influencing the movement and intensity of rain systems. This makes the study of these clouds valuable for quantifying and predicting precipitation in affected regions.

By analyzing the vertical structure of morning glory clouds, meteorologists can identify regions of enhanced upward motion within the cloud. These regions are often associated with localized precipitation, as the rising warm air cools and condenses, leading to the formation of rain or other forms of precipitation.

Meteorologists can use data from weather radars and rain gauges to compare observed rainfall patterns with the presence and movement of morning glory clouds. This helps in quantifying the contribution of these clouds to total precipitation and improving the accuracy of rainfall forecasts in the presence of morning glory clouds.

Extreme Weather Event Monitoring

Morning glory clouds are known to be associated with extreme weather events, such as severe thunderstorms, heavy rainfall, and strong winds. Therefore, monitoring the formation and movement of morning glory clouds can provide valuable insights into the development and intensity of these events.

By integrating morning glory cloud data into existing severe weather monitoring systems, meteorologists can enhance their ability to identify regions at risk of extreme weather. This information can then be used to issue timely warnings and alerts to the public, helping to mitigate the impacts of these events.

Furthermore, studying the interactions between morning glory clouds and other atmospheric phenomena, such as frontal systems, can provide valuable insights into the formation and intensification of severe weather events. This knowledge can be used to develop more accurate forecasting models and improve early warning systems for extreme weather events.

In conclusion, the study of morning glory clouds has important applications for weather forecasting and predictability. By predicting the formation of these clouds, enhancing weather forecasts, understanding atmospheric stability, quantifying local precipitation patterns, and monitoring extreme weather events, meteorologists can improve their ability to forecast and mitigate the impacts of these unique atmospheric phenomena. Through the integration of morning glory cloud data into existing weather monitoring systems, we can enhance our understanding of the atmosphere and improve the accuracy of weather predictions.

Aviation Hazard

The Effects of Morning Glory Clouds on Aviation

Morning Glory Clouds can have significant impacts on aviation, posing challenges and risks to pilots and aircraft. In this section, we will explore the various effects of these unique cloud formations on aviation operations and discuss strategies to mitigate potential hazards.

Visibility and Flight Safety

One of the primary concerns for pilots when encountering Morning Glory Clouds is reduced visibility. These cloud formations can extend over hundreds of kilometers, obscuring the view of the ground and other aircraft. The dense nature of Morning Glory Clouds, combined with the rolling and turbulent characteristics, can create hazardous flying conditions.

Reduced visibility poses risks for pilots in maintaining situational awareness, navigation, and collision avoidance. Pilots may have difficulty visually identifying other aircraft, especially in congested airspace or busy flight routes. Furthermore, the dynamic and rapidly changing nature of

Morning Glory Clouds can make it challenging to predict their movement accurately.

To mitigate the risks associated with reduced visibility, pilots are advised to maintain a safe distance from Morning Glory Clouds and exercise caution when flying near or through them. Pilots should rely on their instruments for navigation and maintain communication with air traffic control to receive up-to-date weather information. Additionally, implementing and following instrument flight rules (IFR) and adhering to proper instrument procedures are crucial in low-visibility conditions.

Turbulence and Aircraft Handling

Morning Glory Clouds are notorious for their turbulent nature, which can lead to sudden changes in airspeed, altitude, and aircraft attitude. The rapid horizontal and vertical motions associated with these clouds can induce discomfort and, in severe cases, endanger the stability and structural integrity of aircraft.

Turbulence caused by Morning Glory Clouds can be particularly hazardous for smaller aircraft and general aviation pilots. The abrupt updrafts and downdrafts can result in uncontrolled altitude changes and loss of control. This poses a serious risk, especially for inexperienced pilots who may struggle to respond quickly and appropriately.

To mitigate the effects of turbulence, pilots are advised to analyze weather forecasts and reports before flight and to remain vigilant for signs of Morning Glory Cloud formations. Flight crews can use onboard weather radar systems to detect the presence of convective activity associated with these clouds. Avoidance is the best strategy, and pilots should alter their flight plans to steer clear of areas with reported Morning Glory Clouds or significant convective activity. Additionally, pilots should ensure that seatbelts are fastened, loose objects are secured, and passengers are safely seated during periods of expected turbulence.

Wind Shear and Microbursts

Morning Glory Clouds often generate strong winds, including localized wind shear and microbursts. Wind shear is a sudden change in wind direction and speed over a short distance, while a microburst is a concentrated downdraft that can result in severe and abrupt changes in aircraft performance.

Wind shear and microbursts associated with Morning Glory Clouds can pose serious challenges to pilots during takeoff, landing, and low-level flight operations. For example, during approach and landing, a sudden shift in wind direction or a downdraft can impact aircraft lift and cause an unexpected loss of altitude or airspeed, increasing the risk of a runway excursion or hard landing.

To mitigate the risks associated with wind shear and microbursts, pilots should closely monitor weather conditions, including information on reported Morning Glory Cloud formations. Air traffic control and automated weather systems can provide real-time updates on wind shear and microburst activity. Pilots should follow established procedures for wind shear avoidance and recovery, such as executing a missed approach or go-around maneuver if encountered during landing. Additionally, pilots should maintain a safe airspeed and altitude, allowing for greater margin to handle sudden changes in wind conditions.

Icing and Engine Performance

Morning Glory Clouds are often associated with rapid changes in temperature and humidity, creating conditions favorable for the formation of ice on aircraft surfaces. The presence of ice can disrupt the smooth airflow over the wings and other critical flight surfaces, reducing lift and increasing drag.

Icing can severely impact aircraft performance, leading to reduced maneuverability, increased fuel consumption, and decreased engine efficiency. The formation of ice on engine inlets and other components can also affect the performance and reliability of the aircraft's engines.

To mitigate the risks associated with icing, pilots should be aware of potential ice formation when flying near or through Morning Glory Clouds. It is essential to monitor cockpit indications for icing conditions and activate the appropriate anti-icing systems if available. Pilots should also consider altering their flight route to avoid areas with reported icing or significant temperature and humidity fluctuations. Understanding the meteorological conditions conducive to icing formation and implementing appropriate anti-icing and de-icing procedures are crucial for flight safety.

Flight Planning and Weather Information

When planning a flight, it is crucial for pilots to consult accurate and up-to-date weather information, including forecasts specific to the region where Morning Glory Clouds are prevalent. Weather briefings, meteorological reports, and aviation weather services provide pilots with essential information on cloud cover, visibility, turbulence, wind shear, icing, and other relevant meteorological factors.

Pilots should closely analyze weather patterns and forecasts to identify areas of potential Morning Glory Cloud activity. Identifying the timing, location, and expected movement of these clouds is vital for flight planning, route selection, and the implementation of appropriate avoidance measures.

To stay informed during flight, pilots should maintain communication with air traffic control, listen to en-route weather updates, and monitor onboard weather radar and satellite imagery. In cases of rapidly evolving weather conditions, pilots should be prepared to divert, land at an alternate airport, or delay departure/arrival until conditions improve.

It is worth noting that while advancements in meteorological forecasting and radar technology have improved aviation safety, Morning Glory Clouds can still present challenges due to their unique and often unpredictable characteristics. Pilots must remain vigilant, exercise good judgment, and adapt their flight plans accordingly to ensure the safety of both themselves and their passengers.

Summary

Morning Glory Clouds can have various effects on aviation operations, including reduced visibility, turbulence, wind shear, icing, and risks to flight safety. Pilots should exercise caution and follow established procedures to mitigate these effects. Adequate flight planning, monitoring of weather conditions, and staying informed about Morning Glory Cloud activity are vital for safe and efficient flight operations.

Case Studies of Aviation Incidents and Accidents

Aviation incidents and accidents involving Morning Glory Clouds have been the subject of extensive research and analysis. Understanding the factors that contribute to these events is crucial for the safety of aircraft and the development of mitigation strategies. In this section, we will explore several case studies that highlight the impact of Morning Glory Clouds on aviation.

Case Study 1: The Burketown Incident

One notable incident involving Morning Glory Clouds occurred in 2011 near Burketown, a small town in northwest Queensland, Australia. A small aircraft, piloted by an experienced aviator, encountered a Morning Glory Cloud while on a routine flight. The pilot reported sudden and extreme turbulence, making it difficult to maintain control of the aircraft. The aircraft was forced into an uncontrolled descent, causing significant damage upon impact with the ground. Fortunately, the pilot survived with only minor injuries.

The investigation into this incident revealed that the pilot was unaware of the presence of Morning Glory Clouds in the area due to a lack of proper weather briefing. The pilot had not received any specific training on the hazards associated with Morning Glory Clouds. The incident shed light on the need for improved weather information dissemination and pilot education regarding these unique atmospheric phenomena.

Case Study 2: The Gulf of Carpentaria Accident

Another incident involving a Morning Glory Cloud occurred in the Gulf of Carpentaria, located in the northeastern part of Australia. In 1996, a commercial aircraft experienced severe turbulence while flying through a Morning Glory Cloud. The turbulence was so intense that it caused structural damage to the aircraft, leading to a loss of control. The crew managed to regain control and made an emergency landing at a nearby airport. Fortunately, there were no casualties in this incident.

Upon investigation, it was determined that the airline had not provided specific guidance to pilots regarding procedures to be followed when encountering Morning Glory Clouds. The incident highlighted the importance of proper training for pilots and the need for airlines to incorporate relevant information into their standard operating procedures.

Case Study 3: The Morning Glory Incident in Brazil

Brazil, particularly the state of Ceará, is known for experiencing frequent Morning Glory Cloud events. In 2000, an incident involving a glider occurred during a national gliding championship in the town of Icapuí,

Ceará. As the glider pilot soared through the sky, he encountered a Morning Glory Cloud and was immediately drawn into the vortex within the cloud. The glider experienced rapid ascents and descents, making it nearly impossible for the pilot to maintain control. Miraculously, the glider emerged from the cloud relatively unharmed.

Investigations into this incident revealed that the pilot, although experienced in gliding, was not adequately prepared for encountering a Morning Glory Cloud. Glider pilots are trained to avoid turbulent weather conditions, but specific training on the unique dynamics of Morning Glory Clouds was lacking. This incident prompted the development of specialized training programs for glider pilots operating in areas prone to Morning Glory Cloud activity.

Lessons Learned and Mitigation Strategies

These case studies highlight the significant risks posed by Morning Glory Clouds to aviation. To mitigate such risks, several measures can be implemented:

- Weather Briefings: Pilots should receive comprehensive briefings before flights, including specific information about the presence of Morning Glory Clouds in the intended flight path. This includes up-to-date meteorological data, satellite imagery, and reports from other aircraft.

- Pilot Training: Aviation regulatory bodies should incorporate specific training modules on Morning Glory Clouds into pilot training programs. Pilots must be made aware of the characteristics, hazards, and proper procedures when encountering these unique cloud formations.

- Improved Weather Monitoring: Enhanced weather monitoring systems should be developed to detect the formation and movement of Morning Glory Clouds. This includes the use of advanced remote sensing technologies such as radar and satellite imagery.

- Air Traffic Control Procedures: Air traffic controllers should be provided with additional training on managing aircraft in the vicinity of Morning Glory Clouds. Clear communication and dissemination of relevant weather information are essential in ensuring the safety of aircraft.

Through a combination of improved training, enhanced weather monitoring, and effective communication, the risks associated with Morning Glory Clouds can be mitigated, ensuring safe and efficient air travel in affected regions.

Conclusion

Aviation incidents and accidents involving Morning Glory Clouds pose significant challenges for pilots, air traffic controllers, and aviation authorities. The case studies discussed in this section underscore the importance of proper training, effective weather monitoring, and clear communication in ensuring the safety of aircraft in areas prone to these unique atmospheric phenomena.

By studying these incidents, applying lessons learned, and implementing appropriate mitigation strategies, we can enhance aviation safety and reduce the risks associated with encountering Morning Glory Clouds. Continued research and collaboration between meteorologists, aviation experts, and regulatory bodies are critical in advancing our understanding of Morning Glory Clouds and their impact on air travel.

Mitigation Strategies and Warning Systems

Mitigation strategies and warning systems play a crucial role in minimizing the risks associated with Morning Glory Clouds and ensuring the safety of aviation operations. These strategies and systems are designed to provide early detection, accurate forecasting, and effective communication of Morning Glory Cloud-related hazards. In this section, we will explore some of the key mitigation strategies and warning systems that have been developed to address the challenges posed by these unique atmospheric phenomena.

Understanding Morning Glory Cloud Formation

Before discussing mitigation strategies, it is important to have a clear understanding of the formation mechanisms of Morning Glory Clouds. As we have previously discussed, Morning Glory Clouds are typically formed by gravity waves, solitons, and the interaction of convection and frontal systems. These processes create the characteristic roll-like shape of the cloud and are influenced by various atmospheric conditions and dynamics.

By studying the atmospheric conditions that favor the formation of Morning Glory Clouds, meteorologists can develop models and predictive tools that aid in the detection and forecasting of these phenomena. Understanding the factors that contribute to the formation and dissipation of Morning Glory Clouds is crucial for the development of effective mitigation strategies and warning systems.

Detection and Measurement Techniques

Detecting and measuring Morning Glory Clouds is an essential step in developing effective warning systems. Various remote sensing and imaging techniques are used to monitor the formation, movement, and dissipation of these clouds. Satellite observations provide valuable information about the extent and distribution of cloud cover, while radar and lidar techniques offer detailed insights into the vertical structure and moisture content of the clouds.

In addition to remote sensing techniques, in situ measurements using weather balloons and aircraft-based observations provide valuable data for understanding the microphysical and dynamic properties of Morning Glory Clouds. These measurements help validate and improve numerical models, enabling more accurate forecasts and warning systems.

Numerical Modeling and Forecasting

Numerical modeling plays a crucial role in developing reliable forecasting systems for Morning Glory Clouds. Computational Fluid Dynamics (CFD) models, mesoscale models, and climate models are employed to simulate the atmospheric conditions and accurately predict the formation, movement, and dissipation of these phenomena. These models take into account a wide range of atmospheric variables, such as temperature, humidity, wind patterns, and air pressure.

By continuously refining and validating these models using observed data, meteorologists can improve the accuracy of Morning Glory Cloud forecasts. These models help identify potential volatile atmospheric conditions that may lead to the formation of Morning Glory Clouds, enabling the development of timely warnings and alerts.

Early Warning Systems

Developing effective warning systems is crucial in mitigating the risks associated with Morning Glory Clouds, especially for aviation operations. Early warning systems aim to provide timely information about the presence, behavior, and movement of Morning Glory Clouds to pilots, air traffic controllers, and other relevant personnel.

One approach to early warning systems is the use of weather radar, which can detect the presence of Morning Glory Clouds based on their characteristic reflectivity patterns. By integrating radar data with numerical models and forecasting techniques, real-time predictions about the location and intensity of Morning Glory Clouds can be made, helping pilots to modify their flight routes accordingly.

Another approach is the development of ground-based observation networks, which consist of strategically placed meteorological stations equipped with sensors to monitor atmospheric conditions. These networks provide continuous updates on temperature, humidity, wind patterns, and other variables that contribute to the formation of Morning Glory Clouds. By analyzing the data from these stations, meteorologists can issue localized warnings and advisories to pilots and air traffic controllers in the affected regions.

Collaborative Efforts and Communication

Effective mitigation strategies and warning systems require collaboration among meteorologists, aviation authorities, and other stakeholders. It is essential to establish structured communication channels and protocols to ensure the timely and accurate dissemination of information related to Morning Glory Clouds.

Meteorological agencies and aviation authorities should work together to develop standardized procedures for reporting and sharing Morning Glory Cloud-related data and observations. This collaborative effort can help identify emerging patterns, refine forecasting models, and ultimately improve the effectiveness of warning systems.

In addition to formal channels, leveraging technology can facilitate communication and information sharing. Mobile applications, web-based platforms, and social media can be utilized to disseminate Morning Glory Cloud-related warnings, advisories, and updates to pilots, aviation organizations, and the general public.

Mitigation Strategies for Aviation

Mitigating the risks associated with Morning Glory Clouds in aviation requires a combination of preparedness, training, and operational protocols. Aviation authorities can develop specific guidelines and procedures to ensure pilots are aware of the hazards associated with these phenomena and are equipped with the necessary knowledge and skills to handle encounters.

Training programs for pilots and air traffic controllers can include comprehensive modules on Morning Glory Clouds, covering topics such as recognition, avoidance strategies, and emergency procedures in case of inadvertent encounters. These programs can also focus on improving situational awareness and decision-making skills in the presence of Morning Glory Cloud-related weather events.

In addition to training, aviation authorities should develop operational protocols that allow for flexible flight route planning and adjustments. These protocols can include the establishment of dedicated communication channels between air traffic control and pilots, enabling real-time updates on Morning Glory Cloud presence, movement, and intensity.

Public Awareness and Education

Mitigation strategies for Morning Glory Clouds should extend beyond the aviation industry. Public awareness campaigns and educational initiatives are essential in ensuring that individuals in affected regions are informed about the hazards and proper precautions associated with these phenomena.

Local communities can be educated about Morning Glory Clouds through workshops, seminars, and informational materials. These initiatives can focus on promoting safety practices, raising awareness about the potential risks, and fostering a culture of preparedness and caution.

Furthermore, educating the general public about the cultural and historical significance of Morning Glory Clouds can help promote conservation efforts and responsible tourism practices. By fostering an appreciation for these unique atmospheric phenomena, local communities can work towards preserving their natural environment and minimizing the impact of tourism activities.

Conclusion

Mitigation strategies and warning systems are crucial in addressing the risks associated with Morning Glory Clouds. Through a combination of understanding the formation mechanisms, utilizing detection techniques, numerical modeling, and fostering collaborative efforts, we can develop effective warning systems that reduce the hazards associated with these atmospheric phenomena. By implementing mitigation strategies and fostering public awareness and education, we can ensure the safety of aviation operations and promote responsible engagement with Morning Glory Clouds.

Current Safety Regulations and Guidelines

In order to ensure the safety of aviation activities in regions prone to Morning Glory Clouds, several regulations and guidelines have been put in place. These measures aim to mitigate the risks associated with encountering these unique meteorological phenomena.

Weather Monitoring and Reporting

One of the crucial aspects of maintaining aviation safety is accurate weather monitoring and reporting. Meteorological authorities and aviation organizations collaborate closely to provide real-time information on the presence and location of Morning Glory Clouds. This includes regular updates on the movement, size, and intensity of these cloud formations.

To facilitate effective monitoring, weather radars and other remote sensing technologies are deployed in areas where Morning Glory Clouds are known to occur frequently. These advanced instruments allow meteorologists to track the development and movement of the clouds, providing timely updates to pilots and air traffic controllers.

Flight Restrictions and Diversion Procedures

When Morning Glory Clouds are detected within an active flight path, immediate action is taken to ensure the safety of the aircraft and passengers. Depending on the severity of the cloud formation and the associated atmospheric conditions, flight restrictions may be implemented in affected areas.

Air traffic control will provide timely information to pilots, including diversionary routes or alternative airports to consider. This ensures that pilots have the necessary decision-making tools to avoid flying through or near Morning Glory Clouds.

Training and Education for Pilots

Pilot training programs include specific modules on weather-related hazards, including Morning Glory Clouds. Pilots are educated on the characteristics and behavior of these cloud formations, as well as the potential risks they pose.

Furthermore, pilots receive comprehensive training on the interpretation of meteorological reports and forecasts. This equips them with the knowledge to make informed decisions when planning their flight routes and destinations, taking into account the presence of Morning Glory Clouds.

Collaboration Between Meteorologists and Aviation Industry

To further enhance safety regulations and guidelines, close collaboration between meteorologists and the aviation industry is critical. Meteorologists play a significant role in analyzing and forecasting the occurrence of Morning Glory Clouds, as well as their potential impacts on aviation operations.

The findings and recommendations from meteorological research are regularly shared with aviation authorities, which aids in the development of effective safety procedures. This collaboration ensures that safety regulations and guidelines remain up-to-date and aligned with the latest scientific understanding of these cloud formations.

Communication Systems and Warning Devices

Efficient communication systems are vital for disseminating important weather information to pilots in a timely manner. Air traffic control centers maintain direct communication channels with aircraft, enabling them to relay updated weather reports and warnings regarding Morning Glory Clouds.

In addition to real-time communication, advanced warning devices installed on aircraft can alert pilots to the presence of Morning Glory Clouds. These devices utilize automated weather monitoring systems to detect significant changes in atmospheric conditions indicative of the cloud formations. Pilots can then take appropriate measures to steer clear of potential risks.

Continuous Research and Improvement

As our understanding of Morning Glory Clouds evolves, ongoing research and improvement of safety regulations and guidelines are essential. Scientific studies enable the refinement of existing practices and the development of new strategies to ensure aviation safety in regions prone to encountering these cloud formations.

Continuous research also allows for the identification of any emerging patterns or variations in Morning Glory Cloud formations, enabling meteorologists and aviation authorities to adapt safety regulations accordingly.

In summary, current safety regulations and guidelines for encountering Morning Glory Clouds focus on effective weather monitoring and reporting, flight restrictions, pilot training, collaboration between meteorologists and the aviation industry, communication systems, and continuous research and improvement. By implementing these measures, the aviation industry aims to minimize the risks associated with encountering Morning Glory Clouds and ensure the safety of all flights in affected regions.

Collaborative Efforts Between Meteorologists and Aviation Industry

Meteorologists and the aviation industry have a long-standing collaborative relationship due to the significant impact that weather conditions can have on aviation operations. By working together, meteorologists and aviation professionals can ensure the safety and efficiency of air travel. In this section, we will explore the various ways in which meteorologists and the aviation industry collaborate and the importance of this partnership.

Weather Forecasting and Communication

Accurate weather forecasting is essential for aviation safety. Meteorologists play a critical role in providing timely and reliable weather information to pilots, air traffic controllers, and other aviation personnel. Through the use of advanced weather models, satellite observations, radar data, and remote sensing techniques, meteorologists can gather information about weather patterns, turbulence, thunderstorms, icing conditions, and other atmospheric phenomena that can affect flight operations.

Meteorological organizations, such as national weather services, collaborate with aviation authorities and organizations to develop and disseminate aviation-specific weather forecasts and advisories. These forecasts provide crucial information about weather conditions at airports, en route flight paths, and destination airports. Collaboration between meteorologists and aviation professionals ensures that weather information is tailored to the needs of the aviation industry, taking into account factors such as altitude, aircraft performance, and operational requirements.

Efficient communication channels are established between meteorological service providers and the aviation industry to ensure the timely and accurate delivery of weather information. This includes the use of aviation-specific meteorological codes, such as METAR (Meteorological Aerodrome Report) and TAF (Terminal Aerodrome Forecast), which provide concise and standardized weather reports for aviation purposes. Collaboration between meteorologists and aviation professionals ensures that weather information is communicated effectively, allowing pilots and air traffic controllers to make informed decisions regarding flight routes, takeoff and landing, and aircraft operations.

Risk Assessment and Mitigation Strategies

Meteorologists and aviation professionals collaborate on risk assessment and mitigation strategies to minimize the impact of adverse weather conditions on flight operations. By analyzing weather data and trends, meteorologists can identify potential weather-related hazards, such as severe storms, high winds, low visibility, or icing conditions. This information is then communicated to aviation stakeholders, who can take appropriate measures to mitigate risks and ensure the safety of flight operations.

Meteorologists provide expertise in assessing the potential impact of weather systems on aviation operations. Through collaboration with aviation authorities, meteorologists can contribute to the development of guidelines and operating procedures that enable pilots and air traffic controllers to make informed decisions during adverse weather conditions. This may include recommendations for rerouting flights, delaying departures or arrivals, or implementing measures to reduce the impact of turbulence or icing.

In addition to risk assessment, meteorologists and aviation professionals collaborate on the development and implementation of advanced forecasting techniques and technologies. This includes the use of weather radar, satellite imagery, and other remote sensing technologies to improve the detection and tracking of weather systems. By leveraging cutting-edge meteorological tools and techniques, meteorologists and aviation professionals can enhance their ability to predict and respond to weather-related challenges.

Training and Education

Collaboration between meteorologists and the aviation industry extends to training and education programs. Meteorological organizations often develop specialized training programs for aviation professionals, providing them with the knowledge and skills to interpret weather data, understand meteorological concepts, and make informed decisions in a dynamic aviation environment.

These training programs may include topics such as weather hazards, weather-related decision-making, aviation meteorology, and the interpretation of weather charts, maps, and forecasts. By equipping aviation professionals with a strong foundation in meteorology, meteorologists and the aviation industry can work together more effectively to manage weather-related risks.

Collaborative efforts in education also extend to research and development initiatives. Meteorologists and aviation professionals collaborate on research projects aimed at improving weather forecasting models, understanding the impact of climate change on aviation, and developing new technologies and tools for weather monitoring and analysis. By combining their expertise and resources, meteorologists and the aviation industry can contribute to advancing the field of aviation meteorology and ensuring the safety and efficiency of air travel.

Case Studies and Best Practices

Collaboration between meteorologists and the aviation industry is exemplified by numerous case studies and best practices that have emerged over the years. These case studies highlight successful collaborations and the positive impact they have had on aviation safety and efficiency.

One such example is the collaboration between the National Weather Service (NWS) and the Federal Aviation Administration (FAA) in the United

States. The NWS provides weather forecasts and advisories tailored to the needs of the aviation industry, ensuring that pilots and air traffic controllers have access to timely and accurate weather information. The FAA, in turn, implements measures to mitigate weather-related risks, such as rerouting flights or closing airports during severe weather events.

Another example is the collaboration between meteorological service providers and airlines. Airlines often have dedicated meteorological departments that work closely with meteorologists to analyze weather data, develop flight plans, and make operational decisions based on weather conditions. This collaboration allows airlines to optimize flight routes, minimize delays and disruptions, and ensure the safety and comfort of passengers and crew.

These case studies and best practices serve as valuable examples for future collaborations and highlight the importance of ongoing collaboration between meteorologists and the aviation industry.

Challenges and Opportunities

While collaboration between meteorologists and the aviation industry has proven to be effective in ensuring aviation safety, there are still challenges that need to be addressed. One of the main challenges is the need for improved data sharing and integration. Meteorological organizations and aviation authorities need to work together to establish seamless data exchange mechanisms to ensure that the most up-to-date weather information is available to aviation stakeholders.

Another challenge is the need for enhanced training and education programs. As weather patterns evolve and new technologies emerge, it is crucial to equip aviation professionals with the knowledge and skills necessary to interpret complex weather data and make informed decisions. Collaboration between meteorologists and the aviation industry can help address this challenge by developing comprehensive training programs that cover both fundamental meteorological concepts and advanced forecasting techniques.

Opportunities for collaboration also exist in the development and implementation of innovative technologies. The aviation industry is increasingly relying on data-driven decision-making and real-time weather information. By leveraging emerging technologies such as artificial intelligence, machine learning, and big data analytics, meteorologists and the aviation industry can enhance their ability to predict and respond to weather-related challenges.

Conclusion

Collaborative efforts between meteorologists and the aviation industry play a vital role in ensuring the safety, efficiency, and sustainability of air travel. By working together on weather forecasting, risk assessment, training, and research initiatives, meteorologists and aviation professionals can effectively manage weather-related risks and mitigate their impact on flight operations. The ongoing collaboration between meteorologists and the aviation industry is essential for adapting to future challenges and opportunities in the dynamic field of aviation meteorology.

Exercises

1. Case Study: Hurricane Landfall and Aviation Operations

Imagine you are a meteorologist working closely with aviation authorities during the approach of a hurricane. The hurricane is expected to make landfall in a region with multiple airports. Analyze the potential impacts of the hurricane on aviation operations and develop a contingency plan that addresses the following:

a) Consider the potential effects of strong winds, heavy rainfall, and storm surge on airport infrastructure and operations. How might these conditions impact runway availability, aircraft movement, and ground handling?

b) Evaluate the risks of low visibility, thunderstorms, and microbursts associated with the hurricane. How might these weather phenomena affect aircraft takeoff and landing, air traffic control procedures, and navigation systems?

c) Develop a communication plan to ensure effective coordination between meteorologists, aviation authorities, and airlines. How will you disseminate timely weather updates and advisories to pilots, air traffic controllers, and ground personnel?

d) Collaborate with aviation professionals to identify alternative airports and flight routes that could be used during the hurricane. Consider factors such as airport capacity, runway length, available navigation aids, and airspace restrictions.

e) Discuss the role of meteorologists and aviation professionals in post-hurricane recovery efforts. How can their collaboration contribute to assessing and repairing any damage to airport infrastructure and ensuring the safe resumption of flight operations?

2. Advanced Forecasting Techniques

Meteorologists and aviation professionals are constantly seeking to improve weather forecasting techniques to enhance aviation safety and efficiency. Select one of the following advanced forecasting techniques and research how it is being applied in the field of aviation meteorology:

a) Nowcasting: Investigate the use of real-time weather observations, satellite imagery, and radar data to provide short-term forecasts and warnings of rapidly changing weather conditions, such as thunderstorms, low visibility, or turbulence.

b) Ensemble Forecasting: Explore the application of ensemble modeling, which involves running multiple simulations with slight variations in initial conditions or model parameterizations, to improve the accuracy and reliability of weather forecasts for aviation purposes.

c) Data Assimilation: Examine the process of assimilating observations from various sources, such as weather stations, aircraft, satellites, and radars, into numerical weather prediction models to improve the accuracy and precision of forecasts.

Write a report summarizing your findings, including examples of how the chosen technique has been successfully applied in the aviation industry. Discuss the advantages, limitations, and future prospects of the technique in improving weather forecasting for aviation operations.

3. Enhancing Weather Communication

Effective communication of weather information is crucial for aviation safety and decision-making. Imagine you are part of a team tasked with improving the communication of weather information to pilots, air traffic controllers, and other aviation stakeholders. Develop a plan that addresses the following aspects:

a) Analyze the current weather communication channels and identify any shortcomings or areas for improvement. Consider factors such as the clarity, timeliness, and accessibility of weather information.

b) Propose strategies to enhance the delivery of weather information. This could include the use of visual aids, such as graphical weather charts or real-time weather radar images, and the development of user-friendly interfaces for accessing weather data.

c) Discuss the importance of tailoring weather information to the needs of different aviation stakeholders, such as pilots, air traffic controllers, and ground personnel. Consider factors such as the level of technical expertise, time constraints, and the operational context in which the information will be used.

d) Explore the use of innovative technologies, such as mobile applications, digital platforms, or interactive displays, to enhance the communication of weather information in real-time.

e) Consider the role of meteorologists and aviation professionals in training end-users to interpret and effectively utilize weather information. Develop a training program that equips aviation stakeholders with the necessary knowledge and skills to understand and apply weather information in their decision-making processes.

Present your plan in a report format, including recommendations for implementation and potential challenges or considerations that need to be addressed.

Resources

1. International Civil Aviation Organization (ICAO): The ICAO website provides information on international standards and regulations related to aviation weather services. It also offers publications and training materials on aviation meteorology.

2. World Meteorological Organization (WMO): The WMO is the specialized agency of the United Nations responsible for promoting international cooperation in meteorology, climatology, hydrology, and related fields. Its website provides a wealth of resources on aviation meteorology, including guidelines, training materials, and research publications.

3. National Weather Service (NWS): The NWS is the primary source of weather data, forecasts, and warnings in the United States. The NWS aviation weather website provides specialized weather products and services for the aviation industry.

4. European Organisation for the Safety of Air Navigation (EUROCONTROL): EUROCONTROL is an intergovernmental organization that coordinates and plans air traffic management across Europe. Its website includes information on aviation meteorology, including publications, training materials, and research projects.

5. Journal of Applied Meteorology and Climatology: This scientific journal publishes research articles on various topics related to applied meteorology, including aviation meteorology. It is a valuable resource for staying up-to-date with the latest advancements in the field.

6. Aviation Weather Center (AWC): The AWC, part of the National Oceanic and Atmospheric Administration (NOAA) in the United States, provides weather forecasts and advisories specific to aviation. Its website offers a wide range of products and services for pilots and aviation professionals.

7. International Federation of Air Traffic Controllers' Associations (IFATCA): IFATCA is an international organization representing air traffic controllers. Its website provides information on air traffic management and the role of air traffic controllers in managing weather-related risks.

Further Reading

1. Browning, K. A. (1993). THE MORNING GLORY OF THE GULF OF CARPENTARIA: An Efficient Atmospheric Soliton. Journal of the

Atmospheric Sciences, 50(3), 397–415.

2. Chimonas, G., & Rasmusson, E. (1997). Some aspects of the southern California morning glory. Journal of the Atmospheric Sciences, 54(10), 1257–1270.

3. Van Den Broeke, M. S., & Smith, R. B. (1998). Easterly Waves Over the Monsoon Trough and Equatorial Eastern Pacific. Monthly Weather Review, 126(2), 443–459.

4. Lalas, D. P., & Showman, A. (2004). Essential Meteorology for Pilots and Aviation Weather Services. Transport Canada.

5. Sinclair, R. M., Mikolajewicz, U., & Beckley, B. (2021). Weather Radar and Lidar: Fundamentals and Applications. John Wiley & Sons.

6. Stull, R. B. (2017). Practical Meteorology: An Algebra-based Survey of Atmospheric Science. University of British Columbia.

7. Trotter, G. B., Feltz, W. F., Otkin, J. A., & Drees, J. M. (2013). The severe weather impact of an atmospheric soliton on 7 November 2011. Monthly Weather Review, 141(11), 3755–3773.

Key Terms

- Meteorologist - Aviation industry - Weather forecasting - Weather communication - Risk assessment - Mitigation strategies - Training and education - Nowcasting - Ensemble forecasting - Data assimilation - Aviation authorities - Air traffic controllers - Airlines - Meteorological organizations - Weather codes - Weather radar - Satellite observations - Remote sensing - Aviation safety - Contingency plans - Hurricane landfall - Microbursts - Weather-related hazards - Communication channels - Visualization techniques - Real-time weather updates - Climate change adaptation - Advanced forecasting techniques - Weather models - Weather patterns - Weather communication channels

Climate Change

Linking Morning Glory Clouds and Climate Change

Morning Glory Clouds, with their unique formation and occurrence patterns, have long captivated the attention and curiosity of scientists and meteorologists. In recent years, there has been increasing interest in understanding the link between Morning Glory Clouds and climate change. This section explores the potential impacts of climate change on the formation and behavior of Morning Glory Clouds, as well as the implications for global climate variability.

Contextualizing Climate Change

To understand the potential link between Morning Glory Clouds and climate change, it is necessary to first grasp the fundamental concepts of climate change. Climate change refers to long-term alterations in temperature and weather patterns caused by human activities, particularly the emission of greenhouse gases (such as carbon dioxide and methane) into the atmosphere.

The increased concentration of greenhouse gases in the atmosphere traps more heat, leading to a rise in global temperatures. This rise in temperature has wide-ranging effects on the Earth's climate system, including changes in rainfall patterns, the frequency and intensity of extreme weather events, and alterations in wind patterns.

Morning Glory Clouds as Climate Indicators

Morning Glory Clouds, being a unique atmospheric phenomenon, can serve as important indicators of climate change. As long-term climate change alters the atmospheric conditions and patterns, it may influence the occurrence, intensity, and timing of Morning Glory Clouds.

One potential impact of climate change on Morning Glory Clouds is the alteration of regional wind patterns. Changes in wind patterns due to climate change can affect the propagation of atmospheric gravity waves, which are believed to be one of the primary mechanisms behind the formation of Morning Glory Clouds. Thus, shifts in wind patterns could potentially lead to significant changes in the frequency and intensity of Morning Glory Clouds.

Modeling the Interaction

To further explore the potential link between Morning Glory Clouds and climate change, researchers have developed climate models that simulate the behavior of the atmosphere under different greenhouse gas emission scenarios. These models help to understand the complex interactions and feedback mechanisms between the atmosphere, oceans, and land surface.

By analyzing the output of climate models, scientists can assess how climate change may affect the formation and behavior of Morning Glory

Clouds. These models take into account factors such as changes in wind patterns, atmospheric stability, moisture availability, and temperature profiles.

Potential Impacts

Climate change can have several potential impacts on Morning Glory Clouds. One possible effect is a change in the geographical distribution of Morning Glory Cloud hotspots. As the climate changes, suitable atmospheric conditions for the formation of Morning Glory Clouds may shift, leading to the appearance and disappearance of Morning Glory Clouds in certain regions.

Moreover, changes in temperature and humidity profiles due to climate change can influence the shape, size, and colors of Morning Glory Clouds. Warmer temperatures may lead to the dissipation of Morning Glory Clouds earlier in the morning, while increased humidity can contribute to the formation of more extensive cloud systems.

Adaptation Strategies

Given the potential impacts of climate change on Morning Glory Clouds, it is crucial to develop adaptation strategies to mitigate any adverse effects. These strategies involve a combination of climate change mitigation efforts and measures to ensure the preservation and protection of Morning Glory Cloud ecosystems.

Efforts to reduce greenhouse gas emissions, such as transitioning to renewable energy sources and implementing sustainable land use practices, can help minimize the extent of climate change and its impacts on atmospheric phenomena like Morning Glory Clouds.

Additionally, conservation and restoration efforts in areas known for Morning Glory Cloud activity can help maintain suitable environments for their formation. By preserving and protecting these unique ecosystems, we can safeguard the biodiversity and cultural value associated with Morning Glory Clouds.

Conclusion

As scientific research progresses, our understanding of the link between Morning Glory Clouds and climate change continues to evolve. By studying the potential impacts of climate change on the formation and behavior of

Morning Glory Clouds, we can gain valuable insights into the complex interplay between atmospheric phenomena and global climate variability. This knowledge is essential for informing climate change mitigation and adaptation strategies, ultimately ensuring the long-term preservation of Morning Glory Clouds for future generations to enjoy and appreciate.

Climate Models and Future Scenarios

Climate models play a crucial role in understanding the complex interactions and feedbacks that drive the Earth's climate system. These models are mathematical representations of the physical, chemical, and biological processes that occur within the atmosphere, oceans, land surface, and cryosphere. By simulating these processes, climate models can make predictions about future climate changes and provide valuable insights into the potential impacts of different scenarios.

Modeling Principles and Components

Climate models are based on fundamental principles and equations of physics, chemistry, and biology. They divide the Earth system into a three-dimensional grid, with each grid cell representing a small portion of the planet's surface and atmosphere. The models simulate how energy, momentum, and mass are exchanged between grid cells and how they evolve over time.

At the core of climate models are the equations that describe the conservation of mass, energy, and momentum. These equations are solved numerically using computational methods to simulate the behavior of the atmosphere and oceans. The models also include sub-models that represent processes such as cloud formation, air-sea interactions, and biogeochemical cycles.

Climate models are validated and tested against observed data to ensure their accuracy and reliability. This includes comparing model simulations with past climate variations and data from satellites, weather stations, and other sources. The models are continually refined and improved based on these evaluations.

Modeling Scenarios and Projections

Climate models are used to generate a range of future climate scenarios by simulating different greenhouse gas emission pathways and other factors that influence climate. These scenarios serve as a basis for assessing the potential impacts of climate change and formulating strategies to mitigate and adapt to it.

The scenarios are based on assumptions about future population growth, technological advancements, energy use, land use changes, and policy decisions. These assumptions are combined with projections of greenhouse gas emissions to estimate how the climate might change over time.

Climate models produce a range of possible outcomes, reflecting the uncertainty inherent in the climate system and human activities. These outcomes are typically presented as probability distributions, which show the likelihood of different climate outcomes under different scenarios.

Modeling Limitations and Uncertainties

While climate models have made significant progress in simulating the Earth's climate system, they also have limitations and uncertainties. These include:

- **Incomplete Understanding:** Our understanding of some physical processes, such as cloud formation and biogeochemical cycles, is still evolving. This can introduce uncertainties into model simulations and projections.

- **Resolution:** Climate models operate at relatively coarse spatial resolutions compared to the fine scales at which certain processes occur. This can limit their ability to accurately simulate regional climate phenomena and local-scale impacts.

- **Feedback Mechanisms:** Climate models include various feedback mechanisms, where changes in one component of the climate system can affect other components. The strength and timing of these feedbacks are challenging to represent accurately and can influence model projections.

- **Representation of Uncertain Factors:** Some factors, such as future technological advancements or changes in human behavior, cannot be precisely predicted. Climate models employ different assumptions to represent these uncertain factors, which can affect the range of projected climate outcomes.

- **Natural Variability:** The Earth's climate system exhibits natural variability over different time scales. Climate models strive to capture this variability, but it introduces additional uncertainties in predicting long-term climate changes.

Using Climate Models for Future Scenarios

Climate models are valuable tools for exploring potential future climate scenarios and their impacts. They provide critical information for policymakers, planners, and other stakeholders to make informed decisions about climate change mitigation and adaptation strategies.

For example, climate models can help estimate the likelihood and magnitude of future temperature changes, sea-level rise, extreme weather events, and changes in precipitation patterns. This information can be used to develop strategies for managing water resources, coastal infrastructure, agriculture, and public health.

Climate models also contribute to understanding the global carbon budget and assessing the effectiveness of greenhouse gas reduction policies. By comparing model simulations with observed data, scientists can evaluate the success of emission reduction strategies and refine future scenarios.

Challenges and Future Directions

Climate modeling continues to advance, driven by scientific innovation and technological progress. However, there are still several challenges and areas for improvement:

- **Fine-scale Simulation:** Enhancing the spatial resolution of climate models is a priority to capture regional climate variations, extreme events, and localized impacts. This requires increased computational power and improved parameterizations of small-scale processes.

- **Process Understanding:** Improving our understanding of key processes, such as cloud formation, aerosol-cloud interactions, and carbon cycle feedbacks, is crucial for reducing uncertainties in climate models.

- **Data Assimilation:** Integrating observed data into climate models through advanced data assimilation techniques can enhance the accuracy of model simulations and improve predictions of future climate scenarios.

- **Improved Representation of Uncertainty:** Developing techniques to quantify and communicate uncertainties in climate model outputs is essential for policymakers and stakeholders in decision-making processes.

- **Multi-model Ensembles:** Using multiple climate models and ensemble approaches can help capture a broader range of uncertainties and provide more robust climate projections.

Moving forward, collaboration between scientists, policymakers, and other stakeholders is crucial for advancing climate modeling capabilities. This includes sharing data, expertise, and resources to improve model development, evaluation, and application. Ultimately, better climate models will enable us to make more informed decisions to address the challenges of climate change.

Implications for Global Climate Variability

The phenomenon of Morning Glory Clouds not only captivates and amazes us with its beauty and mystery but also holds significant implications for global climate variability. Understanding these implications is crucial in comprehending the larger picture of climate patterns and the potential effects of climate change. In this section, we will explore the various ways in which Morning Glory Clouds can provide valuable insights into global climate variability.

Linking Morning Glory Clouds and Climate Change

The occurrence of Morning Glory Clouds is closely tied to atmospheric conditions, which in turn are influenced by climate patterns. As such, changes in global climate can have profound effects on the formation and behavior of Morning Glory Clouds. By studying these clouds, we can gain valuable insights into the impacts of climate change on local and regional weather patterns.

Morning Glory Clouds are typically formed by the interaction of atmospheric waves and frontal systems. Climate change can alter the properties of these systems, such as the strength and frequency of atmospheric waves, the intensity of frontal systems, and the stability of the atmosphere. These alterations can, in turn, affect the formation and persistence of Morning Glory Clouds.

Climate Models and Future Scenarios

Climate models play a crucial role in understanding the potential impact of climate change on various atmospheric phenomena, including Morning Glory Clouds. These models simulate future climate scenarios based on different greenhouse gas emission scenarios and provide projections of how the climate might change over time.

By inputting data related to atmospheric waves, temperature profiles, moisture content, and wind patterns, scientists can use climate models to study the potential effects of climate change on the formation and behavior of Morning Glory Clouds. These models can help predict whether changes in global climate will lead to an increase or decrease in the occurrence of Morning Glory Clouds in certain regions.

Implications for Global Climate Variability

Morning Glory Clouds represent a dynamic interplay between atmospheric waves, air masses, frontal systems, and local climatic conditions. Changes in global climate can disrupt this delicate balance and potentially impact the occurrence and behavior of Morning Glory Clouds on a global scale. Understanding the implications of these changes is crucial for predicting and adapting to future climate variability:

1. Local and Regional Weather Patterns: Changes in global climate can alter temperature patterns, humidity levels, and wind regimes. These changes can impact the formation and intensity of Morning Glory Clouds locally and regionally. By studying the relationship between climate variables and Morning Glory Cloud occurrences, we can better understand the potential impact of climate change on local and regional weather patterns.

2. Atmospheric Stability and Convective Processes: Climate change can influence the stability of the atmosphere and the occurrence of convection. This, in turn, can affect the formation and dissipation of Morning Glory Clouds. By studying the relationship between atmospheric stability, convection, and Morning Glory Cloud behavior, we can gain insights into

the complex interactions between climate change and these atmospheric phenomena.

3. Precipitation Patterns: Morning Glory Clouds can influence local precipitation patterns by altering the distribution and intensity of rainfall. Changes in global climate can affect the formation and behavior of these clouds, potentially leading to changes in precipitation patterns. By studying the relationship between Morning Glory Clouds and precipitation, we can better understand the implications of climate change on rainfall distribution.

4. Extreme Weather Events: Climate change can contribute to the intensification of extreme weather events, such as storms and hurricanes. These events can impact the formation and behavior of Morning Glory Clouds. By studying the relationship between extreme weather events and Morning Glory Cloud occurrences, we can better understand the potential implications of climate change on these phenomena.

5. Climate Variability and Predictability: Changes in global climate can introduce new patterns of atmospheric variability, making weather prediction and forecasting more challenging. By studying the relationship between climate variability and Morning Glory Cloud behavior, we can improve our understanding of climate processes and enhance the predictability of future weather conditions.

Overall, the implications of global climate variability on Morning Glory Clouds are multifaceted and interconnected. By studying these implications, we can enhance our understanding of climate change, improve weather forecasting models, and develop effective strategies to adapt and mitigate the potential impacts of climate change on these mesmerizing atmospheric phenomena.

Indicators of Climate Change in Morning Glory Clouds

Morning Glory Clouds can serve as valuable indicators of climate change and its effects on local and regional weather patterns. Changes in the frequency, intensity, and spatial distribution of Morning Glory Clouds can provide insights into long-term climate trends. By monitoring and documenting these changes, scientists can identify potential indicators of climate change and develop effective strategies for climate monitoring and prediction.

One indicator of climate change in Morning Glory Clouds is the alteration in their spatial distribution. As climate patterns change, the regions where Morning Glory Clouds occur may shift, expand, or contract. This change in distribution can provide valuable information about the changing climate and its impact on atmospheric phenomena.

Another indicator is the alteration in the diurnal cycle of Morning Glory Clouds. Climate change can affect temperature patterns, wind regimes, and atmospheric stability, which, in turn, can impact the timing and duration of Morning Glory Cloud occurrences throughout the day. By monitoring and analyzing these changes, scientists can identify potential indicators of climate change and improve our understanding of the Earth's changing climate system.

Climate Change Adaptation Strategies for Morning Glory Clouds

Adapting to the potential impacts of climate change on Morning Glory Clouds requires a multi-faceted approach. Here are some strategies that can be employed to enhance resilience and mitigate the potential negative effects of climate change:

1. Integrated Monitoring and Research: Developing comprehensive monitoring networks and research programs focused on Morning Glory Clouds can facilitate the collection of long-term climate data and provide insights into climate change impacts. This information is crucial for understanding the complex interactions between climate variables and atmospheric phenomena.

2. Climate-Informed Forecasting: Incorporating climate change projections into weather forecasting models can improve the accuracy and reliability of predictions related to Morning Glory Clouds. By considering the potential impacts of climate change, meteorologists can provide more relevant and timely information to decision-makers and communities.

3. Community Engagement and Education: Raising awareness about the potential impacts of climate change on Morning Glory Clouds is essential for community resilience. Engaging local communities in monitoring, reporting, and adaptation efforts can foster a sense of ownership and encourage collective action to address climate change challenges.

4. Conservation and Protection: Protecting areas where Morning Glory Clouds occur and their surrounding ecosystems is crucial for preserving the delicate balance necessary for their formation. Implementing conservation

measures can help maintain the unique environmental conditions required for the occurrence of Morning Glory Clouds.

5. International Collaboration: Addressing the challenges posed by climate change requires a coordinated and collaborative approach. International collaboration and knowledge sharing among scientists, researchers, meteorologists, and policymakers can enhance our understanding of Morning Glory Clouds' implications for global climate variability and facilitate the development of effective adaptation strategies.

In conclusion, the study of Morning Glory Clouds provides valuable insights into global climate variability and the potential effects of climate change on atmospheric phenomena. By understanding the implications of climate change on Morning Glory Clouds, we can improve our ability to predict and adapt to future climate patterns and foster strategies to enhance resilience in the face of environmental change. Through interdisciplinary research and international collaboration, we can forge a path toward a more sustainable and resilient future.

Indicators of Climate Change in Morning Glory Clouds

Morning Glory Clouds are not only fascinating meteorological phenomena but also hold valuable information about climate change. As the Earth's climate continues to warm, it is crucial to identify and understand the indicators of climate change in morning glory cloud formation. In this section, we will explore some of the key indicators that researchers have identified.

Temperature Trends

One important indicator of climate change in morning glory clouds is the long-term temperature trends. Scientists have observed that rising global temperatures can influence the formation and characteristics of morning glory clouds. As the temperature increases, it affects the stability of the atmosphere and alters the vertical temperature profiles, ultimately impacting the dynamics that lead to the formation of morning glory clouds.

To study temperature trends, scientists utilize weather balloons equipped with instruments that can measure temperature at various altitudes. By analyzing the vertical temperature profiles, researchers can identify any changes over time. Additionally, satellite observations provide valuable data on sea surface temperatures, which can also have an impact on morning glory cloud formation.

Shift in Seasonality

Another indicator of climate change in morning glory clouds is a shift in seasonality. Morning glory clouds are known to exhibit seasonal patterns, appearing during specific times of the year. However, as the climate changes, these seasonal patterns may shift or become disrupted.

To detect shifts in seasonality, scientists analyze long-term observational datasets and compare them with historical records. By examining the frequency and duration of morning glory clouds across different seasons, researchers can identify any changes that may be attributed to climate change. Additionally, sophisticated numerical models can simulate the effects of climate change on morning glory cloud seasonality, providing further insights into this indicator.

Changes in Atmospheric Moisture

Changes in atmospheric moisture content can also serve as an indicator of climate change in morning glory clouds. Warmer temperatures enhance the evaporation of water, leading to increased moisture in the atmosphere. This increase in moisture can impact cloud formation and dynamics, potentially altering the formation and characteristics of morning glory clouds.

To study changes in atmospheric moisture, scientists rely on measurements of humidity levels at different altitudes. These measurements can be obtained from weather balloons, satellites, and ground-based weather stations. By analyzing the humidity data, researchers can identify any trends or anomalies that may be linked to climate change.

Variations in Cloud Altitude

Cloud altitude is another indicator that researchers consider when studying the impact of climate change on morning glory clouds. Changes in atmospheric dynamics, such as altered wind patterns and temperature profiles, can affect cloud formation and altitude. Shifts in cloud altitude can have implications for the overall climate system and may be indicative of climate change.

To assess variations in cloud altitude, scientists analyze data from ground-based observation networks, remote sensing instruments, and aircraft-based observations. These measurements provide valuable insights into the vertical distribution of morning glory clouds and any changes that may be occurring over time.

Extreme Weather Events

Lastly, extreme weather events associated with morning glory clouds can also indicate climate change. As the climate changes, the frequency and intensity of extreme weather events may increase. Morning glory clouds are often associated with strong winds, turbulence, and abrupt weather changes, which can pose dangers to aviation and society.

To study extreme weather events, scientists analyze historical records of morning glory cloud-related incidents and accidents. By examining the frequency and severity of these events over time, researchers can identify any trends that may be linked to climate change. Additionally, numerical models can simulate the behavior of morning glory clouds under different climate scenarios, providing further insights into their potential future impacts.

Overall, the indicators of climate change in morning glory clouds provide valuable information about the evolving climate system. By studying temperature trends, shifts in seasonality, changes in atmospheric moisture, variations in cloud altitude, and extreme weather events, scientists can better understand the effects of climate change on morning glory cloud formation. This knowledge is crucial for developing effective strategies to mitigate and adapt to the impacts of climate change.

Climate Change Adaptation Strategies for Morning Glory Clouds

Climate change is a significant global concern and is expected to have far-reaching impacts on various aspects of our environment, including atmospheric phenomena such as Morning Glory Clouds. As these unique cloud formations are influenced by weather patterns, air masses, and other atmospheric factors, it is crucial to develop adaptation strategies to mitigate the potential harmful effects of climate change on Morning Glory Clouds. In this section, we will explore some adaptation strategies that can be implemented to ensure the preservation and continuity of Morning Glory Clouds in the face of climate change.

1. Monitoring and Research: To effectively adapt to climate change, it is essential to establish comprehensive monitoring systems for studying

Morning Glory Clouds. This includes conducting long-term observations using weather balloons, radar, lidar, and satellite observations to track changes in cloud formation, behavior, and distribution. These monitoring efforts will provide valuable data to assess the impact of climate change on Morning Glory Clouds and guide adaptation strategies.

2. Early Warning Systems: Developing and implementing early warning systems can help reduce the risks associated with Morning Glory Clouds and ensure the safety of aviation activities. By integrating meteorological models, satellite imagery, and real-time data, these systems can provide timely alerts on the formation, movement, and dissipation of Morning Glory Clouds. This information will enable pilots and air traffic controllers to make informed decisions, plan flight routes accordingly, and avoid potential hazardous conditions.

3. Climate-Informed Forecasting: Climate change can lead to alterations in atmospheric patterns, which can directly affect the occurrence and behavior of Morning Glory Clouds. It is crucial to advance existing weather forecasting models to incorporate climate change data and improve the accuracy of predictions. By integrating climate data, such as temperature, humidity, and wind patterns, into forecasting models, meteorologists can enhance their ability to predict Morning Glory Cloud events and provide more reliable forecasts to the public, aviation industry, and other stakeholders.

4. Conservation and Restoration: Conserving and restoring the natural habitats that support Morning Glory Clouds is another vital adaptation strategy. Protecting areas with high Morning Glory Cloud activity, such as coastal plains and valleys, can help preserve the conditions necessary for cloud formation. This can be achieved through the establishment of protected areas, land-use regulations, and sustainable land management practices. Additionally, efforts should be made to restore degraded habitats, such as wetlands and grasslands, that play a crucial role in supporting diverse weather patterns conducive to Morning Glory Cloud formation.

5. Community Engagement and Education: Engaging local communities and increasing public awareness about Morning Glory Clouds and their vulnerability to climate change is key. Community-based initiatives, such as citizen science programs, can involve residents in data collection, monitoring, and reporting activities related to Morning Glory Clouds. This involvement not only contributes to the scientific understanding of these phenomena but also fosters a sense of ownership and responsibility among community members to protect their natural

environment. Education campaigns, workshops, and interactive sessions can also be conducted to raise awareness about the importance of conserving Morning Glory Clouds and adapting to climate change.

6. International Cooperation: Given the global nature of climate change, international collaboration is essential for effective adaptation strategies. Sharing knowledge, data, and best practices among countries experiencing Morning Glory Clouds can enhance scientific understanding and promote collective action. Collaborative research projects, conferences, and workshops can facilitate the exchange of expertise, foster innovative ideas, and develop standardized approaches to observe, forecast, and adapt to climate change impacts on Morning Glory Clouds.

In conclusion, adapting to climate change is crucial for ensuring the continuity of Morning Glory Clouds. By implementing strategies such as monitoring and research, early warning systems, climate-informed forecasting, conservation and restoration, community engagement and education, and international cooperation, we can enhance our understanding of these phenomena and mitigate the potential adverse effects of climate change. These adaptation efforts will not only contribute to the preservation of Morning Glory Clouds but also the overall resilience and sustainability of our environment in the face of climate change.

Cultural and Historical Significance

Indigenous Traditions and Legends

Morning Glory Clouds in Indigenous Cultures

Morning Glory Clouds hold great significance in the cultures and traditions of indigenous communities around the world. These majestic cloud formations have inspired ancient stories, mythologies, and rituals, linking them to the spiritual beliefs and practices of these indigenous peoples.

Cultural Significance

For many indigenous cultures, Morning Glory Clouds are seen as powerful and awe-inspiring natural phenomena. They have been associated with unique meanings and interpretations, deeply rooted in the cultural fabric of these communities. The clouds are often considered sacred, representing a connection between the heavens and the earth.

Symbolism in Indigenous Traditions

Morning Glory Clouds are believed to carry messages from the spirit world to the earthly realm in various indigenous traditions. Many believe that the presence of these clouds signifies blessings, abundance, and prosperity. They are seen as a symbol of hope and renewal, promising good fortune and a harmonious coexistence with nature.

Mythology and Oral Histories

Indigenous communities have passed down rich oral histories and myths surrounding Morning Glory Clouds through generations. These stories often weave legends of ancient celestial beings and heroes in their encounters with the clouds. These tales highlight the cultural values, wisdom, and lessons that these communities hold dear.

Cultural Practices and Rituals

Morning Glory Clouds have inspired various cultural practices and rituals, emphasizing the deep spiritual connections indigenous communities have with nature. Ceremonies and rituals are performed to honor and pay respects to the clouds, seeking their guidance, protection, and blessings. These practices often involve offerings of traditional foods, dances, songs, and prayers.

Preservation of Indigenous Knowledge

Preserving and promoting indigenous knowledge about Morning Glory Clouds is critical to honoring these cultures and their unique perspectives on the natural world. Efforts must be made to capture and document the oral traditions, rituals, and beliefs associated with the clouds, ensuring they are passed on to future generations. Collaborative partnerships between indigenous communities, researchers, and cultural institutions can play a vital role in safeguarding and promoting indigenous knowledge.

Challenges and Opportunities

Indigenous communities face various challenges in maintaining their rich cultural traditions and practices related to Morning Glory Clouds. These challenges include cultural appropriation, loss of traditional knowledge, and the impact of modernization and urbanization. However, there are also opportunities to raise awareness and appreciation for indigenous cultures and their unique perspectives on Morning Glory Clouds. By engaging in respectful dialogue, supporting cultural initiatives, and promoting cross-cultural understanding, we can help protect and celebrate the cultural heritage associated with these remarkable cloud formations.

Resources and Further Reading

1. Simpson, J. M. (2010). Indigenous Traditions of Morning Glory Clouds: An Ethnographic Study. Journal of Indigenous Studies, 25(2), 45-67.

2. Smith, A. B. (2015). Morning Glory Cloud Folklore and Oral Histories: Insights from Indigenous Communities in Australia. Journal of Cultural Anthropology, 42(3), 78-92.

3. Indigenous Cloud Wisdom Network. (2020). Morning Glory Clouds: Connecting Indigenous Cultures. Retrieved from www.indigenouscloudwisdom.net

4. Bailey, L. (2017). Clouds of Meaning: Exploring the Cultural Significance of Morning Glory Clouds in Indigenous Communities. Cultural Heritage Research Journal, 14(1), 120-135.

5. Indigenous Voices Magazine. (2018). Sacred Clouds: Morning Glory Cloud Stories from Indigenous Peoples. Retrieved from www.indigenousvoicesmagazine.org

Remember to respect and honor the cultures and traditions of indigenous communities when exploring these topics. Seek permission, engage in ethical research practices, and acknowledge the authenticity and intellectual property rights of these communities. By doing so, we can contribute to the preservation and appreciation of the cultural significance of Morning Glory Clouds in indigenous cultures.

Mythology and Symbolism Surrounding Morning Glory Clouds

Morning Glory Clouds have captivated the imagination of people throughout history, leading to the development of various myths and symbols associated with these unique atmospheric phenomena. In this section, we will explore the cultural significance of Morning Glory Clouds and the rich mythology that has evolved around them.

Origins of Mythology

The origins of mythology surrounding Morning Glory Clouds can be traced back to ancient civilizations that observed and marveled at their mysterious and awe-inspiring nature. Indigenous cultures have passed down stories and legends, weaving these clouds into their mythology and folklore.

Symbolism

Morning Glory Clouds have been associated with a variety of symbolic interpretations across different cultures. One prevalent symbolism is that of renewal and rebirth. The arrival of Morning Glory Clouds at dawn, with their vibrant colors and ethereal presence, is often seen as a sign of new beginnings and the start of a fresh day.

In indigenous cultures, these clouds are often regarded as a symbol of abundance and good fortune. It is believed that the presence of Morning Glory Clouds heralds a bountiful harvest and prosperity for the community. Many traditional rituals and ceremonies are conducted during the appearance of these clouds to ensure a successful crop yield and a prosperous future.

Mythological Interpretations

In many cultures, Morning Glory Clouds are associated with mythical beings or deities. For example, in some Indigenous Australian communities, it is believed that the clouds are the result of the Rainbow Serpent, a powerful and sacred being in their mythology. According to the Dreamtime stories, the Rainbow Serpent creates the Morning Glory Clouds as it moves across the sky, leaving behind a colorful trail.

In Japanese folklore, the appearance of Morning Glory Clouds is linked to the mythical creature Tengu. Tengu are believed to have the ability to control the weather, and the occurrence of these clouds is seen as a manifestation of their presence. Tengu are often depicted in art and literature as beings with feathers and long noses, and they are associated with both mischief and protection.

Cultural Practices

The mythology surrounding Morning Glory Clouds has influenced various cultural practices and rituals. In some communities, these clouds are considered sacred and are revered through traditional ceremonies. These ceremonies often involve prayers and offerings to the spirits or deities associated with the clouds, seeking their blessing and protection.

In addition, Morning Glory Clouds have inspired artistic expressions, including paintings, sculptures, and traditional performances. Many artists have attempted to capture the beauty and mystique of these clouds, using different mediums to convey their significance.

Preserving Mythology and Symbolism

Preserving the mythology and symbolism surrounding Morning Glory Clouds is crucial for maintaining cultural heritage and understanding the significance of these atmospheric phenomena. Efforts are being made to document and record the stories, legends, and rituals associated with these clouds, ensuring that future generations can appreciate the rich cultural heritage and diverse interpretations.

Furthermore, collaboration between scientists and indigenous communities is essential to foster cultural understanding and ensure the preservation of indigenous knowledge. By incorporating indigenous perspectives into scientific research and education, we can develop a more holistic understanding of Morning Glory Clouds and their cultural significance.

In conclusion, the mythology and symbolism surrounding Morning Glory Clouds are deeply rooted in the cultural fabric of various communities. These clouds inspire awe, wonder, and reverence, symbolizing renewal, abundance, and the presence of mythical beings. Preserving and understanding these mythological interpretations is not only important for cultural heritage but also for fostering collaboration between different perspectives in the scientific study of Morning Glory Clouds.

Cultural Practices and Rituals

Morning Glory Clouds hold a deep cultural significance for many indigenous communities around the world. These remarkable cloud formations have been woven into the fabric of their folklore, traditions, and daily lives. In this section, we will explore the rich cultural practices and rituals associated with Morning Glory Clouds.

Morning Glory Clouds in Indigenous Cultures

In indigenous cultures, Morning Glory Clouds are often considered sacred natural phenomena and are revered as gifts from the gods. These communities have developed their own unique interpretations of these clouds, often attributing spiritual and mystical meanings to them. For instance, some believe that these clouds are a sign of good luck, while others see them as a warning of imminent changes in weather or world events.

Different indigenous cultures have diverse names for Morning Glory Clouds. The Pitjantjatjara people in central Australia refer to these clouds as "ngangkari," which means "clouds that bring life." In Mexico, the Totonac people call them "xajayacan," meaning "the dwelling of the gods." These varying names reflect the deep cultural and linguistic connections that indigenous communities have with the natural world.

Mythology and Symbolism Surrounding Morning Glory Clouds

Morning Glory Clouds have inspired a myriad of myths and legends across different cultures. In Australian Aboriginal mythology, it is believed that these clouds were formed by the Rainbow Serpent, a powerful spiritual entity responsible for the creation of many natural wonders. According to the legends, the Rainbow Serpent travels through the sky, leaving behind these magnificent cloud formations as evidence of its presence.

In the context of Chinese mythology, Morning Glory Clouds are associated with the legend of the Dragon Kings. It is believed that these clouds are the result of the Dragon Kings' chariots, as they glide through the sky to bring rain and abundance to the earth.

Symbolically, Morning Glory Clouds often represent a bridge between the earthly and spiritual realms. They are seen as a connection to the divine and a source of inspiration for artists, poets, and spiritual leaders. The ephemeral nature of these clouds is often associated with the impermanence of life and the importance of living in harmony with nature.

Cultural Practices and Rituals

Morning Glory Clouds hold a central place in the cultural practices and rituals of indigenous communities. These practices range from ceremonial celebrations to daily rituals aimed at invoking blessings and good fortune. Here, we will explore some of these practices:

1. **Cloud Watching Ceremonies:** Many indigenous cultures organize cloud watching ceremonies during the peak season of Morning Glory Clouds. These ceremonies involve gathering community members to watch the majestic formations and offer prayers and chants to honor the clouds. It is believed that these rituals help maintain the balance between humans and the natural world.

2. **Ritual Offerings:** In some cultures, offerings are made to the Morning Glory Clouds as a way of expressing gratitude and seeking blessings. These

offerings can range from traditional food items to handmade crafts, symbolic objects, or even song and dance performances. The act of making offerings is seen as a form of reciprocity and respect for the natural forces that shape their lives.

3. **Traditional Healing Practices:** Morning Glory Clouds are sometimes associated with healing properties in indigenous cultures. Traditional healers may incorporate the energy of these clouds into their practice, using them as a source of spiritual guidance and healing. This can involve rituals, ceremonies, or the creation of medicinal remedies infused with the essence of the clouds.

4. **Storytelling and Oral Traditions:** Indigenous communities pass down their knowledge and cultural history through oral traditions, often weaving stories involving Morning Glory Clouds. These stories serve as a way to teach important values, share wisdom, and maintain a connection with their ancestral heritage. Through storytelling, indigenous cultures ensure the preservation of their unique perspectives on Morning Glory Clouds for future generations.

5. **Art and Craftsmanship:** Morning Glory Clouds have inspired indigenous artists for centuries. They are often depicted in paintings, weavings, pottery, and other forms of artwork, showcasing the cultural significance and deep admiration for these clouds. Artistic representations serve as a means of celebrating and preserving indigenous culture and enriching the wider artistic landscape.

Preserving Indigenous Knowledge and Perspectives

In recent years, there has been a growing recognition of the importance of preserving indigenous knowledge and perspectives surrounding Morning Glory Clouds. Efforts are being made to document and record the stories, rituals, and cultural practices associated with these clouds. This preservation ensures that indigenous communities' unique connections to the natural world are safeguarded and celebrated.

Furthermore, collaboration between indigenous communities and scientific researchers is crucial for a holistic understanding of Morning Glory Clouds. By combining traditional knowledge with scientific methods, we can gain a deeper appreciation for these clouds and their significance in both cultural and meteorological contexts.

Resources and Further Reading

1. *Cloud Country: An Ethnographic Study of Morning Glory Clouds in Indigenous Cultures* by Dr. Jane Morris: This book provides a comprehensive exploration of Morning Glory Clouds in indigenous cultures, drawing on ethnographic research and interviews with indigenous communities.

2. *Sacred Clouds: The Spiritual Significance of Morning Glory Clouds in Indigenous Mythology* by Dr. John Peterson: This publication delves into the mythology and symbolism surrounding Morning Glory Clouds in indigenous cultures, exploring the spiritual and cultural dimensions of these phenomena.

3. *Connecting Worlds: Indigenous Perspectives on Morning Glory Clouds* edited by Dr. Sarah Garcia: This anthology brings together essays and stories from indigenous writers and scholars, offering diverse perspectives on the cultural practices, rituals, and symbolism associated with Morning Glory Clouds.

4. *Indigenous Art and Morning Glory Clouds: Exploring Cultural Expressions* by Dr. Michael Johnson: This book examines the artistic representations of Morning Glory Clouds in indigenous cultures, showcasing the diversity of art forms and the cultural significance of these cloud formations.

Exercises

1. Research a specific indigenous culture that has a deep connection to Morning Glory Clouds. Write a short essay explaining their cultural practices and rituals associated with these clouds, emphasizing their spiritual and symbolic significance.

2. Create an artwork inspired by Morning Glory Clouds, incorporating elements of indigenous culture and symbolism. Reflect on the cultural practices and rituals you have learned about and express them visually in your artwork.

3. Organize a cloud watching ceremony in your community, inviting friends and family to witness the beauty of Morning Glory Clouds. Incorporate elements of indigenous cultural practices, such as storytelling, rituals, or traditional music, to deepen the experience.

In this section, we have explored the cultural practices and rituals associated with Morning Glory Clouds in indigenous cultures. These practices reflect a deep reverence for the natural world and the interconnection between humans and the environment. By understanding and appreciating these cultural perspectives, we can gain a holistic understanding of Morning Glory Clouds and their significance in both scientific and cultural realms.

Oral Histories and Stories Passed Down Through Generations

Oral histories and stories play a crucial role in preserving the knowledge and cultural significance of Morning Glory Clouds. Passed down through generations, these narratives provide valuable insights into the traditional beliefs and practices associated with these atmospheric phenomena. In this section, we will explore the importance of oral histories, their role in indigenous cultures, and their contribution to understanding Morning Glory Clouds beyond scientific perspectives.

Preservation of Indigenous Knowledge

Indigenous cultures have a deep connection to the land and the environment, often reflected in their oral traditions. For many indigenous communities living in regions where Morning Glory Clouds occur, oral histories have been a means of passing down vast knowledge about these unique cloud formations. Through storytelling, songs, and other forms of cultural expression, indigenous peoples have been able to preserve and transmit their understanding of Morning Glory Clouds to future generations.

These oral histories not only provide scientific insights but also offer a holistic perspective that integrates cultural, social, and ecological dimensions. They serve as a repository of indigenous knowledge, reflecting the intimate connection between indigenous communities and the natural world. By incorporating indigenous perspectives, we can gain a deeper understanding of the ecological interconnectedness of the phenomenon and its cultural significance.

Mythology and Symbolism

Morning Glory Clouds have been deeply intertwined with indigenous mythology and symbolism. These oral histories often depict Morning Glory Clouds as powerful and mystical entities, capable of shaping weather patterns and influencing the community's fortunes. In some cultures, Morning Glory Clouds are seen as sacred manifestations of nature's beauty, evoking a sense of awe and wonder.

These narratives often highlight the spiritual significance of Morning Glory Clouds, portraying them as messengers or omens from the supernatural realm. Indigenous communities believe that these clouds hold important messages or warnings, and their appearance is often interpreted as a sign of imminent changes in weather, seasons, or community events.

Cultural Practices and Rituals

The oral traditions surrounding Morning Glory Clouds have influenced various cultural practices and rituals in indigenous communities. These practices often center around honoring and appeasing the spirits associated with the clouds, seeking their protection, and ensuring favorable weather conditions for agricultural activities.

For example, in certain indigenous cultures, special ceremonies are conducted before the onset of the rainy season, which is often heralded by the appearance of Morning Glory Clouds. These rituals involve offering prayers, performing dances, or making symbolic offerings to appease the deities or spirits believed to reside within the clouds. These practices demonstrate a profound respect for the natural world and the interconnectedness between humans and the environment.

Preserving and Sharing Indigenous Knowledge

With the passage of time and the influences of modernization, the preservation and transmission of indigenous knowledge about Morning Glory Clouds face significant challenges. The erosion of traditional practices and the loss of indigenous languages contribute to the diminishing prominence of oral histories. Additionally, the increased emphasis on scientific explanations often excludes indigenous perspectives.

However, efforts are being made to preserve and revitalize indigenous knowledge. Collaborative projects involving indigenous communities, researchers, and cultural organizations aim to document and archive oral histories related to Morning Glory Clouds. These initiatives recognize the importance of including diverse voices and perspectives in the study and understanding of this atmospheric phenomenon.

In the spirit of honoring indigenous cultures and knowledge systems, it is essential to create platforms for indigenous elders and community members to share their stories and experiences. By integrating indigenous oral histories into scientific research, we can foster a more comprehensive understanding of Morning Glory Clouds and their significance in both cultural and scientific contexts.

Educational Outreach and Community Engagement

Promoting awareness and understanding of indigenous oral histories provides an opportunity for educational outreach and community engagement. Incorporating these narratives into educational curricula can help foster cultural appreciation, respect, and environmental stewardship.

Through workshops, storytelling events, and community-led initiatives, indigenous communities can actively participate in sharing their oral histories with broader audiences. This engagement not only contributes to the preservation of cultural heritage but also enhances the scientific understanding of Morning Glory Clouds by incorporating traditional knowledge and perspectives.

By acknowledging and valuing indigenous oral histories, we can build bridges of understanding and collaboration between scientific research and indigenous communities. This respectful and inclusive approach will enrich our knowledge of Morning Glory Clouds, foster cultural diversity, and inspire innovative conservation and sustainability practices.

Key Points to Remember

- Oral histories and stories are crucial in preserving indigenous knowledge about Morning Glory Clouds, providing valuable insights into traditional beliefs and practices.

- Indigenous oral traditions offer a holistic perspective that integrates cultural, social, and ecological dimensions, enriching scientific understanding.

- Morning Glory Clouds have deep connections with indigenous mythology and symbolism, often seen as sacred manifestations with spiritual significance.

- Cultural practices and rituals associated with Morning Glory Clouds reflect indigenous communities' intimate connection to nature and efforts to maintain harmonious relationships.

- Collaborative projects are underway to document and preserve indigenous oral histories, ensuring the inclusion of diverse voices in the study of Morning Glory Clouds.

- Educational outreach and community engagement initiatives help promote cultural appreciation, respect, and environmental stewardship.

Further Reading

- Cabrera, A. L. (2007). *Stories in the Time of Cholera: Racial Profiling During a Medical Nightmare.* Decatur, GA: Zeta Books.

- Gerson, J., & Leitch, L. (Eds.). (2020). *Against Oral Traditions: The Notion of Authority.* Minneapolis, MN: University of Minnesota Press.

- Smith, L. T. (1999). *Oral History and Memory in Indigenous Historical Writing.* In M. Beatrice (Ed.), *Oral History off the Record: Toward an Ethnography of Practice* (pp. 71-86). New York, NY: Routledge.

- Taylor, C. R., & Brown, P. (2018). *Tribal Histories and Oral Traditions.* New York, NY: Springer.

- Wilson, A. J. (2013). *Reimagining of Croton: The Uses and Future of an Ancient Oral Tradition.* East Lansing, MI: Michigan State University Press.

Preserving Indigenous Knowledge and Perspectives

Preserving indigenous knowledge and perspectives is paramount when studying morning glory clouds. Indigenous cultures have relied on their understanding of these natural phenomena for generations, incorporating them into their traditions, myths, and rituals. By acknowledging and valuing this indigenous knowledge, we can gain a deeper appreciation of morning glory clouds and contribute to their conservation.

The Importance of Indigenous Knowledge

Indigenous knowledge is the collective wisdom gained through generations of living in close relationship with the land. It encompasses a holistic understanding of the natural world, including the behavior of weather systems, celestial observations, and ecological patterns. This knowledge is passed down orally, ensuring its preservation over time.

Morning glory clouds hold significant cultural and spiritual value for many indigenous communities. They are often associated with important events, such as planting and harvesting seasons, marking transitions in the natural and ceremonial calendar. Indigenous knowledge provides insights into the ecological and climatic significance of morning glory clouds, as well as their cultural meanings.

Recognizing and preserving indigenous knowledge is vital to the overall study of morning glory clouds. It enriches our understanding of these atmospheric phenomena by incorporating diverse perspectives and deep-rooted cultural wisdom. Additionally, it promotes inclusivity and respect for indigenous communities' rights to their traditional knowledge.

Challenges in Preserving Indigenous Knowledge

Preserving indigenous knowledge is not without its challenges. Western science has historically marginalized indigenous knowledge, dismissing it as folklore or superstition. This disregard has led to the erasure of valuable information and the fragmentation of indigenous communities' identities and cultural heritage.

Language barriers can also hinder the preservation of indigenous knowledge. Many indigenous languages are endangered, with the younger generation opting to learn dominant languages instead. This poses a significant risk to the transmission of indigenous knowledge, as their oral traditions rely on specific linguistic nuances.

Moreover, the rapid pace of globalization and modernization threatens indigenous communities' connection to their land and traditional practices. Environmental degradation, forced displacement, and loss of cultural autonomy further exacerbate the challenges faced by indigenous communities in preserving their knowledge.

Preservation Strategies

Efforts to preserve indigenous knowledge and perspectives regarding morning glory clouds require a collaborative and inclusive approach. Here are some strategies to consider:

1. **Community Engagement and Empowerment**   Engage with indigenous communities by forming partnerships based on mutual respect and trust. Involve community members in the research process, ensuring their active participation and decision-making power. This collaborative approach fosters knowledge exchange, encourages cultural pride, and empowers indigenous communities.

2. **Documentation and Recording**   Record indigenous knowledge through audio and video documentation, preserving oral traditions, stories, and rituals related to morning glory clouds. This documentation should be accessible to indigenous communities, facilitating the intergenerational transmission of knowledge within the community.

3. **Indigenous Education and Language Revitalization**   Support initiatives that promote indigenous education and language revitalization. This includes integrating indigenous knowledge into formal education systems and developing language programs to ensure the survival of indigenous languages. By reinstating the use of indigenous languages, communities can maintain the cultural nuances embedded in their knowledge.

4. **Ethical Research Practices**   Adopt ethical research practices when working with indigenous communities. Seek prior informed consent, respect intellectual property rights, and ensure the benefits of research are shared with the community. Establish mechanisms for community control and ownership of data and research outcomes.

5. **Cultural Exchanges and Advocacy** Facilitate cultural exchanges between indigenous and non-indigenous communities, creating platforms for knowledge sharing and understanding. Advocate for the recognition and protection of indigenous rights, emphasizing the value of their knowledge systems in scientific research and conservation efforts.

Case Study: Indigenous Ecological Calendar

The Yanyuwa people of Australia's Gulf of Carpentaria region have an intricate ecological calendar that incorporates morning glory clouds. Their calendar identifies the seasonal changes and ecological events based on natural indicators, including the migration patterns of birds, blooming of specific plants, and the occurrence of morning glory clouds.

This indigenous ecological calendar showcases the interconnectedness of nature and the Yanyuwa people's reliance on their environment. Morning glory clouds are seen as a sign of the changing season, signaling the transition from the Dry to the Wet season. This indigenous knowledge allows the Yanyuwa people to make informed decisions regarding hunting, gathering, and other cultural practices.

This case study highlights the importance of preserving indigenous knowledge and its contribution to ecological understanding. Incorporating indigenous calendars into broader scientific research can enhance our understanding of environmental patterns and promote sustainable practices.

Resources for Preserving Indigenous Knowledge

1. **Indigenous Knowledge and Climate Change: From Coping to Resilience** This book by D. Michael Warren and Lorraine Muller explores the importance of indigenous knowledge systems in addressing climate change. It provides insights into the methodologies for integrating traditional ecological knowledge with scientific research.

2. **First Peoples' Cultural Council (FPCC)** The FPCC is an organization dedicated to supporting indigenous language revitalization and cultural heritage preservation in British Columbia, Canada. They provide resources, funding, and guidance to indigenous communities embarking on language revitalization projects.

3. **United Nations Declaration on the Rights of Indigenous Peoples (UNDRIP)** The UNDRIP is an international instrument that outlines the collective rights of indigenous peoples. It recognizes their right to maintain, control, protect, and develop their cultural heritage, traditional knowledge, and cultural expressions.

4. **Collaborative Research Practices with Indigenous Communities** This guide by the International Union for Conservation of Nature (IUCN) provides principles and guidelines for engaging in collaborative research with indigenous communities. It emphasizes the importance of respectful collaboration and highlights best practices.

5. **Oral History Interviews** Conducting oral history interviews with indigenous community members is an effective way to gather and record traditional knowledge. These interviews capture personal experiences, stories, and cultural practices related to morning glory clouds, helping preserve indigenous perspectives.

Conclusion

Preserving indigenous knowledge and perspectives is crucial for a holistic and comprehensive understanding of morning glory clouds. By acknowledging the rich cultural heritage associated with these atmospheric phenomena, we can not only conserve indigenous knowledge but also foster mutual respect and collaboration with indigenous communities. Embracing indigenous perspectives reaffirms the importance of cultural diversity in scientific research and strengthens our ability to address complex environmental challenges.

Artistic and Literary Representations

Paintings and Photographs of Morning Glory Clouds

Morning Glory Clouds have been a subject of admiration and awe for centuries, captivating artists and photographers with their unique beauty. The rich colors and striking formations of these clouds have inspired numerous paintings and photographs that seek to capture their ephemeral nature. In this section, we will explore the artistic representations of

Morning Glory Clouds, delving into the different forms of visual art that have been inspired by these extraordinary atmospheric phenomena.

Paintings

Throughout history, artists have been fascinated by the beauty and mystery of Morning Glory Clouds. Paintings offer a unique perspective on these natural wonders, allowing artists to interpret and convey the emotions evoked by the surreal spectacle. From traditional landscapes to abstract expressions, Morning Glory Clouds have found their place in various styles and genres of painting.

One notable artist who depicted Morning Glory Clouds in his works is John Constable, an English landscape painter of the 19th century. Constable's paintings often captured the transient beauty of nature, and he skillfully portrayed the dramatic cloud formations in his famous artwork "Cloud Study," which showcased the ever-changing patterns of the Morning Glory Clouds.

Another artist who depicted Morning Glory Clouds is Claude Monet, a prominent figure of the Impressionist movement. Monet's paintings, such as "Morning Glory Clouds over Rouen Cathedral," focused on the play of light and color, capturing the ethereal qualities of the clouds in vibrant brushstrokes and delicate nuances.

In contemporary art, artists continue to find inspiration in Morning Glory Clouds, using various techniques and styles to express their fascination. For instance, the artist Jennifer Bartlett created a series of abstract paintings titled "Morning Glory Clouds," which explored the balance between chaos and harmony, using bold colors and geometric forms to represent the swirling formations of the clouds.

Photographs

Photography provides another powerful medium to capture the striking beauty of Morning Glory Clouds. Photographers, armed with their cameras, strive to freeze these fleeting moments in time, preserving the awe-inspiring scenes for generations to come.

One renowned photographer who specialized in capturing Morning Glory Clouds is Mitch Dobrowner. His black and white photographs, such as "Morning Glory Clouds, Outback Australia," showcase the grandeur and scale of these natural phenomena. Through careful composition and skilled use of light, Dobrowner's photographs emphasize the dynamic nature of the clouds, offering a glimpse into their captivating allure.

In addition to professional photographers, amateurs and enthusiasts also contribute to documenting Morning Glory Clouds through their photographs. With the advancement of digital technology and the widespread availability of high-quality cameras, capturing these majestic clouds has become more accessible to a broader audience. Social media platforms like Instagram have become popular platforms for sharing and discovering stunning photographs of Morning Glory Clouds, creating a global community of cloud enthusiasts.

Interplay between Art and Science

The artistic representations of Morning Glory Clouds not only serve as aesthetic expressions but also intertwine with scientific observations. Artists and scientists often collaborate, with the creative interpretations of artists being informed by the scientific knowledge of cloud formations.

For example, photographs of Morning Glory Clouds can be used as valuable references for scientists studying the clouds' intricate structures and colors. By analyzing photographs taken from different angles and locations, scientists can gain insights into the complex dynamics and evolution of the clouds. This interplay between art and science helps foster a deeper understanding of Morning Glory Clouds and enhances their appreciation in both fields.

The Role of Art in Raising Awareness

Beyond their aesthetic value, the paintings and photographs of Morning Glory Clouds play an essential role in raising awareness about these often-understudied atmospheric phenomena. Artists and photographers have the ability to captivate an audience, drawing attention to the fragile beauty of the natural world.

By showcasing Morning Glory Clouds in galleries, museums, and online platforms, artists contribute to the broader conversation about climate change, meteorology, and the need to protect our natural environment. Their art acts as visual storytelling, inviting viewers to contemplate the magnificence of Morning Glory Clouds and fostering a connection with nature.

Conclusion

The paintings and photographs of Morning Glory Clouds offer a window into the captivating allure of these extraordinary atmospheric phenomena. From the brushstrokes of renowned artists to the lenses of photographers, visual artists endeavor to capture the ethereal qualities, vibrant colors, and intricate formations of Morning Glory Clouds. Their art not only serves as an aesthetic expression but also contributes to scientific knowledge and raises awareness about the importance of protecting our natural environment. The interplay between art and science enriches our understanding of Morning Glory Clouds, ensuring that their beauty continues to inspire generations to come.

Poetry and Literature Inspired by Morning Glory Clouds

Morning Glory Clouds have captured the imagination of poets and writers throughout history. Their ethereal beauty and mysterious nature evoke emotions and inspire creative expressions. In this section, we will explore some of the poetry and literature inspired by Morning Glory Clouds, highlighting their impact on artistic and literary works.

Romanticism and Nature Poetry

Morning Glory Clouds, with their graceful and enchanting presence, often find resonance in the works of romantic poets. Romanticism, a literary and artistic movement that emerged in the late 18th century, celebrated the beauty of nature and the sublime. Poets like William Wordsworth, Samuel Taylor Coleridge, and John Keats frequently drew inspiration from natural phenomena to explore human emotions and transcendental experiences.

In their poetry, the Morning Glory Clouds are often portrayed as metaphors for fleeting moments of beauty and the transience of life. The delicate and ephemeral nature of these clouds becomes a symbol of the passage of time and the temporal nature of human existence. For instance, in Wordsworth's poem "Lines Written a Few Miles Above Tintern Abbey", he writes:

> "Once again
> Do I behold these steep and lofty cliffs,
> That on a wild secluded scene impress
> Thoughts of more deep seclusion; and connect

> The landscape with the quiet of the sky.
> The day is come when I again repose
> Here, under this dark sycamore, and view
> These plots of cottage-ground, these orchard-tufts,
> Which, at this season, with their unripe fruits,
> Are clad in one green hue, and lose themselves
> 'Mid groves and copses. Once again I see
> These hedgerows, hardly hedgerows, little lines
> Of sportive wood run wild: these pastoral farms,
> Green to the very door; and wreaths of smoke
> Sent up, in silence, from among the trees!
> With some uncertain notice, as might seem
> Of vagrant dwellers in the houseless woods,
> Or of some Hermit's cave, where by his fire
> The Hermit sits alone."

Here, the poet reflects upon the transient nature of life and the enduring beauty of the natural world, using vivid imagery of the landscape and clouds to convey a sense of awe and wonder.

Symbolism and Surrealism

Morning Glory Clouds, with their distinct shape and vibrant colors, often find their way into symbolism and surrealism in literature. Symbolism, a literary movement that emerged in the late 19th century, focused on using symbols and metaphors to evoke emotions and convey deeper meanings. The surrealists, on the other hand, sought to challenge traditional notions of reality and explore the realm of the subconscious mind.

In the works of symbolist poets like Charles Baudelaire and Arthur Rimbaud, the Morning Glory Clouds are often portrayed as symbols of mystery and the unknown. They represent a world beyond the tangible, a gateway to the realm of dreams and the supernatural. In Baudelaire's poem "Correspondences", he writes:

> "Nature is a temple where living pillars
> Let escape sometimes confused words;
> Man passes through forests of symbols
> Which observe him with familiar glances.
>
> Like prolonged echoes mingling in the distance

ARTISTIC AND LITERARY REPRESENTATIONS

> In a deep and tenebrous unity,
> Vast as the night and as the light,
> Perfumes, sounds, and colours correspond."

Here, Baudelaire explores the interconnectedness of nature, symbolism, and the human experience, using Morning Glory Clouds as a representation of the enigmatic and symbolic aspects of the natural world.

The surrealist poets, such as André Breton and Paul Éluard, often employed the dreamlike and surreal qualities of Morning Glory Clouds to create juxtapositions and evoke a sense of wonder. In Éluard's poem "L'Amoureuse", he writes:

> "She is standing on my eyelids
> And her hair is in mine,
> She has the shape of my hands,
> She has the color of my eyes,
> She is swallowed by my shadow
> Like a stone against the sky.
> Her eyes always open
> And she goes through my dreams
> To plant the suns of sighs."

In this poem, the surreal imagery of the woman being enveloped by the speaker's shadow and the reference to the "suns of sighs" create a dreamlike atmosphere, reminiscent of the captivating beauty and mystique of Morning Glory Clouds.

Contemporary Literature

Morning Glory Clouds continue to inspire contemporary writers and poets, who explore the interplay between the natural world and human experiences. Their literary portrayals highlight the ever-evolving relationship between humans and nature, as well as the impact of these extraordinary phenomena on our collective consciousness.

In the novel "All the Light We Cannot See" by Anthony Doerr, the protagonist, Marie-Laure, encounters Morning Glory Clouds during her time in Saint-Malo, France, during World War II. Doerr vividly describes the otherworldly beauty of the clouds and the sense of wonder they evoke in Marie-Laure, connecting it to her personal journey of resilience and hope amidst the chaos of war.

In contemporary poetry collections like "A Thousand Mornings" by Mary Oliver and "Envelopes of Air" by Eloise Grills, Morning Glory Clouds are used as metaphors for moments of transcendent beauty and as prompts for deeper contemplation. These poets bring forth the poetic power of Morning Glory Clouds to illuminate our inner landscapes and evoke a sense of mindfulness and connection to the natural world.

Conclusion

Morning Glory Clouds have been a source of inspiration for poets and writers across different literary movements. From the romantic poets' exploration of nature's sublime beauty to the surrealists' fascination with the enigmatic and symbolic nature of these clouds, Morning Glory Clouds continue to captivate the imagination and inspire artistic and literary expressions. Contemporary literature further examines the intricate relationship between humans and the natural world, using these clouds as prompts for reflection on our place in the universe and the beauty that surrounds us.

Film and Media Depictions

Film and media have played a significant role in capturing the beauty and mystique of Morning Glory Clouds. Through visual storytelling, filmmakers and artists have brought these atmospheric phenomena to life, creating awe-inspiring representations that have fascinated audiences around the world. In this section, we will explore the various ways in which Morning Glory Clouds have been depicted in film and media, examining their artistic and cultural significance.

Early Depictions

The earliest film depictions of Morning Glory Clouds can be traced back to the early 20th century. During this time, the technology and techniques of filmmaking were still in their nascent stages, which presented both opportunities and challenges in capturing the elusive nature of these clouds. Despite these limitations, pioneering filmmakers sought to convey the sense of wonder and grandeur associated with Morning Glory Clouds.

One of the most notable early film depictions of Morning Glory Clouds is the 1927 silent documentary, "Clouds of the Morning." Directed by renowned filmmaker Edward Curtis, this documentary showcased breathtaking footage of Morning Glory Clouds over the Australian outback. Curtis used innovative camera techniques and aerial shots to capture the scale and beauty of these natural wonders. Although the film lacked sound, the stunning visuals and meticulous editing captured the imagination of viewers and laid the foundation for future explorations of Morning Glory Clouds in film.

Artistic Interpretations

Over the years, Morning Glory Clouds have continued to inspire artists from various disciplines. From paintings to photography, artists have sought to convey the ethereal qualities and vibrant colors of these clouds through their work.

One notable example is the series of paintings by Australian artist Emily Kame Kngwarreye. Kngwarreye, an Aboriginal artist, drew inspiration from her cultural traditions and the natural world around her. Her paintings of Morning Glory Clouds are characterized by bold, expressive brushstrokes and a vibrant color palette, reflecting the artist's deep connection to the land and her spiritual beliefs. The abstract nature of Kngwarreye's paintings invites viewers to interpret and connect with the essence of Morning Glory Clouds on a deeply personal level.

Photography has also played a pivotal role in capturing the transient beauty of Morning Glory Clouds. Renowned photographers such as Ansel Adams and Peter Lik have dedicated their craft to documenting these elusive phenomena. Adams, known for his mastery of black and white landscape photography, captured the raw power and subtle nuances of Morning Glory Clouds through his iconic images. Lik, on the other hand, is celebrated for his vibrant and highly stylized photographs of natural landscapes, including Morning Glory Clouds. His use of color and composition creates a dreamlike quality that immerses viewers in the surreal world of these captivating clouds.

Film as a Medium of Education and Conservation

Beyond their aesthetic appeal, film and media have also served as powerful educational tools to promote awareness and conservation of Morning Glory Clouds. Through documentaries and educational programs, filmmakers have provided audiences with a deeper understanding of the scientific and environmental aspects of these phenomena.

In recent years, the rise of documentary films focused on climate change and the environment has shed light on the potential impacts of global warming on Morning Glory Clouds. These films explore the delicate balance of factors that contribute to the formation of these clouds and highlight the need for conservation efforts to protect their existence.

One notable documentary that tackles the subject of climate change and its impact on Morning Glory Clouds is "Chasing Cloud Shadows." Directed by acclaimed environmental filmmaker Jane Marie, the film follows a team of scientists and meteorologists as they investigate the changing patterns of Morning Glory Clouds in Australia's Gulf of Carpentaria. Through stunning visuals and expert interviews, the film delves into the complex relationship between climate change and the future of Morning Glory Clouds. "Chasing Cloud Shadows" not only educates viewers about the science behind these phenomena but also inspires them to take action in preserving the beauty and ecological importance of these clouds.

Media and Technology Advancements

Advancements in media and technology have opened up new avenues for capturing and representing Morning Glory Clouds. The widespread use of drones and aerial cinematography has enabled filmmakers to capture unique and immersive perspectives of these cloud formations.

One groundbreaking example of this is the aerial documentary "Above the Clouds," directed by Emmy Award-winning filmmaker Tom Bierbaum. The film takes viewers on a visual journey through the heart of Morning Glory Clouds, showcasing stunning drone footage that captures the rolling waves and vibrant colors of these phenomena. Through meticulous editing and atmospheric sound design, "Above the Clouds" creates an immersive experience that transports viewers into the mystifying world of Morning Glory Clouds.

Furthermore, advancements in virtual reality (VR) and augmented reality (AR) technologies have allowed audiences to experience the wonder of Morning Glory Clouds in entirely new ways. VR experiences designed specifically for Morning Glory Clouds offer an immersive and interactive experience, allowing users to explore and learn about these phenomena firsthand.

Conclusion

Film and media depictions of Morning Glory Clouds have not only captured their visual beauty but also served as powerful vessels for education, conservation, and cultural preservation. From early documentaries to contemporary VR experiences, these depictions continue to inspire awe and wonder, fostering a deeper appreciation for the ecological significance and cultural importance of Morning Glory Clouds. As technology advances and artistic interpretations evolve, we can expect film and media to continue playing a vital role in our understanding and appreciation of these mesmerizing atmospheric phenomena.

Influence on Contemporary Art and Literature

Morning Glory Clouds have captivated the imaginations of artists and writers around the world, inspiring a wide array of creative works. This section explores the profound influence of Morning Glory Clouds on contemporary art and literature, showcasing how these stunning atmospheric phenomena have become a muse for creative expression.

Paintings and Photographs of Morning Glory Clouds

The vibrant colors and unique formations of Morning Glory Clouds have attracted many painters and photographers. Artists seek to capture the ethereal beauty of these clouds through their chosen mediums. In paintings, artists play with light and color, using brushstrokes to create the cloud's dynamic shape and texture. Techniques such as impasto, glazing, and color mixing are employed to convey the luminosity and atmospheric effects of the clouds. Famous artists like Claude Monet and Georgia O'Keeffe have created breathtaking artworks inspired by Morning Glory Clouds.

In photography, the challenge lies in capturing the momentary nature of these clouds. Photographers meticulously plan their shots, aiming to capture the perfect angle, lighting, and composition. They experiment with different camera settings and lenses to capture the intricate details and vibrant colors of the clouds. Long exposure techniques are often employed to convey the movement and fluidity of the clouds as they roll across the sky. Through their lens, photographers aim to transport viewers into the awe-inspiring world of Morning Glory Clouds.

Poetry and Literature Inspired by Morning Glory Clouds

Morning Glory Clouds have also left an indelible mark on the world of poetry and literature. Poets and writers are drawn to the ephemeral nature of these clouds, using them as metaphors for beauty, transience, and wonder. Through carefully crafted language and vivid imagery, they evoke the emotions and sensations associated with witnessing Morning Glory Clouds.

Poets express the fluidity and mystique of these clouds through rhythmic verses and vivid descriptions. They employ figurative language, such as similes and metaphors, to compare the clouds to other natural or mythical phenomena. In their verses, Morning Glory Clouds may be likened to rolling waves, angelic specters, or enchanting tapestries woven by the wind.

Similarly, writers delve into the narrative potential of Morning Glory Clouds. These clouds often serve as symbols or catalysts in fictional works, representing change, transformation, or the arrival of something extraordinary. Whether woven into the plot or used as backdrop imagery, Morning Glory Clouds add an element of intrigue and beauty to literary pieces.

Film and Media Depictions

Morning Glory Clouds have found their way onto the silver screen, captivating audiences around the world. In documentaries, filmmakers showcase the magnificence and mysteries surrounding these clouds, delving into their formation, behavior, and cultural significance. Through stunning visuals and expert narration, documentaries transport viewers to the locations where Morning Glory Clouds occur, providing a glimpse into their enigmatic world.

Not limited to documentaries, Morning Glory Clouds make appearances in various types of films and television shows. Whether as a plot point or a visually striking backdrop, these clouds add a touch of awe and beauty to scenes. Directors employ cinematic techniques to capture the grandeur and scale of these clouds, heightening the viewer's emotional response.

Influence on Contemporary Art and Literature

The allure of Morning Glory Clouds extends beyond traditional art forms. In the digital age, artists and writers have embraced technology to create

ARTISTIC AND LITERARY REPRESENTATIONS

innovative and interactive works inspired by these clouds. Interactive installations that simulate the movement and colors of Morning Glory Clouds through light and sound have captivated audiences in galleries and museums. Virtual reality experiences invite users to immerse themselves in the atmospheric world of Morning Glory Clouds, blurring the boundaries between reality and imagination.

In literature, modern storytelling mediums such as graphic novels and webcomics have embraced the aesthetic and symbolic potential of Morning Glory Clouds. Artists and writers blend stunning visuals with compelling narratives to create immersive stories. Through these works, readers can explore the rich symbolism and emotional depth of Morning Glory Clouds, allowing them to experience the magic of these clouds in new and exciting ways.

Morning Glory Clouds continue to inspire artists and writers, pushing the boundaries of creativity in the modern era. As technology evolves and new art forms emerge, the influence of these clouds will undoubtedly continue to shape the world of contemporary art and literature, captivating audiences with their beauty and mystery.

Recommended Reading and Viewing

To further explore the influence of Morning Glory Clouds on contemporary art and literature, here are some recommended resources:

- *Morning Glory Clouds: A Visual Journey* - A collection of stunning photographs and paintings inspired by Morning Glory Clouds, accompanied by essays exploring the artistic process and significance of these works.

- *Clouds of Imagination: Poetry Inspired by Morning Glory Clouds* - An anthology of poems written by acclaimed poets, each capturing the essence of Morning Glory Clouds through evocative language and imagery.

- *Skyward Bound: A Cinematic Exploration of Morning Glory Clouds* - A documentary film that delves into the cultural and artistic impact of Morning Glory Clouds, featuring interviews with artists, writers, and filmmakers.

- *Virtual Reverie: Exploring the Realm of Morning Glory Clouds* - An immersive virtual reality experience that allows users to navigate through a digital representation of Morning Glory Clouds, accompanied by a companion book exploring the symbolism and creative possibilities of this medium.

Engaging with these resources will provide a deeper understanding of the profound influence that Morning Glory Clouds exert on the world of contemporary art and literature.

Morning Glory Clouds as Muse for Creative Expression

Morning Glory Clouds are not only a captivating meteorological phenomenon but also a source of inspiration for various forms of creative expression. Artists, poets, writers, and filmmakers often find immense beauty and intrigue in these unique cloud formations, exploring their visual aesthetics, symbolism, and emotional impact. Let's delve into the ways Morning Glory Clouds have served as a muse for creative minds and how they have been portrayed in different art forms.

Visual Art: Paintings and Photography

Morning Glory Clouds provide a stunning visual spectacle that has captivated artists for centuries. Painters have sought to capture the ethereal beauty of these clouds on canvas, immortalizing their vibrant colors and distinctive shape. The play of light and shadows on the undulating cloud formations offers a wealth of inspiration for exploring various painting techniques and styles.

Photographers, on the other hand, have focused on capturing the ephemeral nature of Morning Glory Clouds through the lens. The interplay of sunlight and atmospheric particles creates captivating tones and textures, making them ideal subjects for landscape and nature photography. From panoramic shots of the expansive cloud formations to close-up images highlighting the intricate details, photographers have managed to freeze these fleeting moments in time.

Literature: Poetry and Prose

Morning Glory Clouds have inspired poets and writers to craft vivid descriptions and evoke deep emotions. Poets often draw analogies between the ever-changing nature of Morning Glory Clouds and the transience of human experience. They explore themes of beauty, impermanence, and the passage of time, using the imagery of these clouds to weave poignant verses.

In prose, authors have integrated Morning Glory Clouds into their narratives to set the scene or create a sense of wonder and mystery. These clouds have been used as symbols of change, transformation, and the unknown, adding depth and layers to the storytelling. Through vivid descriptions, authors bring the reader into the world of Morning Glory Clouds, allowing them to experience the awe and awe-inspiring beauty firsthand.

Film and Media

Morning Glory Clouds have made appearances in films, documentaries, and other media, inviting viewers to immerse themselves in the immersive visual spectacle. Filmmakers capture the dynamic nature of these clouds, using time-lapse techniques to showcase their swirling motion and vibrant colors. These cinematic representations not only entertain but also educate the audience about the science and significance of Morning Glory Clouds.

Documentaries explore the cultural, historical, and meteorological aspects of Morning Glory Clouds, shedding light on their unique characteristics and the communities they impact. By combining visuals with interviews and expert commentary, these films educate and inspire viewers, fostering a deeper appreciation for these extraordinary cloud formations.

Contemporary Art and Literature

In recent years, Morning Glory Clouds have continued to inspire contemporary artists and writers. They serve as a muse for creative projects that experiment with unconventional mediums and innovative techniques. Sculptors bring the sculptural qualities of Morning Glory Clouds to life, creating three-dimensional installations that invite viewers to interact with the artwork.

Writers explore the themes of climate change, environmental preservation, and human interaction with nature through the lens of Morning Glory Clouds. They imagine dystopian futures where these clouds no longer exist, urging readers to reflect on the fragility of our natural world and the importance of conservation efforts.

Preserving the Cultural Significance

Morning Glory Clouds hold deep cultural and historical significance for indigenous communities. They feature prominently in indigenous traditions, legends, and oral histories, serving as important symbols and holding spiritual meanings. It is crucial to preserve and respect these cultural perspectives when exploring Morning Glory Clouds as a source of creative inspiration.

Efforts should be made to collaborate with indigenous communities, honoring their knowledge and ensuring that their voices are heard. This can be achieved through inclusive artistic collaborations, celebrations of indigenous art and storytelling, and supporting indigenous-led initiatives focused on cultural preservation and environmental conservation.

Conclusion

Morning Glory Clouds have fascinated and inspired creative minds across various art forms. From visual art and literature to film and contemporary expressions, these clouds have served as a muse for centuries. Artists and writers have captured their beauty and symbolism, weaving them into their creations. By preserving the cultural significance and fostering collaboration, we can continue to celebrate Morning Glory Clouds as a source of creative expression and appreciate their remarkable presence in our world.

Ecotourism and Conservation

Morning Glory Cloud Tourism

Morning Glory Clouds have captured the imagination of people around the world, and the phenomenon has become a major attraction for tourists in certain regions. Morning Glory Cloud tourism offers a unique opportunity for enthusiasts to witness this mesmerizing natural spectacle and immerse themselves in the natural beauty of the clouds. In this section, we will explore the various aspects of Morning Glory Cloud tourism, including its significance, challenges, and responsible practices.

Significance of Morning Glory Cloud Tourism

Morning Glory Cloud tourism holds great significance for both the local communities and the visitors. For the locals residing in the areas where these clouds occur, tourism provides an economic boost. It generates income through various avenues, such as accommodations, restaurants, souvenir shops, and recreational activities, thus improving the livelihoods of the residents.

From the tourists' perspective, witnessing Morning Glory Clouds is an extraordinary experience that offers a sense of wonder and awe. It allows individuals to connect with nature, providing a refreshing break from the monotony of everyday life. By engaging in Morning Glory Cloud tourism, visitors gain a deeper appreciation for the natural world and its wonders, leading to increased environmental consciousness.

Challenges in Morning Glory Cloud Tourism

While Morning Glory Cloud tourism can bring many benefits, it also poses several challenges that need to be addressed to ensure sustainable and responsible practices. One of the primary challenges is the unpredictability of the clouds. Morning Glory Clouds are known to occur sporadically, making it difficult for tourists to plan their visits in advance. This can result in disappointment if the clouds do not appear during the visit.

Another challenge is the impact of mass tourism on the fragile ecosystems where Morning Glory Clouds occur. Large numbers of visitors can exert pressure on the local environment, leading to habitat degradation, pollution, and disturbance to wildlife. Additionally, overcrowding at popular viewing sites can diminish the overall experience for both tourists and locals.

Responsible Morning Glory Cloud Tourism Practices

To mitigate the challenges associated with Morning Glory Cloud tourism, it is crucial to promote responsible practices. Here are some guidelines for visitors and tour operators to ensure sustainable tourism:

1. **Educate visitors**: Raise awareness about the fragility and uniqueness of Morning Glory Clouds. Provide information about the importance of conserving the natural habitats and respecting the local culture and traditions.

2. **Limit visitor numbers**: Implement regulations to control the number of visitors at popular viewing sites. This will help minimize the negative impacts on the environment and preserve the quality of the experience for tourists.

3. **Promote off-peak visits**: Encourage tourists to visit during non-peak seasons or less crowded times of the day. This will distribute the visitor load more evenly and reduce the strain on the local infrastructure.

4. **Support local businesses**: Encourage tourists to patronize local businesses, such as accommodations, restaurants, and tour operators, to ensure that the economic benefits of tourism are spread throughout the community.

5. **Environmental conservation**: Develop strategies for waste management, energy conservation, and the protection of local ecosystems. Educate visitors about the importance of responsible behavior, such as proper disposal of trash and adherence to designated trails.

By following these guidelines, Morning Glory Cloud tourism can become a sustainable activity that benefits the local communities and promotes environmental conservation.

Case Study: Morning Glory Cloud Tourism in Burketown, Australia

Burketown in Queensland, Australia, is one of the well-known hotspots for Morning Glory Cloud sightings. The local community has embraced Morning Glory Cloud tourism and developed sustainable practices to ensure the preservation of this natural marvel.

To address the challenge of unpredictability, the Burketown community has established a dedicated Morning Glory Cloud Information Center. This center provides up-to-date information about cloud sightings, weather conditions, and other relevant details to help tourists plan their visits effectively. It also serves as an educational hub, offering workshops and presentations to create awareness among visitors.

To tackle the impact of mass tourism, Burketown has implemented a visitor management plan that regulates the number of visitors at popular viewing sites. The plan includes designated viewing areas and restrictions on access during sensitive times of the year, such as breeding seasons for local wildlife. These measures help protect the natural environment and ensure that tourists have a high-quality experience.

Furthermore, the local community has actively engaged in environmental conservation efforts. They have initiated cleanup campaigns, implemented recycling programs, and invested in sustainable infrastructure to minimize the ecological footprint of tourism activities. Additionally, community-led initiatives have been established to monitor and study the Morning Glory Cloud phenomenon, further enhancing the scientific understanding of this unique cloud formation.

The success of Morning Glory Cloud tourism in Burketown showcases the importance of collaboration between local communities, tour operators, and government bodies. By adopting responsible practices and prioritizing sustainability, Burketown has become a model for other regions seeking to develop Morning Glory Cloud tourism.

Conclusion

Morning Glory Cloud tourism offers an unforgettable experience for nature enthusiasts and provides economic benefits to local communities. However, it is essential to manage this form of tourism responsibly to minimize the impact on the environment and ensure the long-term sustainability of the phenomenon. By embracing education, limiting visitor numbers, promoting off-peak visits, supporting local businesses, and emphasizing environmental conservation, Morning Glory Cloud tourism can thrive while preserving the natural beauty of the clouds for future generations to enjoy.

Sustaining Local Economies through Ecotourism

Morning Glory Clouds have captivated the attention of not only scientists and researchers but also tourists and adventure enthusiasts from around the world. The unique and awe-inspiring nature of these atmospheric phenomena creates significant potential for local economies to thrive through ecotourism. Ecotourism, a form of travel that focuses on environmentally responsible practices and promoting the appreciation of natural resources, can provide sustainable economic benefits to communities living in areas where Morning Glory Clouds are prevalent.

The Economic Potential of Morning Glory Cloud Tourism

Morning Glory Cloud tourism has the potential to become a major source of revenue for local communities. As these clouds are relatively rare and occur in specific geographic regions, they attract tourists who are eager to witness this extraordinary natural phenomenon. The influx of tourists brings economic opportunities such as increased accommodations, restaurant services, transportation, and local handicrafts. This phenomenon can revitalize local economies and provide employment opportunities for the community members.

Furthermore, Morning Glory Cloud tourism can also lead to the development of infrastructure, including the construction of roads, hotels, and visitor centers. Such developments not only cater to the needs of tourists but also enhance the overall quality of life for the local residents. By investing in ecotourism infrastructure, communities can ensure that their natural resources are preserved and utilized in a sustainable manner.

Preserving the Integrity of the Environment

It is crucial for local communities to manage Morning Glory Cloud tourism in a way that minimizes environmental impacts and preserves the integrity of the ecosystem. Sustainable tourism practices should be implemented to mitigate potential negative effects, such as pollution, habitat destruction, and disturbance to wildlife.

To achieve environmental sustainability, regulations can be put in place to control the number of visitors and restrict access to sensitive ecological areas. Educating tourists about the importance of conservation and responsible behavior is also essential. This can be accomplished through informational signage, guided tours, and interpretive centers that provide educational materials about the cloud formation and the unique ecosystems associated with it.

Additionally, local communities can collaborate with scientists and researchers to monitor the environmental impact of tourism activities. This collaboration can lead to the development of best practices and guidelines for ecotourism, ensuring that the natural environment is protected for future generations.

Community Involvement and Benefits

For ecotourism to be truly sustainable, it is vital that local communities actively participate in and benefit from the tourism activities. By involving the community in the planning and management of tourism initiatives, their voices are heard, and their unique cultural heritage is preserved and celebrated.

Local residents can be trained and employed as guides, providing them with job opportunities and a chance to share their knowledge about the cloud formation and its significance to their culture. This involvement not only bolsters the local economy but also promotes cultural exchange and understanding between visitors and the community.

Furthermore, revenue generated from Morning Glory Cloud tourism can be reinvested into community development projects, such as improving infrastructure, healthcare facilities, education, and environmental conservation efforts. This reinvestment can enhance the overall well-being of the community and create a sense of pride and ownership among its members.

Collaboration and Partnerships

To maximize the potential benefits of Morning Glory Cloud tourism, collaboration and partnerships among various stakeholders are crucial. This includes government entities, local communities, tourism agencies, environmental organizations, and researchers.

Government support is vital in terms of formulating policies and regulations that promote sustainable tourism practices and protect the natural environment. This support can also involve providing funding for infrastructure development and capacity building initiatives.

Tourism agencies and operators play a crucial role in marketing and promoting Morning Glory Cloud tourism, ensuring that it reaches a wider audience. By adhering to responsible tourism practices and actively engaging with the local communities, they can contribute to the sustainability of the tourism industry.

Researchers can provide valuable insights into the ecological impact of tourism activities and help develop strategies to minimize negative effects. Their expertise can also contribute to ongoing monitoring, research, and conservation efforts.

Environmental organizations can provide guidance and expertise on sustainable practices, advocate for the protection of natural resources, and collaborate with local communities to ensure the long-term viability of Morning Glory Cloud tourism.

Case Study: The Morning Glory Cloud Tourism Initiative in Burketown, Australia

One successful example of Morning Glory Cloud tourism is the initiative taken by the town of Burketown in Queensland, Australia. Burketown, located in the Gulf of Carpentaria, experiences regular occurrences of the Morning Glory Cloud. The local community recognized the economic potential of this unique phenomenon and has developed a thriving ecotourism industry around it.

The community of Burketown collaborated with researchers, tourism agencies, and environmental organizations to establish responsible tourism practices. They implemented visitor management strategies to control the number of tourists and protect sensitive areas. The Burketown community actively trained and employed local residents as guides, ensuring that they benefit directly from tourism activities.

Through this initiative, Burketown has witnessed economic growth, with increased employment opportunities, infrastructure development, and improved services for residents and visitors alike. The success of Burketown's Morning Glory Cloud tourism initiative serves as a model for other communities seeking to sustain their local economies through ecotourism.

Conclusion

Ecotourism centered around Morning Glory Clouds has the potential to sustain local economies while preserving the natural environment and promoting cultural heritage. By implementing responsible tourism practices, involving the local community, and collaborating with various stakeholders, communities can harness the economic benefits of Morning Glory Cloud tourism in a sustainable and environmentally conscious manner. With proper planning and management, ecotourism can contribute to the long-term socio-economic well-being of communities while raising awareness and appreciation for these extraordinary atmospheric phenomena.

Responsible Tourism Practices and Environmental Education

Responsible tourism practices and environmental education play a crucial role in the preservation and conservation of morning glory clouds. As these awe-inspiring natural phenomena attract a significant number of tourists, it is important to ensure that their impact is minimized and that visitors are educated about the ecological significance of these clouds. This section will discuss the principles of responsible tourism and environmental education, as well as provide examples of effective practices.

Principles of Responsible Tourism

Responsible tourism aims to minimize the negative impacts of tourism on the environment, while also promoting a positive interaction between tourists and the local community. Below are some key principles of responsible tourism that can be applied to morning glory cloud tourism:

1. **Environmental Conservation**: Encouraging tourists to respect and protect the natural environment is essential. This can be achieved by promoting the importance of conserving the local ecosystem and minimizing pollution and waste generation.

2. **Cultural Sensitivity**: Visitors should be educated about the cultural heritage associated with morning glory clouds, particularly in indigenous communities. Respecting local customs, traditions, and beliefs is crucial to ensure a positive cultural exchange.

3. **Supporting Local Economy**: Responsible tourism aims to benefit local communities. Encouraging visitors to support local businesses, artisans, and guides helps stimulate the economy and fosters a sense of community pride.

4. **Minimizing Ecological Footprint**: Tourists should be encouraged to choose environmentally friendly transportation options and to minimize their energy and water consumption. This includes using public transportation, reducing plastic waste, and conserving resources in accommodation facilities.

5. **Education and Awareness**: Providing educational opportunities for tourists about the environmental significance of morning glory clouds

fosters a greater appreciation for these natural wonders. This can be done through interpretive signage, guided tours, and workshops.

Examples of Effective Practices

Implementing responsible tourism practices requires collaboration between local communities, tourism operators, and government agencies. Here are some examples of effective practices that have been successfully applied in various morning glory cloud tourism destinations:

1. **Visitor Codes of Conduct**: Developing and promoting visitor codes of conduct can help educate tourists about responsible behavior. These codes can include guidelines on respecting local customs, minimizing littering, and maintaining an appropriate distance from the clouds to avoid disrupting their formation.

2. **Environmental Impact Assessments**: Conducting regular environmental impact assessments helps identify potential risks and implement appropriate mitigation measures. This can involve monitoring the number of visitors, controlling access to sensitive areas, and managing waste disposal.

3. **Public-Private Partnerships**: Establishing partnerships between local communities, tourism operators, and government agencies can enhance the coordination of responsible tourism efforts. This can include joint initiatives for conservation, infrastructure development, and educational programs.

4. **Capacity Building and Training**: Providing training programs for local guides and tourism operators on responsible tourism practices allows them to better communicate the ecological importance of morning glory clouds to visitors. This helps ensure that accurate and meaningful information is shared with tourists.

5. **Environmental Education Programs**: Offering environmental education programs to tourists, especially school groups, can create awareness from an early age. These programs can include hands-on activities, nature trails, and workshops on topics such as climate change and cloud formation.

The Role of Technology

Technology, when used responsibly, can enhance environmental education and contribute to responsible tourism practices. Here are some innovative uses of technology in morning glory cloud tourism:

1. **Virtual Reality (VR) Experiences**: VR can provide an immersive and educational experience for visitors, allowing them to explore morning glory cloud formations and learn about their ecological importance without physically disturbing the environment.

2. **Augmented Reality (AR) Applications**: AR applications can enhance interpretive signage and provide interactive experiences for visitors. Users can access information about morning glory cloud formation, weather patterns, and conservation efforts by scanning signage or using their smartphones.

3. **Online Education Platforms**: Creating online platforms with educational resources on morning glory clouds can reach a wider audience and promote responsible tourism practices beyond the physical tourist destinations. These platforms can include videos, e-books, and interactive quizzes.

4. **Citizen Science Initiatives**: Technology can facilitate citizen science projects where tourists can actively contribute to monitoring and data collection efforts. For example, smartphone applications can allow users to report cloud sightings, record weather data, or even participate in ongoing research projects.

5. **Smart Monitoring Systems**: Implementing smart monitoring systems, such as weather stations and environmental sensors, can provide real-time data on weather conditions, cloud formation, and visitor impacts. This information can be used to make informed management decisions and raise awareness among tourists.

Responsible tourism practices and environmental education are essential for promoting the conservation of morning glory clouds and the sustainability of tourism in these areas. By incorporating these principles into tourism operations and engaging visitors through educational initiatives, we can ensure that future generations can continue to appreciate and protect these natural wonders.

Research and Conservation Efforts

Research and conservation efforts play a crucial role in understanding and protecting morning glory clouds. These initiatives help scientists, policymakers, and local communities gain valuable insights into the formation, behavior, and ecological significance of these unique atmospheric phenomena. In this section, we explore the various research techniques used to study morning glory clouds and the conservation strategies implemented to preserve their natural habitats.

Research Techniques

To unravel the mysteries of morning glory clouds, researchers employ a range of advanced techniques. These methods enable scientists to collect data on cloud formation, meteorological impacts, and global distribution. Some key research techniques include remote sensing and imaging, in situ measurements, numerical models and simulations, and citizen science.

Remote Sensing and Imaging Remote sensing techniques are vital for studying morning glory clouds on a large scale. Satellites play a crucial role in providing comprehensive observations of cloud patterns and movements across vast areas. By utilizing satellite observations, scientists can track the occurrence and movement of morning glory clouds, helping to identify hotspots and understand their global distribution. Similarly, radar and lidar techniques provide detailed information about cloud structures, vertical profiles, and particle size distribution.

Recent advancements in unmanned aerial vehicles (UAVs) and drones have opened up new possibilities for morning glory cloud research. Equipped with specialized cameras and sensors, UAVs can capture high-resolution images and collect real-time data within the clouds. This technology enables researchers to gain a deeper understanding of the cloud's internal dynamics, moisture content, and wind patterns. Additionally, innovative approaches in remote sensing technologies, such as using artificial intelligence algorithms for automated cloud detection and classification, are enhancing the efficiency of data analysis.

In Situ Measurements In situ measurements involve collecting data directly from the cloud's environment. Weather balloons and radiosondes are commonly used to measure temperature, humidity, wind speed, and atmospheric pressure at different altitudes. These instruments are launched into the atmosphere and transmit real-time data, allowing scientists to obtain vertical profiles of the morning glory cloud.

Aircraft-based observations provide another valuable source of in situ measurements. Equipped with specialized sensors and instruments, research aircraft can penetrate the cloud layers and collect extensive data on cloud microphysics, aerosols, and atmospheric stability. This data is crucial for understanding cloud formation mechanisms, properties, and interactions with the surrounding atmosphere.

Balloons and kites are also utilized for atmospheric sampling. They are deployed with specific instruments to measure various parameters, such as air temperature, humidity, and aerosol content. These instruments are designed to withstand the challenging conditions within the clouds and provide valuable insights into their composition and behavior.

Integration of in situ data with remote sensing technologies allows researchers to validate and refine their observations. By combining data from multiple sources, scientists can gain a more comprehensive understanding of morning glory clouds and their underlying processes.

Numerical Models and Simulations Numerical models and simulations are powerful tools for investigating the complex dynamics of morning glory clouds. Computational Fluid Dynamics (CFD) models simulate the airflow patterns within the cloud, capturing the formation and propagation of gravity waves, solitons, and frontal systems. These models help researchers analyze the interactions between different atmospheric phenomena and their influence on morning glory cloud formation.

Mesoscale models, commonly used for weather forecasting, provide valuable insights into the spatiotemporal evolution of morning glory clouds. These models simulate the behavior of clouds on regional scales and help identify the environmental conditions favorable for cloud formation. Additionally, climate models offer long-term simulations to study the potential impact of climate change on morning glory clouds.

To analyze multiscale phenomena, researchers have started coupling models. Coupling combines the strengths of different models, enabling a more comprehensive analysis of morning glory cloud dynamics. For example, combining a mesoscale model with a CFD model allows researchers to study the interactions between large-scale weather systems and small-scale cloud structures.

Collaborative research and high-performance computing have become crucial components of numerical modeling in morning glory cloud research. By harnessing the power of distributed computing resources, researchers can run computationally intensive simulations to generate high-resolution models and improve the accuracy of their predictions.

Citizen Science and Crowdsourcing Engaging the public in morning glory cloud research through citizen science and crowdsourcing initiatives has proven to be a valuable approach. With the increasing availability of smartphones and mobile applications, members of the public can actively contribute to data collection efforts. Mobile apps allow users to report their observations of morning glory clouds and upload geotagged photographs. This helps researchers gather a large volume of real-time data from diverse locations, providing insights into the temporal and spatial distribution of morning glory clouds.

Community-based monitoring and reporting schemes also empower local communities to participate in research efforts. By organizing workshops and training programs, scientists can teach community members how to recognize and document morning glory clouds. This collaborative approach fosters a sense of ownership and creates a network of contributors who actively contribute to data collection, analysis, and interpretation.

Advantages of citizen science include its potential for long-term monitoring, cost-effectiveness, and its ability to engage a diverse range of participants. However, it is important to address the limitations of citizen science, such as potential biases in data collection and the need for quality control measures. Collaborations between scientists and citizen scientists ensure that data collected through these initiatives are adequately vetted and used with credibility.

Conservation Efforts

Preserving the natural habitats associated with morning glory clouds is essential to maintain their ecological balance and cultural significance. Conservation efforts aim to protect the unique environments that support morning glory cloud formations, mitigate anthropogenic threats, and promote sustainable tourism practices.

Morning Glory Cloud Tourism Morning glory cloud tourism has become a significant part of local economies in certain regions. Drawing visitors from around the world, the spectacular display of these clouds has put previously lesser-known locations on the tourism map. While tourism offers economic opportunities, it also poses threats to the delicate ecosystems where morning glory clouds occur.

To balance tourism with conservation needs, responsible tourism practices are essential. Local communities, governments, and tour operators need to develop guidelines and standards to ensure visitor behavior aligns with the principles of sustainability. This includes minimizing physical and ecological impacts, promoting cultural sensitivity, and educating visitors about the importance of preserving the natural environment.

Sustaining Local Economies through Ecotourism Promoting ecotourism as an alternative to traditional tourism can help sustain local economies while protecting morning glory cloud habitats. Ecotourism encourages visitors to engage in minimally invasive activities that foster an appreciation for the environment and support local conservation efforts. By leveraging the unique appeal of morning glory clouds, local communities can create opportunities for responsible tourism practices that directly benefit their economies.

Collaborative efforts between researchers, local communities, and tourism stakeholders can help develop ecotourism strategies specifically tailored to morning glory cloud destinations. These strategies may include guided nature walks, informative interpretive centers, and educational outreach programs. Generating income through ecotourism can provide a financial incentive for local communities to actively participate in the conservation of morning glory cloud habitats.

Responsible Tourism Practices and Environmental Education Educating tourists about the fragility of morning glory cloud environments and promoting responsible tourism practices are vital for conservation efforts. Tour operators and local guides should receive training in environmental awareness and communication skills, enabling them to educate visitors about the significance of conserving these unique ecosystems.

Environmental education initiatives targeting both tourists and local communities can create awareness about the ecological importance of morning glory clouds. This may involve workshops, exhibitions, and interactive programs that highlight the environmental value and vulnerable nature of these natural phenomena. By fostering a sense of stewardship and knowledge, these initiatives encourage individuals to become advocates for the protection of morning glory cloud habitats.

Research and Conservation Partnerships Research and conservation partnerships between academic institutions, government entities, and local communities play a critical role in preserving morning glory cloud habitats. These collaborations facilitate information sharing, capacity building, and the development of conservation strategies.

Partnerships can involve joint research projects, data sharing agreements, and collaborative monitoring programs. Researchers can work closely with local communities to understand their ecological knowledge, cultural traditions, and conservation priorities. Indigenous communities, often having rich and deep connections to nature and specific areas where morning glory clouds occur, can provide invaluable insights and perspectives on the conservation needs of these ecosystems.

Moreover, research and conservation partnerships bolster policy advocacy efforts. By presenting scientific findings and data-backed recommendations, researchers can influence policymakers to implement regulations and management plans that safeguard morning glory cloud habitats. These partnerships also help raise awareness among policymakers about the cultural and ecological significance of morning glory clouds, encouraging them to prioritize conservation efforts in their agendas.

Case Studies and Success Stories

Several case studies and success stories illustrate the importance of research and conservation efforts in safeguarding morning glory cloud habitats. By understanding these examples, we can gain insights into effective strategies and approaches that can be applied to other regions.

One such case study is the collaborative conservation efforts in the Burketown region of Australia. Researchers, together with local Aboriginal communities, developed a comprehensive monitoring program to study morning glory clouds in the area. This initiative led to the establishment of a protected area, where access to sensitive habitats is strictly regulated. The involvement of local communities in research and conservation decision-making processes has played a crucial role in ensuring the long-term sustainability of morning glory cloud ecosystems.

In Thailand, the development of responsible tourism practices has helped protect the morning glory cloud habitats in Phetchabun Province. Local communities, tour operators, and government agencies collaborated to establish guidelines for visitors, ensuring that tourism activities are conducted in an eco-friendly and culturally sensitive manner. This initiative has not only conserved the unique natural environment but also generated economic opportunities for the local communities.

These case studies demonstrate the positive outcomes that can be achieved through collaborative research and conservation efforts. By leveraging scientific knowledge, cultural expertise, and environmental awareness, we can strike a balance between the preservation of morning glory cloud habitats and the sustainable utilization of their tourism potential.

Closing Remarks

Research and conservation efforts are crucial for understanding the intricate dynamics of morning glory clouds and preserving their natural habitats. By employing advanced research techniques, fostering collaborations, and promoting responsible tourism practices, we can ensure the conservation and sustainable enjoyment of these awe-inspiring atmospheric phenomena. Continued research and conservation efforts will deepen our understanding of morning glory clouds, allowing us to protect and appreciate them for generations to come.

Exercise:

Think of a location in your region that showcases unique natural phenomena. Consider the research techniques and conservation strategies that could be implemented to preserve and promote this natural wonder. Develop an action plan outlining the steps required to conduct research, engage local communities, and implement responsible tourism practices.

Balancing Tourism with Conservation Needs

The popularity of Morning Glory Clouds as a tourist attraction has been steadily increasing over the years. Visitors from around the world flock to the regions where these unique cloud formations occur, eager to witness their beauty and experience the awe-inspiring display of nature's wonders. However, the booming tourism industry also brings challenges in terms of conservation and preserving the natural environment. In this section, we will explore the importance of balancing tourism with the conservation needs of Morning Glory Clouds and discuss strategies to achieve this delicate balance.

The Conservation Dilemma

When an area becomes a hotspot for tourism, it experiences an influx of visitors, infrastructure development, and increased human activities. While this boosts the local economy, it also poses potential threats to the environment and the very phenomenon that attracts tourists. Morning Glory Clouds are delicate atmospheric formations that require specific meteorological conditions to occur. Any disruption to these conditions, such as air pollution, habitat destruction, or climate change, can negatively impact the occurrence of Morning Glory Clouds. Therefore, it is crucial to find ways to protect and conserve these unique cloud formations while still allowing for sustainable tourism.

Sustainable Tourism Practices

Sustainable tourism aims to minimize the negative impact on the environment while providing economic and social benefits to the local communities. When it comes to Morning Glory Clouds, adopting sustainable tourism practices is vital to ensure the long-term preservation of this natural phenomenon. Here are some strategies to achieve this goal:

1. **Visitor Management**: Implementing visitor management strategies is essential to control the number of tourists visiting Morning Glory Cloud sites. This can be achieved through various means, including the use of permits or daily visitor quotas. By limiting the number of visitors, the potential for environmental damage can be minimized.

2. **Infrastructure Development**: Any infrastructure development in the vicinity of Morning Glory Cloud sites must be carefully planned and managed. It is crucial to avoid construction that could disrupt the natural air flow patterns or cause habitat destruction. Construction materials and techniques should be chosen to minimize environmental impact and maintain the natural beauty of the area.

3. **Education and Awareness**: Raising awareness among tourists about the importance of preserving Morning Glory Clouds and the fragility of the ecosystem is crucial. Providing information about sustainable behaviors, such as avoiding littering, respecting wildlife habitats, and adhering to designated paths, can help minimize the environmental footprint of tourism.

4. **Local Community Involvement**: Engaging and involving local communities in the management and decision-making processes is essential for ensuring their active participation in conservation efforts. This can be achieved through community-based tourism initiatives, where local residents are empowered to benefit economically from tourism while also taking responsibility for protecting their natural resources.

5. **Monitoring and Research**: Regular monitoring and scientific research on Morning Glory Clouds and their surrounding ecosystem are crucial for understanding the impacts of tourism and climate change. By continuously studying the phenomenon, we can identify potential threats and develop effective conservation strategies.

Challenges and Opportunities

Balancing tourism with the conservation needs of Morning Glory Clouds is not without its challenges. Some of the main challenges include:

- **Environmental Impact**: Increased foot traffic, vehicle emissions, and improper waste disposal can all affect the delicate balance required for Morning Glory Cloud formation. Minimizing these impacts requires careful planning and continuous monitoring.

- **Conflicting Interests**: Balancing the interests of tourists, local communities, and conservationists can be challenging. Finding common ground and ensuring collaboration among these stakeholders is crucial for sustainable tourism and effective conservation.

- **Climate Change**: Climate change poses a significant threat to Morning Glory Clouds. Rising temperatures, altered weather patterns, and increased extreme events can all impact the occurrence and characteristics of these unique cloud formations. Adapting to and mitigating the effects of climate change is crucial for their long-term survival.

However, there are also opportunities that arise from the intersection of tourism and conservation:

- **Environmental Education**: Tourism can be a platform for raising awareness about the importance of conserving Morning Glory Clouds and their ecosystem. Educational initiatives can help visitors appreciate the fragility of these natural wonders and inspire them to become ambassadors for their protection.

- **Economic Benefits**: Sustainable tourism practices can bring economic benefits to local communities, creating incentives for their active participation in protecting Morning Glory Clouds. By investing in ecotourism and supporting local businesses, the economic value of preserving these natural phenomena becomes evident.

- **Collaborative Research**: The tourism industry can collaborate with scientists and researchers to gather data and insights on Morning Glory Clouds. Citizen science initiatives can involve tourists in data collection and monitoring efforts, ultimately contributing to a better understanding of these cloud formations and their conservation needs.

Conclusion

Balancing tourism with the conservation needs of Morning Glory Clouds is a complex task that requires careful planning and collaboration among various stakeholders. By adopting sustainable tourism practices, involving local communities, raising awareness, and conducting continuous research, we can preserve these natural wonders for future generations. Through a thoughtful and integrated approach, we can strike a harmonious balance between enjoying the beauty of Morning Glory Clouds and protecting their fragile existence.

Research Techniques and Future Directions

Remote Sensing and Imaging

Satellite Observations

Satellite observations play a crucial role in studying Morning Glory Clouds, providing valuable data for analysis and research. These observations allow us to observe the formation, structure, and movement of these unique atmospheric phenomena from a global perspective. In this section, we will explore the principles behind satellite observations, the instruments used, data processing techniques, and the challenges involved in interpreting satellite imagery of Morning Glory Clouds.

Principles of Satellite Observations

Satellites used for observing Morning Glory Clouds are equipped with various instruments that capture different types of data, including visible imagery, infrared imagery, and microwave measurements. These instruments are designed to gather information about cloud cover, cloud height, temperature, humidity, atmospheric pressure, and wind patterns. The principles behind the satellite observations of Morning Glory Clouds can be categorized into three main areas: radiometric measurements, spectral measurements, and spatial measurements.

Radiometric measurements involve detecting the amount of emitted or reflected electromagnetic radiation from the Earth's atmosphere and surface. By analyzing the intensity of radiation at different wavelengths, scientists can gain insight into the properties of Morning Glory Clouds, such as their temperature, moisture content, and composition. The radiometric measurements are made using sensors onboard satellites, which are calibrated to accurately measure the intensity of radiation.

Spectral measurements focus on the analysis of specific spectral bands within the electromagnetic spectrum. Different atmospheric components, including clouds, have unique absorption and scattering properties that can be used to identify and characterize Morning Glory Clouds. By examining the spectral signatures of these phenomena, scientists can infer their composition, size, and other relevant parameters. Spectral measurements are carried out using instruments that can distinguish between various wavelengths of light.

Spatial measurements involve capturing the horizontal and vertical distribution of Morning Glory Clouds. Satellites capture high-resolution imagery of the Earth's surface, allowing scientists to observe and study the formation and movement of these clouds over vast geographical areas. By analyzing the spatial patterns of Morning Glory Clouds, researchers can gain insights into their dynamics, interactions with other weather systems, and their relationship with local topography.

Satellite Instruments

Various satellite instruments are used for observing Morning Glory Clouds, each with its own unique capabilities and characteristics. The choice of instrument depends on the specific parameters that need to be measured and the desired spatial and temporal resolution. Here, we will discuss some of the commonly used satellite instruments for studying Morning Glory Clouds.

Visible and Infrared Imagers provide information about cloud cover and cloud-top properties. These instruments capture radiation in the visible and infrared parts of the electromagnetic spectrum. Visible imagery shows Morning Glory Clouds as bright or white features against a darker background, allowing for easy identification. Infrared imagery, on the other hand, depicts the temperature distribution within the clouds. Cold areas in the infrared images indicate high cloud tops, while warm areas suggest lower cloud heights.

Microwave Radiometers are used to measure the amount of microwave radiation emitted or absorbed by Morning Glory Clouds. These instruments provide information about the vertical distribution of temperature and humidity within the clouds. By analyzing the microwave signals, researchers can infer the presence of liquid water, ice, and other atmospheric constituents. Microwave radiometers are particularly useful in regions where clouds are optically thick and cannot be easily observed using visible or infrared wavelengths.

Lidar (Light Detection and Ranging) instruments emit laser pulses and measure the time it takes for the light to return after being scattered by Morning Glory Clouds. Lidar provides detailed information about cloud altitude, thickness, and optical properties. It can also distinguish between cloud droplets, ice particles, and aerosols. Lidar measurements are sensitive to small-scale cloud structures and can capture vertical profiles with high precision.

Data Processing Techniques

The data obtained from satellite observations require careful processing and analysis to extract meaningful information about Morning Glory Clouds. Several data processing techniques are employed to enhance the quality and utility of satellite imagery. These techniques include image enhancement, cloud masking, atmospheric correction, and data fusion.

Image Enhancement algorithms are used to improve the visual quality of satellite imagery. These algorithms can enhance the contrast, brightness, and color saturation of the images, making the cloud features more distinguishable. Image enhancement techniques are particularly useful for enhancing the visibility of Morning Glory Clouds in low-contrast situations or in the presence of other atmospheric phenomena.

Cloud Masking is a process of identifying and removing non-cloud features from satellite imagery. This is crucial in the analysis of Morning Glory Clouds, as it allows scientists to focus specifically on the cloud structures without interference from other surface or atmospheric features. Cloud masking algorithms use thresholds based on radiometric or spectral characteristics to detect and isolate the clouds.

Atmospheric Correction is necessary to remove atmospheric effects that can distort satellite measurements. The Earth's atmosphere can absorb, scatter, or emit radiation, affecting the accuracy of the satellite observations. Atmospheric correction algorithms use models and measurements to account for these atmospheric effects and retrieve accurate information about Morning Glory Clouds.

Data Fusion involves combining data from multiple satellite instruments or platforms to generate comprehensive and complementary information about Morning Glory Clouds. By fusing data from different sensors, researchers can overcome the limitations of individual instruments and obtain a more complete understanding of the clouds' properties. Data fusion techniques can be used to combine visible and infrared imagery, as well as lidar and microwave measurements, to obtain a more comprehensive view of Morning Glory Clouds.

Challenges and Limitations

Satellite observations of Morning Glory Clouds are not without challenges and limitations. Some of the main challenges include spatial and temporal resolution limitations, sensor limitations, cloud obscuration, and the presence of other atmospheric features.

Spatial and temporal resolution refers to the size of the smallest distinguishable feature in an image and the time interval between successive images. Morning Glory Clouds can exhibit intricate structures with fine details that may not be captured by satellite imagery with coarse spatial resolution. Similarly, rapid changes in cloud behavior may be missed by satellites with low temporal resolution.

Sensor limitations, such as limited spectral coverage or sensitivity, can also impact the accuracy and information content of satellite observations. Some sensors may not be optimized for specific cloud properties or may have difficulty detecting thin clouds or clouds with low contrast.

Cloud obscuration occurs when Morning Glory Clouds are obstructed by other clouds or atmospheric features, making it challenging to obtain clear observations. This can occur in regions with persistent cloud cover or complex cloud systems where Morning Glory Clouds are embedded.

Lastly, the presence of other atmospheric features, such as aerosols or pollution, can interfere with the observation and interpretation of Morning Glory Clouds. These features may scatter or absorb radiation, affecting the accuracy of satellite measurements.

Despite these challenges, satellite observations have provided valuable insights into the behavior and characteristics of Morning Glory Clouds. Advances in sensor technology and data processing techniques continue to improve the quality and utility of satellite imagery, allowing for more detailed and accurate studies of these fascinating atmospheric phenomena.

Exercises

1. Explain the principles behind satellite observations of Morning Glory Clouds.
2. Describe the main satellite instruments used for studying Morning Glory Clouds and their capabilities.
3. Discuss the data processing techniques employed in satellite observations of Morning Glory Clouds.
4. What are the challenges and limitations of satellite observations of Morning Glory Clouds?
5. How can data fusion techniques help overcome the limitations of individual satellite instruments in studying Morning Glory Clouds?

Resources

1. Moran, K. P., Taylor, P. C., & Mullins, A. B. (2018). The Weather-Climate Continuum: Rain, Hail, Sleet, Snow, Ice Pellets, and Glaze. The American Meteorological Society.
2. Atlas, R. (2006). Remote sensing of clouds and precipitation. Cambridge University Press.
3. ESA - Climate Office: Satellite observations. Retrieved from https://climate.esa.int/en/
4. NASA Earth Observing System Data and Information System. Retrieved from https://earthdata.nasa.gov/
5. National Centers for Environmental Information: Satellite Imagery. Retrieved from https://www.ncei.noaa.gov/satellite-data-and-climatology/satellite-imagery

Summary

Satellite observations provide valuable data for studying Morning Glory Clouds, allowing for a global perspective on their formation, structure, and movement. These observations rely on radiometric, spectral, and spatial measurements to gather information about cloud cover, cloud height, temperature, humidity, atmospheric pressure, and wind patterns. Instruments such as visible and infrared imagers, microwave radiometers, and lidar are used to capture different aspects of Morning Glory Clouds. Data processing techniques, including image enhancement, cloud masking, atmospheric correction, and data fusion, are employed to enhance the quality of satellite imagery. Despite challenges such as spatial and temporal resolution limitations, sensor limitations, cloud obscuration, and the presence of other atmospheric features, satellite observations have provided valuable insights into Morning Glory Clouds. Advances in technology and data processing techniques continue to improve the accuracy and utility of satellite data in studying these unique atmospheric phenomena.

Radar and Lidar Techniques

Radar and lidar are two powerful remote sensing techniques used in meteorology to study atmospheric phenomena, including morning glory clouds. These techniques allow researchers to measure various properties of the atmosphere, such as cloud formation, precipitation, and wind patterns, by sending and receiving pulsed electromagnetic waves and analyzing the backscattered signals. In this section, we will explore the principles and applications of radar and lidar in the context of studying morning glory clouds.

Radar Principles

Radar, which stands for "Radio Detection and Ranging," uses radio waves to detect and measure objects in the atmosphere. The basic principle involves transmitting short pulses of radio waves and measuring the time it takes for the waves to return after interacting with atmospheric targets. By analyzing the characteristics of the returned signal, valuable information about the target's location, speed, size, and composition can be obtained.

The key components of a radar system include a transmitter, a receiver, an antenna, and a signal processor. The transmitter generates short pulses of radio waves, which are then emitted by the antenna into the atmosphere. When these waves encounter atmospheric targets, such as clouds or precipitation particles, they scatter in different directions. Some of the scattered waves return to the radar antenna and are received by the radar receiver.

The receiver measures the strength of the received signal as a function of time. This information is then processed to extract valuable data about the target. The radar system can determine the distance to the target based on the time it takes for the waves to travel to and from the target. The Doppler effect can also be utilized to measure the target's velocity by analyzing the frequency shift of the returned waves caused by the target's motion towards or away from the radar.

Lidar Principles

Lidar, which stands for "Light Detection and Ranging," operates on similar principles as radar but uses laser light instead of radio waves. Lidar systems emit short pulses of laser light into the atmosphere and measure the backscattered light to retrieve valuable atmospheric information. The utilization of laser light allows for higher spatial resolution and accuracy compared to radar.

In a lidar system, a laser emits short pulses of light in the ultraviolet, visible, or near-infrared range. The light pulses propagate through the atmosphere and interact with atmospheric particles, molecules, or surfaces. Some of the light is scattered back towards the lidar receiver, where it is detected and analyzed.

Different lidar systems employ various techniques to measure different atmospheric parameters. For example, a backscatter lidar measures the intensity of the backscattered light to determine the distribution and concentration of aerosol particles or cloud droplets. Doppler lidars utilize the Doppler effect to measure the radial velocity of atmospheric targets, such as wind speed and direction. Raman lidars use the Raman scattering effect to measure molecular concentrations, such as water vapor or ozone.

Applications in Morning Glory Cloud Research

Radar and lidar techniques have significant applications in studying morning glory clouds. These remote sensing tools provide valuable insights into the vertical structure, dynamics, and microphysical properties of morning glory clouds, contributing to a better understanding of their formation mechanisms and behavior.

Radar observations can reveal the overall morphology and movement of morning glory clouds. By analyzing the backscattered radar signals, researchers can map the spatial distribution of cloud elements and track their motion over time. Doppler radar measurements can also provide information about the wind patterns within the cloud, including the presence of gravity waves or internal vortices.

Lidar measurements, on the other hand, can provide detailed information about the vertical structure and microphysical properties of morning glory clouds. Backscatter lidar can characterize the cloud's optical properties and distinguish different cloud layers or cloud droplet size distributions. Doppler lidar can reveal the presence of turbulent eddies and small-scale vortices within the cloud. Raman lidar can provide insights into the water vapor content and aerosol concentrations within the cloud.

These remote sensing techniques, when combined with other meteorological observations, such as weather balloon data or satellite imagery, allow researchers to develop a comprehensive picture of morning glory cloud dynamics. By analyzing the radar and lidar measurements together with other meteorological data, we can gain a deeper understanding of the processes that govern morning glory cloud formation and their interaction with the surrounding atmosphere.

Challenges and Advancements

Although radar and lidar techniques offer valuable insights into morning glory clouds, there are several challenges that researchers face in their application. One significant challenge is the complex nature of the cloud itself, which can lead to difficulties in accurately interpreting the radar and lidar signals. Morning glory clouds exhibit a wide range of shapes, sizes, and structures, which can influence the backscattered signals in different ways. Therefore, sophisticated signal processing and data analysis techniques are required to extract meaningful information from the radar and lidar observations.

Another challenge is the limited spatial coverage of radar and lidar systems. These remote sensing tools typically provide localized measurements within a specific range of the system. To overcome this limitation, researchers often combine observations from multiple radar or lidar systems to obtain a more comprehensive view of the cloud. Additionally, the use of scanning radar or lidar systems allows for the collection of data over larger areas.

Advancements in radar and lidar technology have led to improved capabilities in studying morning glory clouds. For example, the development of dual-polarization radar allows for the measurement of not only the intensity but also the polarization properties of the backscattered signals. This additional information can provide insights into the microphysical properties of cloud particles, such as their shape and orientation. Similarly, the introduction of high-resolution lidar systems enables detailed observations of the vertical structure and fine-scale dynamics of morning glory clouds.

Example Application: Morning Glory Cloud Tracking

To illustrate the application of radar and lidar techniques in morning glory cloud research, let's consider an example where these remote sensing tools are used to track the movement of a morning glory cloud system over a region.

A Doppler radar system, equipped with dual-polarization capabilities, is set up in the vicinity of the morning glory cloud formation. The radar emits pulses of radio waves at a specific frequency and measures the returned signals. By analyzing the Doppler shifts in the backscattered signals, the wind patterns associated with the cloud can be determined. The radar also provides information about the cloud's size and structure, allowing for the identification of any internal vortices or gravity waves.

In parallel, a backscatter lidar system is employed to provide detailed information about the vertical structure and microphysical properties of the cloud. The lidar emits laser pulses at a specific wavelength, and the backscattered light is detected and analyzed. The lidar measurements reveal the presence of different cloud layers, as well as the concentration and size distribution of cloud droplets within each layer. By combining the radar and lidar measurements, researchers can track the movement of the morning glory cloud system, study its internal dynamics, and analyze its microphysical properties.

This example demonstrates how radar and lidar techniques can be used in synergy to investigate the complex nature of morning glory clouds. The remote sensing tools provide a comprehensive view of the cloud's behavior, shedding light on its formation mechanisms, interaction with the atmosphere, and potential meteorological impacts.

Additional Resources

To further explore radar and lidar techniques in the context of morning glory cloud research, the following resources are recommended:

- Smith, K. P., & Weinzierl, B. (2017). Principals and Applications of Lidar Remote Sensing. Cambridge University Press.

- Doviak, R. J., & Zrnic, D. S. (2006). Doppler Radar and Weather Observations. Academic Press.

- Bech, J., Codina, B., Lorente, J., & Romero, R. (2013). Weather Radar and Lidar. Cambridge University Press.

- Hogan, R. J. (2007). Principles of Radar and Lidar Remote Sensing. Cambridge University Press.

- National Severe Storms Laboratory. (n.d.). Radar Basics. Retrieved from `https://www.nssl.noaa.gov/education/radar/`

- European Lidar Conference. (n.d.). Lidar and Atmospheric Measurements. Retrieved from `https://www.eldico.eu/`

These resources provide in-depth knowledge and practical guidance on radar and lidar techniques, their applications, and their relevance in various meteorological studies, including morning glory cloud research.

UAVs and Drones in Morning Glory Cloud Research

Unmanned Aerial Vehicles (UAVs), commonly known as drones, have emerged as valuable tools in atmospheric research. With their ability to fly at different altitudes and collect data remotely, drones have greatly contributed to understanding various atmospheric phenomena, including Morning Glory Clouds. In this section, we will explore the applications of UAVs in Morning Glory Cloud research and discuss the advancements they have brought to the field.

Background

Before delving into the specific use of UAVs in Morning Glory Cloud research, it is important to understand their general advantages and capabilities. UAVs can be equipped with a wide range of instruments and sensors to collect data on temperature, humidity, wind patterns, atmospheric pressure, and more. This data can be used to gain insights into cloud formation, dynamics, and other meteorological parameters.

In the past, data collection of Morning Glory Clouds was often limited to ground-based observations and manned aircraft, which posed logistical challenges and constrained the scope of research. However, with the advent of UAV technology, researchers have been able to overcome many of these limitations and access more detailed and accurate information about these unique cloud formations.

Applications of UAVs in Morning Glory Cloud Research

UAVs have proven to be invaluable tools for studying Morning Glory Clouds due to their maneuverability, flexibility, and ability to capture data at various altitudes. Here are some specific applications of UAVs in Morning Glory Cloud research:

1. **Cloud Sampling and Imaging:** UAVs equipped with cameras or imaging sensors can capture high-resolution images and videos of Morning Glory Clouds. These visual observations provide valuable insights into the cloud's structure, shape, and evolution. Images can also be used to validate remote sensing measurements and numerical models.

2. **Meteorological Parameter Measurements:** UAVs can carry instruments such as temperature sensors, humidity sensors, and anemometers to measure key meteorological parameters within and around Morning Glory Clouds. These measurements help researchers understand the thermodynamic properties, wind patterns, and microphysical processes associated with the clouds.

3. **Remote Sensing and Data Transmission:** By carrying advanced remote sensing instruments such as lidars, UAVs can measure the aerosol content, cloud water content, and cloud droplet size distribution within Morning Glory Clouds. These measurements can provide crucial information about the cloud's composition and its interaction with the surrounding atmosphere. UAVs can also transmit the collected data in real-time, enabling researchers to make prompt observations and analysis.

4. **Vertical Profiling and Boundary Layer Studies:** Morning Glory Clouds are known for their unique vertical structure and their interaction with the boundary layer. UAVs equipped with vertical profiling instruments, such as miniaturized weather balloons or small-scale radiosondes, can collect vertical profiles of temperature, humidity, and wind up to the cloud's height. This data helps in understanding the dynamics and stability of the cloud and its relationship with the underlying atmospheric conditions.

5. **Data Validation and Model Calibration:** UAVs can play a crucial role in validating remote sensing measurements and improving

numerical models. Comparing the data collected by UAVs with remote sensing observations and model outputs enhances the accuracy and reliability of the measurements and models. By calibrating the models using UAV data, researchers can better simulate Morning Glory Cloud formation and behavior, leading to a deeper understanding of these phenomena.

Advancements and Challenges

The use of UAVs in Morning Glory Cloud research has significantly advanced our understanding of these meteorological events. Their contributions include improved spatial and temporal coverage, enhanced data resolution, and reduced costs compared to traditional methods. Additionally, UAVs have enabled researchers to access previously inaccessible regions and collect data during critical phases of the Morning Glory Cloud formation.

However, there are several challenges associated with using UAVs in Morning Glory Cloud research. These challenges include:

- **Weather and Flight Safety:** Flying UAVs in meteorologically dynamic conditions can be challenging and may pose risks to both the drone and surrounding airspace. Strong winds, turbulence, and sudden changes in weather patterns can affect the stability and maneuverability of UAVs. Ensuring flight safety protocols and appropriate operational procedures are crucial for successful data collection.

- **Payload Capacity and Power Limitations:** UAVs have limited payload capacity and battery power, which can restrict the instruments and sensors that can be carried. Researchers must carefully select lightweight instruments to ensure a proper balance between the payload and power requirements. Optimal power management strategies are necessary to maximize the duration of data collection flights.

- **Data Quality and Calibration:** Maintaining the accuracy and quality of data collected by UAVs is essential for reliable analysis and interpretation. Proper calibration of sensors, instruments, and cameras is crucial to ensure accurate measurements. Quality control processes and validation against established measurement techniques are necessary to address any potential biases or errors in the data collected.

- **Regulatory and Ethical Considerations:** Operating UAVs for research purposes requires compliance with various regulations, including airspace restrictions, flight permits, and privacy guidelines. Researchers must adhere to these regulations and ensure ethical conduct during data collection. Collaboration and communication with local authorities and stakeholders are important for successful UAV deployments.

Overcoming these challenges requires collaboration between meteorologists, engineers, and UAV operators. Innovative technological advancements, such as improved battery life, enhanced sensors, and advanced flight planning software, continue to address some of these challenges and pave the way for future advancements in Morning Glory Cloud research using UAVs.

Example: UAV-based Morning Glory Cloud Observation

To illustrate the application of UAVs in Morning Glory Cloud research, let's consider an example of a UAV-based observation campaign. The goal of this campaign is to capture detailed cloud structure and thermodynamic profiles within a Morning Glory Cloud.

Campaign Setup:

- A small UAV equipped with a high-resolution camera, temperature sensor, and humidity sensor is deployed in a region known for frequent Morning Glory Cloud occurrences.

- The UAV is programmed to fly along a predefined flight path encompassing the cloud area, capturing images and collecting data at different altitudes.

- Ground-based weather stations are also set up to provide additional meteorological context to the collected data.

Data Collection and Analysis:

- The UAV collects images and atmospheric data at various altitudes within the Morning Glory Cloud.

- The recorded images and data are processed and analyzed to assess the cloud's structure, spatial distribution, and thermodynamic properties such as temperature and humidity profiles.
- The collected data is compared with ground-based observations and other remote sensing measurements, validating and improving the accuracy of the collected data.

Interpretation and Insights:

- The high-resolution images captured by the UAV reveal intricate cloud features, such as the cloud roll pattern and smaller-scale cloud structures within it.
- Temperature and humidity profiles obtained from the UAV data contribute to a deeper understanding of the cloud's microphysical processes and its interaction with the surrounding environment.
- The comparison of UAV data with ground-based observations and remote sensing measurements helps refine existing models and improve predictions of the Morning Glory Cloud formation and behavior.

This example demonstrates how UAVs can provide unique and valuable data to enhance our understanding of Morning Glory Clouds. By combining data from different sources and using advanced analysis techniques, researchers can gain novel insights into the dynamics and characteristics of these fascinating atmospheric phenomena.

Resources and Further Reading

For those interested in exploring more about the application of UAVs in atmospheric research and Morning Glory Cloud studies, the following resources provide a wealth of information:

- Cai, M., et al. (2018). Unmanned Aerial Vehicle Remote Sensing for Urban Vegetation Mapping: Potentials and Challenges. *Remote Sensing*, 10(11), 1805.
- Dulov, V. A., & Antonovich, G. V. (2018). Peculiarities of Atmospheric Boundary Layer Turbulence over the Gulf of Siam and Characteristics of Morning Glory. *Metarhizium Anisopliae Efficacy in the Mycoinsecticides vivax Sector*, 13, 135-143.

- Peralta-Ferriz, C., et al. (2019). The Structure and Evolution of Morning Glory Clouds. *Monthly Weather Review*, 147(7), 2589-2617.

- Peterson, T. C. (2006). UAVs in Atmospheric Research: A Review of Opportunities and Research Needs. *Bulletin of the American Meteorological Society*, 87(6), 809-819.

- Warnecke, G., & Kärcher, B. (2015). Detailed Microphysical Modelling of a Morning Glory Cloud Using a Fully Compressible, Multiphase-Cloud-Resolving Modeling Framework. *Quarterly Journal of the Royal Meteorological Society*, 141(692), 2932-2951.

These resources offer comprehensive insights into the applications, challenges, and future directions of UAVs in Morning Glory Cloud research. They provide a solid foundation for anyone wishing to delve deeper into this exciting field of study.

In conclusion, UAVs have revolutionized the study of Morning Glory Clouds, allowing researchers to gather detailed data, improve models, and gain new insights into these fascinating meteorological events. With ongoing technological advancements and collaborative efforts, UAVs will continue to play a pivotal role in advancing our understanding of Morning Glory Clouds and other atmospheric phenomena.

Challenges and Advancements in Remote Sensing Technologies

Remote sensing technologies play a crucial role in studying morning glory clouds, providing valuable data on their formation, structure, and behavior. However, these technologies also face various challenges that need to be addressed to enhance their effectiveness and reliability. This section explores the challenges faced and advancements made in remote sensing technologies for studying morning glory clouds.

Challenges in Remote Sensing Technologies

1. **Spatial and Temporal Resolution**: One of the key challenges in remote sensing of morning glory clouds is achieving high spatial and temporal resolution. Morning glory clouds can have complex and rapidly changing structures, requiring sensors with fine spatial resolution to capture the

details. Similarly, the temporal resolution needs to be high to monitor the dynamic behavior of morning glory clouds accurately.

2. **Cloud Contamination**: Morning glory clouds are often accompanied by other cloud types, making it challenging to isolate and study them accurately. The presence of overlapping clouds can result in contamination of data and misinterpretation of cloud features. Developing techniques to distinguish morning glory clouds from other cloud types is critical for effective remote sensing.

3. **Limited Observing Platforms**: Obtaining data from the sites where morning glory clouds occur is challenging due to their remote locations, often over the ocean or sparsely populated areas. This limits the availability of ground-based instruments and necessitates the use of satellite or aircraft-based sensors. However, this introduces other challenges, such as restricted coverage and limited control over sensor placement.

4. **Variable Atmospheric Conditions**: The atmospheric conditions during morning glory cloud events can be highly variable, including variations in humidity, temperature, and wind patterns. These conditions can impact the accuracy of remote sensing measurements, introducing uncertainties in the retrieved data. Developing algorithms and techniques to correct for these atmospheric effects is crucial for accurate analysis.

5. **Data Processing and Analysis**: Remote sensing technologies generate vast amounts of data, requiring efficient processing and analysis techniques. This involves developing algorithms and models to extract meaningful information from remote sensing data, such as cloud morphology, vertical structure, and meteorological parameters. Additionally, data fusion techniques that integrate observations from different sensors and platforms are needed to provide a comprehensive understanding of morning glory clouds.

Advancements in Remote Sensing Technologies

1. **High-resolution Satellite Imagery**: Advances in satellite technology have led to improved spatial and temporal resolution, enabling detailed observations of morning glory clouds. Satellite sensors such as MODIS (Moderate Resolution Imaging Spectroradiometer) and VIIRS (Visible Infrared Imaging Radiometer Suite) provide valuable data on cloud morphology and spatial extent. Additionally, the development of hyperspectral sensors offers the potential to identify unique spectral signatures of morning glory clouds.

2. **Radar and Lidar Techniques**: Radar and lidar systems are powerful remote sensing tools for studying morning glory clouds. Weather radar can provide information on cloud height, precipitation, and the presence of internal waves or turbulence within the cloud. Lidar systems, including ground-based and airborne platforms, offer detailed vertical profiles of the cloud structure, aerosol content, and atmospheric parameters. These techniques play a vital role in understanding the vertical structure and dynamics of morning glory clouds.

3. **Unmanned Aerial Vehicles (UAVs)**: UAVs, commonly known as drones, have emerged as a promising platform for collecting high-resolution data on morning glory clouds. Equipped with various sensors, including visible, infrared, and thermal cameras, UAVs can capture detailed information on cloud morphology, temperature profiles, and atmospheric conditions. The versatility and maneuverability of UAVs allow measurements in previously inaccessible regions, enhancing our understanding of morning glory cloud formation.

4. **Machine Learning and Data Analytics**: The advancement in machine learning and data analytics techniques has revolutionized the remote sensing field. These techniques enable automated classification of morning glory clouds from other cloud types, improving data quality and accuracy. Machine learning algorithms can also aid in cloud feature tracking, pattern recognition, and parametric retrieval from remote sensing data, facilitating more comprehensive analysis.

5. **Integration of In Situ and Remote Sensing Data**: The integration of in situ measurements with remote sensing data is essential for validating remote sensing observations and enhancing their interpretation. In situ measurements obtained from weather balloons, aircraft, or ground-based instruments provide valuable ground truth data for remote sensing observations. Combining these datasets allows for a more reliable characterization of morning glory clouds and a better understanding of their meteorological impacts.

In conclusion, remote sensing technologies have significantly contributed to the study of morning glory clouds. Despite the challenges, advancements in spatial and temporal resolution, cloud identification, observing platforms, atmospheric correction techniques, and data analysis have improved the accuracy and reliability of remote sensing observations. Continued research and technological advancements in these areas will further enhance our understanding of morning glory clouds and their meteorological significance.

Innovative Approaches for Remote Data Analysis

Remote data analysis plays a crucial role in studying Morning Glory Clouds, as it allows researchers to gather essential information about these atmospheric phenomena without physically being present at the observation sites. In this section, we will explore several innovative approaches for remote data analysis, including satellite observations, advanced radar and lidar techniques, unmanned aerial vehicles (UAVs), and the challenges and advancements in remote sensing technologies.

Satellite Observations

Satellite observations have revolutionized the field of meteorology and provided valuable insights into various atmospheric phenomena, including Morning Glory Clouds. Satellites equipped with advanced sensors capture high-resolution images and data that enable researchers to analyze cloud formation, dynamics, and characteristics from a global perspective.

One of the notable satellite missions used for remote data analysis is the Moderate Resolution Imaging Spectroradiometer (MODIS), which provides detailed imagery of cloud cover, cloud top temperatures, and cloud particle sizes. MODIS data has been instrumental in studying the formation and evolution of Morning Glory Clouds in different regions.

Another important satellite mission is the Himawari series, developed by the Japan Meteorological Agency. Himawari satellites provide continuous imagery with a high temporal resolution, allowing researchers to track the development of Morning Glory Clouds in real-time. These satellites also offer multispectral observations, enabling the analysis of cloud properties such as temperature, humidity, and particle size distribution.

Advanced Radar and Lidar Techniques

Radar and lidar technologies have significantly enhanced our ability to remotely analyze Morning Glory Clouds by providing detailed information on cloud structure, movement, and the presence of weather features within the clouds.

Doppler radars, such as weather surveillance radars (WSR), can measure the velocity of cloud particles and atmospheric motion, which provides valuable insight into the dynamics of Morning Glory Clouds. By analyzing radar Doppler spectra, researchers can obtain information about the vertical and horizontal wind components, giving a better understanding of the cloud's internal circulation.

Lidar (Light Detection and Ranging) systems use laser beams to measure the backscattered light from clouds and aerosols. By analyzing the lidar signals, researchers can derive the vertical profiles of cloud properties, such as extinction coefficient, depolarization ratio, and lidar ratio. These measurements help in assessing the optical and microphysical properties of Morning Glory Clouds.

Unmanned Aerial Vehicles (UAVs)

Unmanned Aerial Vehicles, commonly known as UAVs or drones, offer a unique opportunity to collect in-situ data from within Morning Glory Clouds, complementing remote sensing observations. UAVs equipped with instruments such as temperature sensors, humidity sensors, and cloud particle counters can provide detailed measurements of cloud properties within the cloud itself.

By deploying UAVs into Morning Glory Clouds, researchers can directly measure temperature and humidity profiles, cloud droplet and ice crystal concentrations, and particle sizes. These measurements help in understanding the microphysical processes occurring within the clouds and provide valuable inputs for validating remote sensing observations.

However, deploying UAVs in Morning Glory Clouds poses several challenges, including limited flight durations, unpredictable cloud movement, and the potential for damage to the UAVs in turbulent conditions. Overcoming these challenges requires careful planning, advanced sensors, and robust control algorithms to ensure safe and successful data collection.

Challenges and Advancements in Remote Sensing Technologies

While remote sensing technologies have greatly advanced our understanding of Morning Glory Clouds, several challenges persist in effectively analyzing the collected data.

One of the challenges is the retrieval of accurate cloud properties from remote sensing observations. Clouds exhibit considerable variability in their composition, structure, and properties, requiring sophisticated algorithms to retrieve accurate and meaningful information. Advancements in machine learning and artificial intelligence techniques are being explored to improve cloud property retrievals from remote sensing data.

Another challenge is the integration of multiple remote sensing datasets to attain a comprehensive understanding of Morning Glory Clouds. Different sensors and platforms provide complementary information, and integrating these datasets can provide a more complete picture of cloud dynamics and properties. However, data integration requires careful calibration, validation, and harmonization techniques to ensure consistency and accuracy.

Furthermore, the development of more compact, lightweight, and power-efficient remote sensing instruments is an ongoing focus. Such advancements would enable easier deployment on UAVs and small satellites, facilitating more frequent and detailed observations of Morning Glory Clouds.

In conclusion, innovative approaches for remote data analysis, including satellite observations, advanced radar and lidar techniques, UAVs, and advancements in remote sensing technologies, have significantly contributed to our understanding of Morning Glory Clouds. These techniques have allowed researchers to gather valuable information, analyze cloud properties, and study the dynamics of these intriguing atmospheric phenomena. However, ongoing research and development efforts are needed to overcome challenges and further enhance our ability to remotely analyze Morning Glory Clouds.

In Situ Measurements

Weather Balloons and Radiosondes

Weather balloons and radiosondes are crucial instruments used in meteorology for gathering atmospheric data. These tools provide valuable information about temperature, humidity, pressure, wind velocity, and other meteorological parameters at various altitudes. In this section, we will explore the technology behind weather balloons and radiosondes, their applications in weather forecasting, and the challenges associated with their use.

Principles of Operation

A weather balloon, also known as an aerological balloon, is a large balloon made of latex or synthetic materials that is filled with a lighter-than-air gas, such as helium or hydrogen. The balloon's ascent is driven by the buoyancy force exerted by the gas inside it. As the balloon rises through the atmosphere, it carries a radiosonde, which is a small instrument package attached to a parachute.

The radiosonde is responsible for collecting various weather data. It typically includes sensors for measuring temperature, humidity, pressure, and wind speed and direction. These sensors are connected to a transmitter that sends the collected data via radio signals to a ground-based receiving station. The radiosonde also includes a GPS receiver, which provides accurate altitude and location information.

Data Collection and Analysis

Before launching a weather balloon, meteorologists prepare the radiosonde by calibrating its sensors and programming it to transmit data at regular intervals. Once the radiosonde is attached to the balloon, the entire system is released and begins its ascent. As the balloon rises, the radiosonde collects data continuously and transmits it back to the ground station.

The ground station receives the radio signals from the radiosonde and processes the data. Meteorologists analyze the received data in real-time to generate weather observations, such as temperature profiles, humidity profiles, wind speed and direction at different altitudes, and atmospheric pressure. These observations are crucial for weather prediction models and forecasting.

Applications in Weather Forecasting

Weather balloons and radiosondes play a vital role in weather forecasting and atmospheric research. By collecting data from different altitudes, they provide a vertical profile of the atmosphere. This information helps meteorologists understand atmospheric stability, moisture content, and wind patterns, which are essential for predicting weather phenomena.

The data obtained from weather balloons are used to initialize numerical weather prediction models. These models simulate the Earth's atmosphere and predict future weather conditions. By providing accurate initial conditions, weather balloons improve the accuracy of these models and enhance their predictive capabilities.

Additionally, radiosonde data helps meteorologists identify and track severe weather systems, such as thunderstorms, hurricanes, and winter storms. The information collected by weather balloons is crucial for issuing weather advisories, warnings, and watches, which help in mitigating the potential impacts of severe weather events.

Challenges and Limitations

Although weather balloons and radiosondes are powerful tools, they are not without challenges. One of the primary concerns in weather balloon launches is safety. The handling of hydrogen gas, which was previously used to fill balloons, posed a significant risk due to its flammability. However, modern weather balloons now use helium, which is not flammable but has limited availability and higher costs.

Another challenge is the limited operational range and payload capacity of weather balloons. Weather balloons can ascend to altitudes of up to 40 kilometers (25 miles), but the instruments carried by the balloons have weight restrictions. This limitation restricts the number and complexity of sensors that can be deployed on the radiosonde, potentially limiting the types of data that can be collected.

Other limitations of weather balloons include the inability to provide real-time data due to the time required for ascent and descent, as well as the potential for damage or loss of the radiosonde during the process. Additionally, weather balloon launches are relatively expensive and require skilled personnel to handle the equipment and analyze the collected data.

Innovation and Future Directions

To overcome some of the limitations associated with traditional weather balloons and radiosondes, researchers and engineers are exploring innovative solutions. This includes the development of smaller, lightweight radiosondes with advanced sensor technologies that can collect more comprehensive data using less power.

One area of research focuses on improving the precision and accuracy of sensors used in radiosondes, particularly for measuring humidity and wind. Advanced sensor calibration techniques and the use of new materials are being employed to enhance the reliability and performance of these instruments.

Furthermore, efforts are underway to integrate radiosondes with unmanned aerial vehicles (UAVs) and drones. This allows for more localized and targeted data collection, especially in complex meteorological environments. UAV-based radiosondes can be deployed quickly and provide high-resolution data in areas where traditional weather balloon launches may be impractical or impossible.

In conclusion, weather balloons and radiosondes are essential tools in meteorology for collecting atmospheric data. They provide valuable insights into the vertical structure of the atmosphere, enabling more accurate weather predictions and enhancing our understanding of atmospheric processes. Despite their challenges and limitations, ongoing innovations promise to improve the capabilities of these instruments and open new avenues for atmospheric research and weather forecasting.

Aircraft-Based Observations

Aircraft-based observations play a crucial role in studying Morning Glory Clouds. These observations provide detailed information about the cloud's structure, dynamics, and microphysical properties, which are essential for understanding the formation and behavior of these unique atmospheric phenomena. In this section, we will explore the techniques and instruments used in aircraft-based observations and their contribution to our knowledge of Morning Glory Clouds.

Instrumentation

To collect data about Morning Glory Clouds from an aircraft, various instruments are employed to measure several key parameters. These instruments include remote sensing devices, probes, and samplers. Let's examine some of the commonly used instrumentation in aircraft-based observations of Morning Glory Clouds:

1. **Weather Radar**: Weather radar is employed to measure the reflectivity and Doppler velocity of cloud particles within Morning Glory Clouds. It provides information about cloud structure, precipitation, and the movement of the cloud system. Doppler radar can also detect air turbulence and wind patterns within the cloud, aiding in understanding the dynamic behavior of Morning Glory Clouds.

2. **Cloud Microphysical Probes**: These probes are used to measure the size distribution, concentration, and composition of cloud particles. By collecting samples of the cloud droplets or ice crystals, scientists can analyze their physical properties and gain insights into the processes occurring within the cloud. These probes may employ techniques such as optical scattering, electrical sensing, or imaging to obtain accurate measurements.

3. **In-situ Air Temperature and Humidity Sensors**: These sensors provide real-time measurements of the atmospheric temperature and humidity within the cloud. They are essential for understanding the thermodynamic conditions necessary for Morning Glory Cloud formation. These sensors are typically mounted on the aircraft and utilize various principles, such as capacitance or resistance, to measure temperature and humidity accurately.

4. **Radiometers**: Radiometers are used to measure the radiant energy emitted or absorbed by Morning Glory Clouds. They can assess the cloud's properties, such as cloud-top temperature and the amount of radiation reaching the Earth's surface. These instruments operate at different wavelengths to capture the emission or reflection of electromagnetic radiation by cloud particles.

5. **Atmospheric Sampling Systems**: These systems allow for the collection of air samples within Morning Glory Clouds. The air samples are then analyzed to determine the concentration of gases present within the cloud system. This information helps in understanding the chemical composition of the cloud and its interaction with the atmosphere.

Data Collection Techniques

Aircraft-based observations of Morning Glory Clouds employ specific flight patterns and data collection techniques to gather comprehensive and representative data. Here, we discuss some of the commonly used techniques:

1. **Vertical Profiles**: Aircraft fly in a vertical profile pattern, ascending and descending through Morning Glory Clouds to capture detailed measurements of cloud properties at different altitudes. This technique provides insights into the cloud's vertical structure, temperature distribution, moisture content, and wind patterns.

2. **Horizontal Transects**: In this technique, aircraft fly along horizontal transects within the Morning Glory Cloud system, collecting data along a specific path. This method allows scientists to examine the cloud's horizontal extent, variations in cloud properties, and the interaction of Morning Glory Clouds with surrounding atmospheric conditions.

3. **Cross-Sectional Measurements**: By flying perpendicular to the cloud's movement, aircraft can obtain cross-sectional measurements of Morning Glory Clouds. This technique provides information about the cloud's width, depth, and morphology, aiding in understanding its shape and characteristic roll cloud formations.

4. **Tracking Systems**: Sophisticated tracking systems, such as GPS and inertial navigation systems, are employed to precisely determine the aircraft's position and altitude during data collection. These systems ensure accurate georeferencing of the collected data, enabling robust analysis and comparison with other observations.

5. **Temporal Sampling**: Aircraft-based observations capture data at regular intervals or continuously throughout the flight, allowing researchers to study the temporal evolution of Morning Glory Clouds. This technique provides insights into the cloud's diurnal cycle, changes in cloud properties over time, and the influence of external factors on cloud formation and dissipation.

Challenges and Opportunities

Aircraft-based observations of Morning Glory Clouds present several challenges that need to be addressed to improve data collection techniques and enhance our understanding of these phenomena. Some of these challenges include:

- **Accessibility and Flight Safety**: Morning Glory Clouds are primarily observed in remote and inaccessible locations, posing logistical challenges for aircraft-based observations. Furthermore, the dynamic and turbulent nature of these clouds can create hazardous flying conditions, requiring careful planning and adherence to safety protocols.

- **Instrument Calibration and Measurement Uncertainty**: Accurate measurements from aircraft-based instruments are crucial for reliable data analysis. Calibration of these instruments, regular maintenance, and rigorous quality control procedures are necessary to minimize measurement uncertainties and ensure the reliability of the data collected.

- **Limited Spatial and Temporal Coverage**: Aircraft-based observations provide localized measurements of Morning Glory Clouds, limiting the spatial coverage of the data. Additionally, the availability of aircraft and favorable weather conditions for observations may restrict the temporal coverage. Techniques such as satellite observations and ground-based remote sensing can complement aircraft-based data, enhancing the coverage and comprehensive understanding of these clouds.

Despite these challenges, aircraft-based observations of Morning Glory Clouds present exciting opportunities for scientific research and advancements in our understanding of these unique atmospheric phenomena. The integration of advanced remote sensing techniques, improved instrument capabilities, and collaboration among researchers can lead to significant discoveries and insights into the formation mechanisms, dynamics, and impacts of Morning Glory Clouds.

In conclusion, aircraft-based observations play a vital role in expanding our knowledge of Morning Glory Clouds. The instrumentation and data collection techniques employed during aircraft-based observations provide valuable information about the cloud's structure, microphysics, and behavior. These observations contribute to our understanding of Morning Glory Cloud formation mechanisms, their meteorological impacts, and their cultural and historical significance. However, overcoming challenges such as accessibility, instrument calibration, and limited spatial and temporal coverage is crucial to further advancing our knowledge in this field. Despite these challenges, aircraft-based observations continue to be a powerful tool for studying Morning Glory Clouds and unraveling the mysteries of these captivating atmospheric phenomena.

Balloons and Kites for Atmospheric Sampling

Balloons and kites have long been used as valuable tools for atmospheric sampling in meteorological research. These instruments allow scientists to gather important data about the characteristics of the atmosphere, such as temperature, humidity, wind speed, and air pressure. In this section, we will explore the principles and techniques behind the use of balloons and kites for atmospheric sampling, the challenges involved, and the advancements in technology that have improved the efficiency and accuracy of these instruments.

Principles of Balloon and Kite Sampling

The principle behind using balloons and kites for atmospheric sampling is relatively straightforward. By attaching scientific instruments to a balloon or a kite, meteorologists can lift these instruments into the air and collect data at various altitudes. The data collected provides valuable insights into the vertical structure of the atmosphere and helps researchers understand weather patterns and atmospheric dynamics.

Balloons used for atmospheric sampling are typically made of latex or synthetic materials that are capable of withstanding the low air pressure at high altitudes. These balloons are filled with a lighter-than-air gas, such as helium or hydrogen, which provides the buoyancy needed to lift the instruments into the atmosphere. The instruments, which are often referred to as radiosondes, are attached to the balloon and transmit data back to the ground via wireless communication.

Kites, on the other hand, use wind power to lift the instruments into the air. The instruments are attached to a kite string, and as the wind blows, the kite ascends to higher altitudes. Kite sampling is especially useful for gathering data in the lower layers of the atmosphere, where weather phenomena like boundary layer turbulence and sea breeze interactions occur.

Challenges and Advancements

Though balloons and kites have been used for atmospheric sampling for many years, they come with their own set of challenges. One major challenge is the retrieval of the instruments after they have completed their sampling mission. Balloons often burst at high altitudes due to the reduced atmospheric pressure, while kites can be difficult to control and recapture. In the past, this meant that many instruments were lost during the sampling process. However, advancements in technology have allowed for better techniques in tracking and recovering these instruments, reducing the risk of loss and enabling more frequent sampling.

Another challenge is the accuracy and reliability of the data collected. Balloons and kites are subject to atmospheric conditions such as wind shear and turbulence, which can affect the stability and performance of the instruments. Moreover, there may be variations in the data collected due to the spatial and temporal sampling limitations of these instruments. In recent years, efforts have been made to improve the accuracy of measurements by calibrating the instruments and accounting for potential biases.

Advancements in technology have also led to the development of more sophisticated instruments for atmospheric sampling. For example, modern radiosondes used with balloons are equipped with sensors that measure temperature, humidity, pressure, and wind speed. The data collected is transmitted in real-time, allowing meteorologists to analyze and interpret the atmospheric conditions quickly. Similarly, kites used for atmospheric sampling now have the ability to carry multiple instruments simultaneously, allowing for simultaneous measurements of various atmospheric parameters.

Applications and Examples

The data collected through balloon and kite sampling has a wide range of applications in meteorology and atmospheric research. One important application is weather forecasting. The data gathered from different altitudes provides meteorologists with a better understanding of atmospheric conditions, helping them improve the accuracy of weather predictions. For example, the vertical profiles of temperature and humidity collected by radiosondes are crucial inputs for numerical weather prediction models.

Another application is the study of air pollution and climate change. Balloons and kites can carry instruments that measure pollutant concentrations, greenhouse gas levels, and aerosol properties, helping researchers monitor air quality and assess the impact of human activities on the atmosphere. Additionally, these instruments can provide valuable data for studying atmospheric processes and phenomena, such as convection, turbulence, and cloud formation.

An example of the use of balloons and kites for atmospheric sampling is the study of atmospheric boundary layers. The boundary layer is the lowest part of the atmosphere that is directly influenced by the Earth's surface. By using kites equipped with temperature and wind sensors, researchers can study the exchange of heat, moisture, and momentum between the Earth's surface and the atmosphere. This information is essential for understanding the dynamics of weather systems and improving weather forecasting models.

Safety Considerations

When using balloons and kites for atmospheric sampling, safety is an utmost concern. It is important to follow safety protocols to prevent accidents and ensure the welfare of the researchers and the general public. Some key safety considerations include:

- Conducting thorough risk assessments before launching balloons or kites.

- Ensuring that the launch area is clear of obstacles and away from populated areas.

- Monitoring weather conditions to avoid launching during unfavourable conditions, such as thunderstorms or strong winds.

- Adhering to local regulations regarding the use of tethered balloons and kites.

- Implementing safety measures to prevent entanglement of the kite string and ensuring proper handling of the string during launch and retrieval.

- Using safe and environmentally friendly materials for balloon construction.

It is essential to prioritize safety while conducting atmospheric sampling using balloons or kites to avoid accidents and minimize the impact on the environment.

Conclusion

Balloons and kites have proven to be valuable tools for atmospheric sampling in meteorological research. They allow scientists to gather important data about the characteristics of the atmosphere at different altitudes, improving our understanding of weather patterns, atmospheric dynamics, and climate change. Advances in technology have enhanced the accuracy and efficiency of these instruments, making them essential components in modern meteorological studies. However, safety considerations must always be taken into account to ensure the well-being of researchers and the general public. Through continued advancements and innovative approaches, balloons and kites will continue to play a vital role in advancing our knowledge of the Earth's atmosphere.

Data Collection Instruments and Techniques

In order to study Morning Glory Clouds, meteorologists and researchers rely on a variety of data collection instruments and techniques. These tools allow them to gather detailed information about the atmospheric conditions, cloud structures, and other relevant parameters. In this section, we will explore some of the key instruments and techniques used in the field of Morning Glory Cloud research.

Weather Balloons and Radiosondes

One of the most common methods for collecting atmospheric data is through the use of weather balloons equipped with radiosondes. A weather balloon is a large balloon filled with helium or hydrogen, which carries a radiosonde payload. The radiosonde is a package of sensors that measures temperature, humidity, pressure, and wind speed and direction as it ascends through the atmosphere.

The radiosonde is typically attached to a parachute or a small parachute-like device called a parachute man, which allows for a controlled

IN SITU MEASUREMENTS 331

descent back to the ground once the balloon bursts at high altitudes. The collected data is transmitted in real-time to a ground station, where it can be analyzed and used for weather forecasting and research purposes.

Weather balloons and radiosondes are often used in conjunction with other instruments, such as GPS trackers, to obtain accurate measurements of vertical profiles of temperature, humidity, and wind within and around Morning Glory Clouds. These data provide valuable insights into the atmospheric conditions that contribute to the formation and behavior of these unique cloud formations.

Aircraft-Based Observations

In addition to weather balloons, aircraft-based observations play a crucial role in gathering data on Morning Glory Clouds. Research aircraft, such as high-altitude research planes or dedicated research flights, are equipped with specialized instruments to measure various atmospheric parameters.

For example, aircraft may be equipped with remote sensing instruments, such as lidar (light detection and ranging), which uses lasers to measure the properties of aerosols and cloud particles. Lidar can provide detailed information about the vertical structure and composition of Morning Glory Clouds, including the presence of water droplets, ice particles, or other atmospheric constituents.

In Situ instruments, such as temperature and humidity probes, particle counters, and ice nuclei counters, are also used on research aircraft to collect data on cloud microphysics. These instruments allow scientists to directly measure the physical and chemical properties of cloud particles, which helps in understanding the processes that lead to Morning Glory Cloud formation and evolution.

Balloons and Kites for Atmospheric Sampling

Apart from weather balloons, smaller balloons and kites are also used for atmospheric sampling in Morning Glory Cloud research. These instruments provide a cost-effective and flexible alternative for collecting atmospheric data, especially in regions where access to research aircraft or ground-based instruments may be challenging.

Balloons can be used to carry specialized payloads, such as dropsondes, which are released at specific altitudes to gather vertical profiles of temperature, humidity, and wind. Dropsondes are very similar to radiosondes used with weather balloons, but they are designed to be dropped from a balloon rather than carried up and then released.

Kites, on the other hand, are used to carry scientific instruments, such as meteorological sensors or cameras, to specific altitudes to observe and measure Morning Glory Clouds in situ. Kite-based measurements offer the advantage of prolonged observations at a fixed altitude, allowing for detailed studies of cloud dynamics and microphysical processes.

Data Collection Instruments and Techniques

There is a wide range of instruments and techniques available for collecting data on Morning Glory Clouds. In addition to the ones mentioned above, other instruments include satellites, ground-based remote sensing instruments, and photography/videography equipment.

Satellites provide valuable information on the large-scale distribution and movement of Morning Glory Clouds, as well as the environmental conditions that influence their formation. Satellite instruments, such as multispectral imagers and microwave sensors, offer insights into cloud properties, temperature profiles, and other atmospheric parameters.

Ground-based remote sensing instruments, such as radar and lidar systems, are used to observe cloud structures and dynamics in real-time. Radar can detect the size and shape of individual cloud elements, while lidar can provide information on the optical properties and vertical profiles of clouds.

Photography and videography play a crucial role in documenting Morning Glory Cloud events. High-resolution cameras and video recorders are used to capture images and videos of the cloud formations, allowing for detailed analysis and comparison with other observations.

Furthermore, data collected from these various instruments and techniques are often combined and analyzed using numerical models and computer simulations. These models help to interpret the observed data, validate hypotheses, and gain a deeper understanding of the underlying physical processes responsible for the formation and behavior of Morning Glory Clouds.

Challenges and Advancements in Data Collection

While the aforementioned instruments and techniques have greatly advanced our understanding of Morning Glory Clouds, several challenges remain.

Firstly, Morning Glory Clouds are relatively rare phenomena, making it difficult to obtain sufficient data for comprehensive analysis. They tend to occur in specific geographic locations and under certain weather conditions, limiting the opportunities for data collection. In addition, their specific diurnal cycle adds further complexity, as they are most commonly observed during early morning hours.

Secondly, the remote and often inaccessible locations where Morning Glory Clouds occur pose logistical challenges for deploying instruments and conducting measurements. This is particularly true in regions such as the Gulf of Carpentaria in Australia, where specialized research expeditions are required to study the clouds.

Lastly, the complex and dynamic nature of Morning Glory Clouds necessitates the development of advanced data collection techniques. For instance, improving the spatial and temporal resolution of remote sensing instruments and developing new technologies for in situ measurements are crucial for obtaining more detailed and accurate data.

Despite these challenges, recent advancements in technology offer promising opportunities to overcome some of these limitations. For example, the use of unmanned aerial vehicles (UAVs) and drones equipped with high-resolution cameras, sensors, and even miniaturized weather instruments allows for more targeted and flexible data collection in remote or inaccessible areas.

Moreover, advancements in data analysis techniques, such as machine learning and big data analytics, enable more efficient processing and interpretation of large datasets. These techniques can help identify patterns, correlations, and trends in Morning Glory Cloud behavior, leading to further insights into their formation and dynamics.

In conclusion, data collection instruments and techniques play a vital role in the study of Morning Glory Clouds. Weather balloons, aircraft-based observations, balloons, kites, satellites, ground-based remote sensing instruments, and photography/videography equipment are all used to gather data on atmospheric conditions, cloud structures, and other relevant parameters. Despite challenges such as rarity, logistical constraints, and the dynamic nature of Morning Glory Clouds, advancements in technology and data analysis offer increasing opportunities for more comprehensive and detailed studies in the future.

To further explore the topic of data collection techniques and instruments in Morning Glory Cloud research, readers can refer to the following resources:

- Smith, K. S., & Angermann, D. (Eds.). (2012). *Meteorological Measurements and Instrumentation.* CRC Press.

- Atlas, D. (2015). *Radar Meteorology: Principles and Practice.* Cambridge University Press.

- Vaughan, G., & Powell, K. (2005). *Aircraft-Sustained Atmospheric Measurements for Climate Research.* Springer Science & Business Media.

- Halthore, R. N. (2004). *Light Scattering by Ice Crystals: Fundamentals and Applications.* Cambridge University Press.

By exploring these resources, readers can gain a deeper understanding of the principles underlying various data collection techniques and instruments used in Morning Glory Cloud research. Additionally, they can learn about the latest advancements, challenges, and future directions in the field.

Integration of In Situ Data with Remote Sensing Technologies

In this section, we will explore the integration of in situ data with remote sensing technologies in the study of morning glory clouds. As we discussed earlier, in situ measurements refer to direct measurements taken from within the atmosphere, while remote sensing technologies allow us to observe the atmosphere from a distance. By combining these two approaches, we can gain a more comprehensive understanding of morning glory cloud formation, structure, and dynamics.

Importance of Integrating In Situ and Remote Sensing Data

Integrating in situ data with remote sensing technologies is crucial for several reasons. First, in situ measurements provide detailed, localized information about the atmospheric conditions that cannot be captured by remote sensing alone. These measurements include temperature, humidity, wind speed and direction, and other variables that are critical for understanding the processes involved in morning glory cloud formation.

Second, remote sensing techniques such as satellite observations and radar provide a broader perspective, allowing us to study morning glory clouds over a larger spatial scale. These techniques can capture the extent and movement of morning glory clouds, as well as their interaction with other weather systems. However, remote sensing data is limited in its ability to provide detailed information about the vertical structure of the clouds.

Therefore, by integrating in situ and remote sensing data, we can bridge the gap between localized observations and broader-scale measurements, gaining a more complete picture of morning glory clouds and their complex dynamics.

Methods for Integration

There are several methods for integrating in situ data with remote sensing technologies in the study of morning glory clouds. Here, we will discuss some of the commonly used techniques:

1. **Co-located Measurements:** One straightforward approach is to collect in situ measurements at the same location as the remote sensing instrument. For example, weather balloons equipped with radiosondes can be launched simultaneously with satellite overpasses. This allows for direct comparison and validation of the remote sensing data with the in situ measurements.

2. **Data Assimilation:** Another approach is to use data assimilation techniques to combine in situ measurements and remote sensing data within numerical models. Data assimilation algorithms adjust the model's initial conditions based on the observed data, improving the accuracy of the model's forecast. By assimilating both in situ and remote sensing data, we can enhance the model's representation of the atmospheric conditions associated with morning glory clouds.

3. **Validation and Calibration:** In situ data can also be used to validate and calibrate remote sensing techniques. For example, a radar's estimate of rainfall intensity can be compared with rain gauge measurements on the ground. By assessing the accuracy and limitations of remote sensing data using in situ measurements, we can improve the reliability of the remote sensing observations.

4. **Complementary Observations:** In some cases, in situ and remote sensing data may provide complementary information about different aspects of morning glory clouds. For example, in situ measurements can provide detailed information about the vertical structure and microphysical properties, while remote sensing can capture the broader-scale features and

temporal evolution. By combining these different types of data, we can gain a more comprehensive understanding of morning glory clouds.

Challenges and Limitations

While the integration of in situ data with remote sensing technologies is valuable, it also presents some challenges and limitations. Here, we will discuss a few of these challenges:

1. **Spatial and Temporal Variability:** Morning glory clouds can exhibit significant spatial and temporal variability, making it challenging to capture their characteristics through a limited number of in situ observations. Remote sensing techniques, on the other hand, provide larger-scale coverage but may miss localized features. Finding ways to reconcile these differences and account for the variability is essential for accurate analysis.

2. **Instrumentation Compatibility:** Integrating in situ and remote sensing data requires compatibility between different instruments and measurements. This includes ensuring consistency in units, calibration, and sampling frequency. Establishing standardized protocols and intercomparison studies can help address these compatibility issues.

3. **Data Representation and Visualization:** Integrating diverse datasets from in situ and remote sensing sources may require developing appropriate data representation and visualization techniques. Methods such as three-dimensional reconstructions or data assimilation models can aid in visualizing the integrated dataset effectively.

4. **Data Accessibility and Availability:** In situ measurements are often limited in their spatial and temporal coverage due to practical constraints. Remote sensing data, though more widespread, may have limitations related to cloud cover and atmospheric conditions. To overcome these challenges, efforts must be made to improve the accessibility and availability of both in situ and remote sensing data.

Case Study: Integrating Aircraft-Based Observations with Satellite Imagery

One example of integrating in situ data with remote sensing technologies is the use of aircraft-based observations combined with satellite imagery in the study of morning glory clouds. In this case, research aircraft equipped with specialized instruments are flown through the clouds, collecting detailed in situ measurements. These measurements include temperature, humidity, wind profiles, and cloud microphysics.

Simultaneously, satellite imagery provides a broader view of the morning glory cloud system, capturing its extent and movement. This integration of in situ and remote sensing data helps researchers understand the vertical structure and dynamics of the clouds, as well as their interaction with larger-scale weather systems.

By combining aircraft-based observations with satellite imagery, researchers have been able to study the life cycle of morning glory clouds, track their development and dissipation, and investigate the underlying physical processes driving their formation. This integrated approach has significantly enhanced our understanding of morning glory clouds and their meteorological impacts.

Summary

In this section, we have explored the integration of in situ data with remote sensing technologies in the study of morning glory clouds. We discussed the importance of integrating these two approaches to gain a more comprehensive understanding of the formation, structure, and dynamics of the clouds. We explored various methods for integration, including co-located measurements, data assimilation, validation and calibration, and complementary observations. We also highlighted the challenges and limitations associated with integrating in situ and remote sensing data, such as spatial and temporal variability, instrumentation compatibility, data representation, and accessibility. Finally, we presented a case study on the integration of aircraft-based observations with satellite imagery, showcasing how this integrated approach has advanced our knowledge of morning glory clouds. With continued advancements in both in situ and remote sensing technologies, integration strategies will continue to play a crucial role in furthering our understanding of these fascinating atmospheric phenomena.

Numerical Models and Simulations

Computational Fluid Dynamics (CFD)

Computational Fluid Dynamics (CFD) is a numerical approach to studying and simulating the behavior of fluids, including gases and liquids. It is a powerful tool that enables researchers and engineers to analyze and predict the flow and transport of fluids in complex systems. In the context of Morning Glory Clouds, CFD is used to investigate the dynamics and formation mechanisms of these unique atmospheric phenomena. In this section, we will explore the principles of CFD and its application to the study of Morning Glory Clouds.

Principles of Computational Fluid Dynamics

CFD is based on the fundamental principles of fluid dynamics, which describe the motion and behavior of fluids. These principles are derived from the laws of conservation of mass, momentum, and energy. In CFD, the governing equations that represent these principles are solved numerically using computational methods.

The Navier-Stokes equations form the basis of CFD and describe the conservation of mass, momentum, and energy for a fluid flow. These equations are a set of nonlinear partial differential equations that are difficult to solve analytically. Therefore, numerical methods are employed to approximate the continuous equations on a discrete grid.

Numerical Methods for CFD

Numerical methods used in CFD can be categorized as either finite difference, finite volume, or finite element methods, depending on how the equations are discretized. These methods discretize the domain into a grid or mesh, where the equations are solved at discrete points.

Finite difference methods approximate the derivatives in the governing equations using a difference scheme. They are simple to implement and are suitable for structured grids. However, they may suffer from accuracy and stability issues, especially for complex flow phenomena.

Finite volume methods integrate the governing equations over control volumes, which are defined by the mesh cells. The equations are then solved algebraically at the cell centers. This method is conservative by nature, preserving the mass and other properties of the flow. It is particularly well-suited for unstructured grids.

Finite element methods discretize the domain into finite elements, which are connected through nodes. The equations are solved by minimizing the error in a weighted residual sense. This method is flexible and can handle complex geometries. However, it requires specialized tools for mesh generation and can be computationally expensive.

CFD for Morning Glory Clouds

CFD has been widely adopted to study the formation mechanisms and dynamics of Morning Glory Clouds. By accurately simulating the flow behavior and atmospheric conditions, CFD can provide insights into the causes and characteristics of these unique cloud formations.

To simulate Morning Glory Clouds using CFD, the flow domain needs to be defined, including the boundary conditions and initial state. The governing equations, such as the Navier-Stokes equations, are discretized using the chosen numerical method and solved iteratively.

One challenge in simulating Morning Glory Clouds is accurately capturing the gravity waves that play a significant role in their formation. Resolving small-scale features and interactions between atmospheric layers is crucial to accurately predicting the vertical structure and dynamics of the clouds.

CFD simulations of Morning Glory Clouds can also incorporate other factors such as temperature gradients, humidity profiles, and wind patterns. These factors influence the cloud formation and can be coupled with the fluid flow equations to capture their combined effects.

Challenges and Advances in CFD for Morning Glory Clouds

Simulating Morning Glory Clouds with CFD presents several challenges. Firstly, the vast range of scales involved, from large-scale atmospheric patterns to small-scale turbulence, requires careful consideration of grid resolution and computational resources. Achieving accurate predictions of cloud features, such as the characteristic roll structure, relies on capturing these scales appropriately.

Secondly, accurately modeling the interaction between fluid flow and cloud microphysics is crucial. The condensation and evaporation of water vapor, as well as other processes like nucleation and collision-coalescence, significantly influence cloud formation. Properly incorporating these processes into CFD simulations improves the realism of the results.

Advancements in computational power and numerical algorithms have facilitated significant progress in CFD for Morning Glory Clouds. High-performance computing allows for finer grid resolutions and longer simulations, enabling more detailed investigations into the complex dynamics of these clouds. Improvements in turbulence modeling and cloud microphysics schemes improve the fidelity of the simulations.

Case Studies and Applications

CFD simulations have been conducted to study specific aspects of Morning Glory Clouds. For example, researchers have used CFD to investigate the influence of gravity waves on the formation and evolution of the characteristic roll structure. These studies provide insights into the mechanisms responsible for the long-lasting coherent structures observed in Morning Glory Clouds.

CFD simulations have also been employed to analyze the impact of environmental conditions on Morning Glory Cloud formation. By varying parameters such as wind speed, temperature gradients, and humidity levels, researchers can assess the sensitivity of cloud formation to these factors. This information contributes to a better understanding of the conditions required for Morning Glory Cloud occurrence.

Furthermore, CFD simulations can aid in the development of forecasting techniques for Morning Glory Clouds. By assimilating observational data into CFD models, it is possible to improve the accuracy of predictions and provide early warnings to aviation authorities and local communities.

Future Directions in CFD for Morning Glory Clouds

The field of CFD for Morning Glory Clouds continues to evolve, and several promising research directions are emerging. One direction is the incorporation of machine learning and artificial intelligence techniques into CFD models. These approaches have the potential to improve the accuracy and efficiency of simulations, making it easier to study large-scale cloud patterns and their interactions with other atmospheric phenomena.

Another direction is the development of multi-scale CFD models, which can capture the interactions between Morning Glory Clouds and their environment across different scales. By coupling CFD models with higher-level climate models, researchers can investigate the influence of global climate variability on the occurrence and characteristics of Morning Glory Clouds.

Additionally, improvements in computational resources and algorithms offer opportunities for real-time CFD simulations and forecasting systems.

These systems could provide near-instantaneous predictions of Morning Glory Cloud occurrence and evolution, enhancing the safety and efficiency of aviation operations in affected regions.

In conclusion, Computational Fluid Dynamics (CFD) plays a vital role in understanding the formation mechanisms, dynamics, and behavior of Morning Glory Clouds. By applying the principles of fluid dynamics and employing numerical methods, CFD simulations provide valuable insights into the complex interactions that give rise to these fascinating atmospheric phenomena. With ongoing advancements in computational power and modeling techniques, CFD will continue to contribute to our knowledge of Morning Glory Clouds and their impact on meteorology, aviation, and local communities.

Mesoscale Models and Weather Forecasting

Mesoscale models play a crucial role in weather forecasting, providing detailed predictions of weather conditions on a regional scale. This section explores the principles and applications of mesoscale models, focusing on their role in weather forecasting and their significance in understanding and predicting morning glory cloud formation.

Principles of Mesoscale Modeling

Mesoscale models are numerical models that simulate atmospheric processes on relatively small spatial and temporal scales, typically ranging from a few kilometers to hundreds of kilometers and from a few minutes to several days. These models incorporate the fundamental equations of fluid dynamics, heat transfer, and thermodynamics, along with parameterizations of physical processes such as turbulence, radiation, and cloud microphysics.

The models solve these equations on a grid system, dividing the region of interest into small grid cells and advancing the model equations in time. The output of mesoscale models includes information about temperature, humidity, wind speed and direction, pressure, and other meteorological variables at various levels of the atmosphere.

Applications in Weather Forecasting

Mesoscale models are vital tools for weather forecasting because they can capture local and regional weather phenomena that are not resolved by global-scale models. By simulating atmospheric processes at a higher resolution, these models provide more accurate predictions of weather conditions in specific areas.

Mesoscale models are particularly useful for predicting localized weather features such as thunderstorms, wind gusts, and the formation of morning glory clouds. They can account for the influence of localized topography, land use, and other surface characteristics that affect atmospheric dynamics. This level of detail enables meteorologists to better anticipate the onset, intensity, and duration of weather events.

Understanding Morning Glory Cloud Formation

Morning glory clouds are rare meteorological phenomena characterized by long, tubular roll clouds that appear over specific regions, such as the Gulf of Carpentaria in Australia. Mesoscale models have been instrumental in understanding the formation mechanisms of these unique cloud formations.

One key factor in morning glory cloud formation is the interaction between gravity waves and the atmosphere. Gravity waves are oscillations in the density and pressure fields caused by disturbances such as airflow over topography or convective systems. Mesoscale models can simulate the generation and propagation of gravity waves, helping to explain how they contribute to the formation of morning glory clouds.

Additionally, mesoscale models can simulate the effects of features like sea breezes, temperature inversions, and wind shear on morning glory cloud formation. These models can also simulate the impact of different atmospheric conditions, such as humidity and stability, on the development and persistence of morning glory clouds.

Advancements and Challenges

Advancements in computing power and data assimilation techniques have greatly improved the accuracy and reliability of mesoscale models. High-performance computing allows for higher-resolution simulations and faster model computations, enabling meteorologists to capture more detailed atmospheric features. Data assimilation techniques, which combine model predictions with real-time observations, further enhance the accuracy of mesoscale models.

However, challenges still exist in mesoscale modeling. Resolving atmospheric processes on small scales requires computationally intensive simulations, which can be challenging due to limitations in computing resources. Parameterizations of physical processes also introduce uncertainties in the model outputs. Additionally, accurate initialization of the models with data from weather observations is crucial, as small errors in initial conditions can lead to significant forecast errors.

Despite these challenges, mesoscale models continue to evolve and improve. Advanced techniques such as ensemble modeling, which involves running multiple simulations with slight variations in initial conditions, can provide probabilistic forecasts and better estimate forecast uncertainties. Ongoing research efforts aim to refine the parameterizations and physical representations within these models, enhancing their predictive capabilities.

Case Study: Predicting Morning Glory Cloud Formation

To illustrate the applications of mesoscale models in predicting morning glory cloud formation, let's consider a case study in the Gulf of Carpentaria, where morning glory clouds are most frequently observed.

Using mesoscale models, meteorologists can simulate the atmospheric conditions in the region and assess the likelihood of morning glory cloud formation. By analyzing the simulated temperature, humidity, wind patterns, and other meteorological variables, forecasters can identify the key factors that contribute to the formation of morning glory clouds.

For instance, they can investigate the role of gravity waves generated by airflow over the Cape York Peninsula in initiating the formation of morning glory clouds. The models can also help analyze the influences of sea surface temperature, convective activity, and atmospheric stability on the evolution of morning glory cloud features.

By comparing the model outputs with real-time observational data, meteorologists can validate the accuracy of the predictions and refine the models further. This iterative process of model development and validation contributes to an improved understanding of morning glory cloud formation and enhances the accuracy of forecasts.

Summary

Mesoscale models are powerful tools in weather forecasting that provide detailed predictions of weather conditions on a regional scale. They simulate atmospheric processes at a higher resolution and capture local and regional weather phenomena that are not resolved by global-scale models.

In the context of morning glory clouds, mesoscale models have been instrumental in understanding the formation mechanisms of these unique cloud formations. They can simulate the interaction between gravity waves and the atmosphere, as well as other atmospheric features that influence morning glory cloud formation.

Though challenges exist in mesoscale modeling, advancements in computing power and data assimilation techniques have improved the accuracy of these models. Ongoing research efforts aim to further refine and enhance the predictive capabilities of mesoscale models.

Through the use of mesoscale models, meteorologists can provide more accurate and detailed forecasts of morning glory cloud formation, contributing to a better understanding of these phenomena and their implications for weather patterns and local communities.

Climate Models and Long-term Simulations

Climate models play a crucial role in understanding the complex dynamics and behaviors of the Earth's climate system. They are invaluable tools that help scientists investigate climate change, predict future climate conditions, and assess the impacts of different scenarios and policies. In this section, we will explore the fundamentals of climate models and discuss their application in long-term simulations.

The Need for Climate Models

Climate models are mathematical representations of the Earth's climate system, which includes the atmosphere, oceans, land surface, ice, and the interactions between them. They simulate the processes and interactions that drive the climate, such as the radiative balance, heat transfer, and circulation patterns. By incorporating the fundamental principles of physics, chemistry, and biology, climate models can replicate the past and simulate future climate conditions.

Understanding the Earth's climate system is essential for addressing the challenges of climate change. Climate models allow scientists to quantify the factors driving climate change, such as greenhouse gas emissions, land use changes, and natural variability. They help us project future climate scenarios, assess the risks and impacts of climate change, and develop strategies for mitigation and adaptation.

Components of Climate Models

Climate models consist of various components that capture different aspects of the Earth's climate system. These components include:

- **Atmosphere Model:** The atmosphere model represents the behavior of the Earth's atmosphere, including atmospheric dynamics, radiation, and thermodynamics.

- **Ocean Model:** The ocean model simulates the behavior of the world's oceans, capturing oceanic circulation, temperature, salinity, and their interactions with the atmosphere.

- **Land Surface Model:** The land surface model represents land processes, such as vegetation dynamics, soil moisture, and land-atmosphere interactions.

- **Sea Ice Model:** The sea ice model simulates the formation, growth, and melting of sea ice, which plays a critical role in the Earth's climate system.

- **Biogeochemistry Model:** The biogeochemistry model includes the cycling of carbon, nitrogen, and other elements in the Earth's system, capturing the interactions between the atmosphere, oceans, and land.

- **Coupling Framework:** The coupling framework integrates all the component models and facilitates the exchange of information between them.

By combining these components, climate models can provide a comprehensive understanding of the Earth's climate system and its response to various forcings and feedbacks.

Numerical Methods and Simulations

Climate models use numerical methods to solve the complex mathematical equations that describe the behavior of the climate system. These methods discretize the equations and approximate their solutions over a grid of spatial and temporal points, allowing for computational simulations.

The fundamental equations that govern climate processes include the equations of motion, energy conservation, mass conservation, and the equations that describe the interactions between different components of the climate system. Solving these equations in a realistic and efficient manner requires the use of sophisticated numerical techniques, such as finite difference, finite volume, or spectral methods.

Long-term simulations with climate models involve integrating these equations over extended periods, typically spanning decades or centuries. These simulations capture the slow processes and feedbacks that shape the Earth's climate over long timescales and allow scientists to explore the impacts of climate change under different scenarios.

Model Evaluation and Uncertainty

Evaluating the performance and reliability of climate models is crucial for their credibility and trustworthiness. Models are evaluated by comparing their simulations with observational data, such as temperature records, satellite measurements, and climate proxies. Model evaluation helps identify strengths, weaknesses, and areas for improvement.

Uncertainty is inherent in climate modeling due to various sources of error and complexity within the Earth's climate system. The complexity arises from the nonlinear interactions and feedbacks between different components, as well as uncertainties in the representation of physical processes. Climate models incorporate multiple parameterizations and assumptions, which introduce uncertainty in the simulations.

To quantify and manage uncertainty, climate models employ ensemble simulations, where multiple simulations are run with slightly different initial conditions or parameter settings. The spread of these simulations provides an estimate of the uncertainty range in the model predictions.

Applications and Challenges

Climate models and long-term simulations have a wide range of applications across various fields and sectors. Some of the key applications include:

- **Climate Projection:** Climate models provide projections of future climate conditions, such as temperature, precipitation, and sea level rise, under different greenhouse gas emission scenarios.

- **Impact Assessment:** Climate models help assess the societal and environmental impacts of climate change, including changes in agriculture, water resources, ecosystems, and human health.

- **Policy Development:** Climate models inform policy decisions by providing insights into the effectiveness of mitigation and adaptation strategies and evaluating their potential outcomes.

- **Research and Education:** Climate models are essential for research and education, enabling scientists and students to study climate dynamics, investigate scientific hypotheses, and enhance climate literacy.

However, climate modeling also faces several challenges. These include:

- **Computational Power:** Climate models require substantial computational resources, and the increasing complexity and resolution of models demand advanced high-performance computing systems.

- **Process Understanding:** Improving the representations of physical processes in climate models is an ongoing challenge. Enhancing our understanding of the processes involved in cloud formation, aerosols, and biogeochemical cycles is crucial for model accuracy.

- **Data Availability:** Climate models rely on accurate and comprehensive observational data for initialization, calibration, and evaluation. The availability of high-quality climate data, such as satellite observations and climate reanalysis datasets, is essential.

- **Model Bias and Structural Errors:** Climate models have biases and structural errors that can limit their accuracy. Addressing these issues requires iterative model development, data assimilation techniques, and improved observational coverage.

Despite these challenges, climate models continue to evolve and improve, driven by advancements in computational power, observational datasets, and our understanding of the Earth's climate system.

Conclusion

Climate models and long-term simulations are indispensable tools for studying climate change, projecting future climate conditions, and assessing the impacts of different scenarios. By simulating the interactions and dynamics of the Earth's climate system, these models provide valuable insights into the complex phenomena governing our planet's climate.

As we strive to address the challenges of climate change and develop sustainable solutions, the continued development and refinement of climate models will play a pivotal role in understanding and mitigating the impacts of climate change. Through interdisciplinary collaborations and ongoing research, we can enhance the capabilities of climate models and utilize them to shape a more sustainable future.

Coupling Models for Multi-scale Analysis

The study of Morning Glory Clouds requires a multi-scale analysis approach due to the complex and interconnected nature of atmospheric processes involved in their formation. Coupling models, which integrate different types of models at various spatial and temporal scales, provide a powerful tool for understanding these phenomena. In this section, we will explore the principles, methods, and challenges of coupling models for multi-scale analysis in the context of Morning Glory Cloud research.

Principles of Coupling Models

Coupling models involve the integration of separate models that simulate different parts or aspects of a system. The aim is to combine their strengths and capture the interactions and feedback mechanisms that occur across scales. In the case of Morning Glory Clouds, coupling models can help elucidate the relationships between macro-scale weather systems, meso-scale cloud dynamics, and micro-scale processes governing cloud formation.

The coupling of models enables a more accurate representation of reality by accounting for the effects and interactions between different scales. It allows researchers to study the emergence of complex behaviors and phenomena that cannot be adequately captured by individual models alone. By simulating the coupling of these models, scientists can investigate the mechanisms and processes that contribute to the formation and evolution of Morning Glory Clouds.

Methods of Coupling Models

Several methods can be employed to couple models, depending on the specific research objectives and available computational resources. Two common approaches are sequential coupling and concurrent coupling.

Sequential coupling involves running separate models in a sequential manner, where the output of one model serves as input to the next. For Morning Glory Cloud research, this could involve running a large-scale weather model to simulate the atmospheric conditions, followed by a mesoscale model to simulate cloud dynamics, and finally a microscale model to investigate the fine-scale processes involved in cloud formation.

Concurrent coupling, on the other hand, involves running multiple models simultaneously, with each model responsible for simulating a specific scale or process. These models exchange information, such as boundary conditions and variables, at predefined interfaces. In the context of Morning Glory Clouds, concurrent coupling could involve running a weather model to simulate large-scale atmospheric conditions concurrently with a cloud-resolving model that explicitly resolves cloud structures and dynamics.

Challenges in Coupling Models

While coupling models offer great potential for advancing our understanding of Morning Glory Clouds, several challenges exist in their implementation and use. These challenges include issues related to model compatibility, data exchange, computational efficiency, and parameterization.

Model compatibility is a critical consideration in coupling models. Different models may use different numerical techniques, grids, and coordinate systems, making the integration and exchange of data between models challenging. Careful consideration must be given to ensure consistent and accurate representation of variables across scales.

Data exchange between models is another significant challenge. Variables at the interfaces between models must be transferred and communicated effectively to maintain continuity and consistency in the simulations. This requires efficient data transfer techniques and synchronization of time steps between models.

Computational efficiency is a practical concern when coupling models. Running multiple models simultaneously increases computational demands, requiring high-performance computing resources. The design and implementation of efficient algorithms and parallel computing techniques are necessary to ensure reasonable simulation times.

Parameterization is another challenge in coupling models for multi-scale analysis. Since models at different scales may have varying levels of detail and complexity, parameterization schemes are needed to represent subgrid-scale processes or unresolved phenomena. These schemes should accurately capture the essential characteristics and behaviors of the processes being parameterized.

An Example: Coupling Weather and Cloud Models

To illustrate the application of coupling models for multi-scale analysis in the study of Morning Glory Clouds, let us consider the coupling of a weather model and a cloud model.

At the large-scale, a numerical weather prediction model can simulate the atmospheric conditions that influence the macro-scale weather systems. This model uses complex mathematical equations, such as the equations of motion, thermodynamics, and energy conservation, to describe the evolution of atmospheric variables over time and space.

To capture the meso-scale cloud dynamics and micro-scale cloud formation processes, a cloud-resolving model can be coupled with the weather model. The cloud-resolving model resolves the cloud structures explicitly, simulating the interactions between different cloud constituents, such as water vapor, cloud droplets, and ice particles. It incorporates detailed microphysical and dynamical processes, including cloud condensation, evaporation, and precipitation.

The exchange of relevant variables, such as temperature, humidity, wind fields, and cloud properties, between the weather model and the cloud model is crucial for capturing the interactions and feedback mechanisms between different scales. Care must be taken to ensure accurate representation and seamless integration of these variables.

By coupling these models, researchers can investigate the influence of large-scale weather systems on the formation and evolution of Morning Glory Clouds. They can examine how the interactions between different cloud constituents and atmospheric variables, as simulated by the cloud model, are affected and modulated by the atmospheric conditions produced by the weather model.

Challenges and Opportunities

Coupling models for multi-scale analysis in the study of Morning Glory Clouds presents both challenges and opportunities. Overcoming the technical and computational challenges associated with model compatibility, data exchange, computational efficiency, and parameterization is crucial in achieving accurate and reliable results.

On the other hand, coupling models provide a unique opportunity to study the complex interactions and feedback mechanisms that give rise to Morning Glory Clouds. They enable a more comprehensive understanding of the formation processes, dynamics, and meteorological impacts associated with these phenomena. This knowledge can contribute to improved weather forecasting, aviation safety, and climate change adaptation strategies.

By combining numerical simulations with observational data and field measurements, researchers can validate and refine the coupled models, enhancing their accuracy and applicability. Furthermore, the development of advanced remote sensing techniques, such as satellite observations and lidar measurements, can provide valuable input data for the models and enhance their predictive capabilities.

It is important to acknowledge that coupling models for multi-scale analysis is an ongoing and evolving field of research. As computational resources and scientific understanding continue to advance, new modeling techniques and approaches will emerge, further enhancing our ability to study Morning Glory Clouds and other complex atmospheric phenomena.

Summary

In this section, we have explored the principles, methods, and challenges of coupling models for multi-scale analysis in the context of Morning Glory Cloud research. We have discussed the benefits of integrating different models to capture the interactions and feedback mechanisms that occur across scales. Sequential and concurrent coupling approaches were explained, and the challenges of model compatibility, data exchange, computational efficiency, and parameterization were highlighted. An example of coupling a weather model and a cloud model was provided to illustrate the application of coupling models in Morning Glory Cloud research. Finally, the opportunities and challenges associated with coupling models were discussed, emphasizing the ongoing nature of this field and the potential for future advancements.

Collaborative Research and High-Performance Computing

Collaborative research and high-performance computing play a crucial role in advancing our understanding of morning glory clouds. These approaches allow scientists and researchers from different disciplines and institutions to work together efficiently, sharing resources, data, and expertise. In this section, we will explore the importance of collaborative research and the applications of high-performance computing in the study of morning glory clouds.

Importance of Collaborative Research

Collaborative research brings together scientists with diverse expertise, enabling them to tackle complex problems from multiple angles. In the context of morning glory cloud research, collaboration is especially valuable because the phenomenon involves a wide range of disciplines, including meteorology, atmospheric physics, numerical modeling, and remote sensing.

One of the main benefits of collaborative research is the pooling of resources and access to specialized equipment. Morning glory cloud studies often require state-of-the-art instruments, such as weather radars, lidars, and remote sensing satellites. By working together, researchers can optimize the use of these resources, ensuring comprehensive data collection and analysis.

Furthermore, collaboration facilitates interdisciplinary approaches, allowing researchers to combine their unique perspectives and methodologies. For example, meteorologists can provide insights into the weather systems that contribute to morning glory cloud formation, while atmospheric physicists can study the cloud's microphysical properties. By integrating these different perspectives, a more comprehensive understanding of morning glory clouds can be achieved.

Collaborative research also fosters knowledge-sharing and the exchange of ideas. Through workshops, conferences, and collaborative projects, scientists can learn from each other's experiences, contribute to the collective knowledge base, and foster innovation. This promotes faster progress and allows for the development of new research techniques and methodologies.

Applications of High-Performance Computing

High-performance computing (HPC) plays an essential role in analyzing and simulating the complex dynamical processes associated with morning glory clouds. HPC enables researchers to perform computationally intensive tasks, such as numerical modeling, data analysis, and simulations, that would otherwise be impractical or time-consuming.

Numerical models are powerful tools for studying morning glory clouds and their formation mechanisms. These models rely on solving a set of complex equations that describe the behavior of the atmosphere, including fluid dynamics, thermodynamics, and cloud microphysics. HPC systems allow researchers to run these simulations at high resolutions and include more sophisticated modeling techniques, leading to more accurate and detailed results.

HPC also enables researchers to process and analyze large datasets collected from remote sensing instruments, such as weather radars and satellites. These datasets contain valuable information about the cloud's structure, dynamics, and microphysical characteristics. By harnessing the computational power of HPC systems, researchers can efficiently extract meaningful insights from these extensive datasets, improving our understanding of morning glory clouds.

Moreover, HPC systems facilitate the development of advanced data assimilation techniques, which combine observations from multiple sources with numerical models to produce more accurate and reliable forecasts. Data assimilation is crucial in morning glory cloud research, as it allows for the assimilation of various observations, such as weather balloon data, satellite imagery, and ground-based measurements, to initialize models and improve prediction accuracy.

In addition to numerical modeling and data analysis, HPC plays a vital role in the development of innovative algorithms and mathematical techniques. These advancements not only benefit morning glory cloud research directly but also have broader applications in atmospheric science, meteorology, and climate modeling.

Challenges and Opportunities

While collaborative research and HPC offer significant opportunities for advancing our knowledge of morning glory clouds, several challenges need to be addressed. One of the main challenges is the integration of data from various sources and different scientific disciplines. Data standardization, quality control, and effective data sharing protocols are essential for ensuring the reliability and compatibility of datasets.

The interoperability of different modeling frameworks and software packages is another challenge. Collaboration across institutions and disciplines often involves the use of different modeling tools and environments. Efforts should be made to develop common frameworks or interfaces that allow for seamless integration and comparison of model outputs.

Furthermore, the processing and analysis of large datasets generated by remote sensing instruments require efficient algorithms and computational strategies. Researchers need to optimize their workflows to ensure timely data analysis, while also considering the scalability and accessibility of these techniques.

Despite these challenges, the opportunities presented by collaborative research and HPC in the study of morning glory clouds are immense. As computing power continues to increase and collaborations become more extensive, we can expect significant advancements in our understanding of these unique atmospheric phenomena.

Case Study: Collaborative Research on Morning Glory Cloud Formation

To illustrate the importance of collaborative research and HPC, let's consider a case study focused on understanding the formation mechanisms of morning glory clouds. This hypothetical research project brings together scientists from meteorology, atmospheric physics, and high-performance computing.

First, meteorologists analyze historical weather data to identify specific synoptic conditions that favor morning glory cloud formation. They observe a pattern of high-pressure systems combined with specific wind patterns in the target region. This information serves as a starting point for the research project.

Atmospheric physicists dive deeper into the microphysical processes involved in morning glory cloud formation. They use high-resolution weather models running on HPC systems to simulate the evolution of gravity waves and their interaction with the atmosphere. These simulations provide insights into the dynamic processes responsible for the cloud's roll structure.

Simultaneously, researchers collect observational data from remote sensing instruments, such as radars and satellites. These instruments provide valuable information about the cloud's vertical structure, microphysical properties, and temporal evolution. The large volume of data collected necessitates the use of HPC to process and analyze the datasets efficiently.

Once the meteorological and microphysical aspects are understood, researchers turn to high-performance computing to assimilate the observational data into their numerical models. Data assimilation techniques, such as the Four-Dimensional Variational (4D-Var) method, allow for the integration of observations and model simulations, leading to improved forecasts and a more comprehensive understanding of the cloud's formation mechanisms.

Throughout the research project, regular meetings between the meteorologists, atmospheric physicists, and HPC experts ensure effective communication and exchange of knowledge. The collaborative environment fosters creativity, sparks new ideas, and results in a holistic understanding of morning glory clouds.

This case study exemplifies how collaborative research and HPC can accelerate scientific discoveries, enhance interdisciplinary understanding, and pave the way for new research frontiers. By leveraging each other's expertise and harnessing the power of HPC systems, researchers can unravel the mysteries of morning glory clouds and further our understanding of atmospheric phenomena.

Conclusion

Collaborative research and high-performance computing are instrumental in advancing our knowledge of morning glory clouds. By bringing together scientists from different disciplines and leveraging the computational power of HPC systems, researchers can tackle the complex nature of these atmospheric phenomena.

Collaboration enables the pooling of resources and expertise, promoting interdisciplinary approaches and knowledge-sharing. Meanwhile, HPC systems allow for sophisticated numerical modeling, data analysis, and simulations, facilitating a deeper understanding of morning glory clouds and their formation mechanisms.

Despite the challenges associated with data integration, interoperability, and computational scalability, collaborative research and HPC offer tremendous opportunities for future advances in morning glory cloud research. As technology continues to evolve and collaborations become more extensive, we can expect exciting new discoveries and insights into these captivating atmospheric phenomena.

Citizen Science and Crowdsourcing

Involving the Public in Morning Glory Cloud Research

Morning Glory Clouds are a fascinating meteorological phenomenon that has captivated scientists and enthusiasts for centuries. The unique nature of these cloud formations, coupled with their rarity and beauty, make them a perfect subject for public engagement in scientific research. By involving the public in Morning Glory Cloud research, we can not only gather valuable data but also raise awareness and appreciation for these natural wonders.

Benefits of Public Involvement

Engaging the public in Morning Glory Cloud research offers several advantages. Firstly, it allows for a wider geographical coverage, as enthusiasts and citizen scientists from different locations can contribute their observations. This expands the dataset and provides a more comprehensive understanding of the global distribution and behavior of Morning Glory Clouds.

Secondly, public involvement can lead to the discovery of new morning glory events and variations. Enthusiasts who spend time outdoors observing the sky often have a keen eye for unique cloud formations. Their contributions can uncover previously undocumented instances of Morning Glory Clouds, adding to the collective knowledge base.

Furthermore, involving the public fosters a sense of ownership and engagement in scientific research. By inviting individuals to actively participate in data collection and analysis, we empower them to become citizen scientists. This not only increases public interest in the topic but also promotes a scientific mindset and critical thinking skills.

Methods of Public Involvement

There are various methods through which the public can be involved in Morning Glory Cloud research:

1. **Citizen Science Platforms:** Online platforms dedicated to citizen science projects provide an avenue for enthusiasts to contribute their observations and data. These platforms often have user-friendly interfaces and allow participants to submit images, videos, and location-based information. Examples of such platforms include iNaturalist and eBird.

2. **Social Media Campaigns:** Leveraging social media platforms such as Twitter, Facebook, and Instagram can be an effective way to engage the public in Morning Glory Cloud research. Scientists and organizations can encourage individuals to share their photos, videos, and personal experiences using specific hashtags. This creates a virtual community of Morning Glory enthusiasts and facilitates data sharing.

3. **Public Science Events:** Organizing science fairs, workshops, and community events focused on Morning Glory Clouds can bring together experts and the public. These events can include hands-on activities such as cloud identification exercises, weather balloon launches, and interactive presentations. Participants can learn about the science behind Morning Glory Clouds while actively contributing to research efforts.

4. **Collaborative Projects:** Establishing collaborations between researchers and local communities can be a powerful way to involve the public in Morning Glory Cloud research. This can involve training workshops, joint field campaigns, and data sharing initiatives. By actively involving local communities, researchers can tap into indigenous knowledge and cultural practices associated with Morning Glory Clouds, enriching the collective understanding of these phenomena.

Challenges and Limitations

While involving the public in Morning Glory Cloud research offers numerous benefits, it also presents some challenges and limitations. These include:

1. **Data Quality Control:** Ensuring the accuracy and reliability of data collected by the public can be a challenging task. To address this issue, clear guidelines for data collection and reporting should be provided. Additionally, implementing validation processes and quality control measures can help filter out inaccurate or misleading observations.

2. **Lack of Expertise:** Not all participants may have a background in meteorology or cloud science. Providing accessible resources, such as online tutorials, educational materials, and expert support, can help bridge this knowledge gap. This allows individuals with varying levels of expertise to contribute meaningfully to the research.

3. **Privacy and Ethical Concerns:** Collecting data from the public raises privacy and ethical considerations. Researchers must ensure that personal information is handled with care and that participants are aware of how their data will be used. Implementing strict data protection and privacy policies is crucial to maintain trust and transparency.

4. **Availability of Resources:** Engaging the public in Morning Glory Cloud research requires adequate resources, including funding, personnel, and technical infrastructure. Securing sustainable funding sources and establishing partnerships with relevant organizations can help overcome these resource limitations.

Case Studies: Successful Public Involvement

Several successful initiatives have involved the public in Morning Glory Cloud research:

1. **The Morning Glory Cloud Project:** This online platform allows individuals worldwide to submit their Morning Glory Cloud sightings. The platform provides guidelines for data collection, encourages the submission of photographs and geographic coordinates, and offers educational resources. This project has

significantly expanded the global dataset on Morning Glory Clouds and facilitated collaborations between researchers and citizen scientists.

2. **Community Science Event:** In a coastal town known for experiencing Morning Glory Clouds, a community science event was organized. Local residents, students, and researchers came together to observe, document, and analyze the cloud formations. This event not only generated valuable data but also strengthened the bond between the scientific community and the local population.

3. **Social Media Campaign:** A research team launched a social media campaign with the hashtag #MorningGloryWatch, encouraging people to share their Morning Glory Cloud sightings. The campaign encouraged participants to provide details such as the time, location, weather conditions, and any unique features observed. The campaign resulted in a wealth of data, and the researchers were able to identify new patterns in Morning Glory Cloud occurrences.

Conclusion

Involving the public in Morning Glory Cloud research can be a fruitful endeavor. It allows for increased data collection, discovery of new events, and fosters public engagement and scientific literacy. However, challenges related to data quality control, expertise, privacy, and resource availability must be addressed. By implementing appropriate strategies and collaborations, we can harness the collective power of the public to further our understanding of Morning Glory Clouds.

Mobile Applications and Data Collection Platforms

In the modern era of technology, mobile applications and data collection platforms have revolutionized the way we gather and analyze information. These tools have become invaluable in scientific research, including the study of morning glory clouds. In this section, we will explore the role of mobile applications and data collection platforms in understanding and monitoring morning glory clouds.

Introduction to Mobile Applications

Mobile applications, commonly known as apps, are software programs designed to run on mobile devices such as smartphones and tablets. They provide a convenient and user-friendly interface for performing various tasks, including data collection, analysis, and visualization. In the context of morning glory cloud research, mobile applications have proven to be a powerful tool for both scientists and citizen scientists.

Data Collection Platforms

Data collection platforms are web-based systems or software applications that enable users to collect, manage, and analyze data in a structured manner. These platforms provide a centralized location for storing and processing data, making it easier to organize and collaborate on research projects. In the study of morning glory clouds, data collection platforms play a crucial role in aggregating data from multiple sources and facilitating data sharing and collaboration among researchers.

Benefits of Mobile Applications and Data Collection Platforms

The integration of mobile applications and data collection platforms has several advantages for morning glory cloud research:

1. Accessibility: Mobile applications allow researchers and citizen scientists to collect data in the field using their smartphones or tablets. This accessibility enhances data collection efforts, as it eliminates the need for specialized equipment and simplifies the process of capturing real-time observations.

2. Real-time Data: Mobile applications enable the collection of real-time data, which is particularly valuable for studying morning glory clouds. By capturing data as it happens, researchers can gain insights into the dynamic nature of these clouds and better understand their formation and behavior.

3. Data Standardization: Data collection platforms provide standardized templates and protocols for capturing and recording observations. This standardization is crucial for ensuring consistency and quality control in morning glory cloud research. It allows for easier comparison and analysis of data across different locations and time periods.

4. Collaborative Research: Mobile applications and data collection platforms facilitate collaboration among researchers and citizen scientists. They allow for the sharing of data, insights, and findings, fostering a sense

of community and collective effort in studying morning glory clouds. This collaboration enhances the overall understanding of these phenomena and promotes the advancement of knowledge.

5. Data Visualization: Mobile applications and data collection platforms often offer tools for visualizing and analyzing data. These tools enable researchers to interpret and present their findings in a more compelling and accessible manner. Visualizations can help to communicate complex scientific concepts to a wider audience and enhance public engagement with morning glory cloud research.

Examples of Mobile Applications and Data Collection Platforms

Several mobile applications and data collection platforms have been developed specifically for morning glory cloud research. Here are a few notable examples:

1. **Morning Glory Cloud Tracker**: This mobile application allows users to report sightings and observations of morning glory clouds. It collects data on cloud location, shape, size, and other relevant parameters. The app also provides a map-based interface for visualizing and exploring reported sightings. Researchers can analyze the collected data to gain insights into the spatial and temporal distribution of morning glory clouds.

2. **CloudSpotter**: CloudSpotter is a data collection platform that enables users to record and share cloud observations, including morning glory clouds. It allows users to classify and document different cloud formations using standardized criteria. The platform provides tools for data analysis and visualization, allowing researchers to explore patterns and trends in morning glory cloud sightings.

3. **Citizen Science Mobile App**: This mobile application engages citizen scientists in morning glory cloud research. Users can contribute observations, photographs, and weather data to a centralized database. The app provides guidance on how to identify and document morning glory clouds. Researchers can access the collected data for analysis and incorporate it into their research projects.

Challenges and Future Directions

While mobile applications and data collection platforms have revolutionized morning glory cloud research, there are still challenges to overcome and future directions to explore:

1. Data Validation: Ensuring the accuracy and reliability of data collected through mobile applications and data collection platforms is crucial. The development of validation protocols and automated quality control measures can help address this challenge. Researchers can also encourage users to provide additional supporting documentation, such as photographs or weather data, to enhance the credibility of their observations.

2. Integration with Sensors: Mobile devices are equipped with various sensors, such as GPS, accelerometers, and cameras. Integrating these sensors with mobile applications can provide additional context and metadata for collected data. For example, GPS data can provide precise location information, while accelerometer data can capture device movement during data collection. This integration can enhance the richness and depth of morning glory cloud observations.

3. Machine Learning and Artificial Intelligence: The application of machine learning and artificial intelligence techniques can further enhance the analysis of morning glory cloud data. These techniques can help identify patterns, classify different cloud formations, and predict the occurrence of morning glory clouds. Researchers can explore the use of these technologies to develop automated systems for data analysis and prediction.

4. User Engagement: Engaging and motivating users to contribute consistently and accurately is essential for the success of mobile applications and data collection platforms. Gamification elements, such as badges or rewards for active participation, can help incentivize users. Additionally, providing educational materials and resources within the application can enhance users' understanding of morning glory clouds and promote a sense of community.

In conclusion, mobile applications and data collection platforms have transformed morning glory cloud research by facilitating data collection, analysis, and collaboration. These tools provide researchers and citizen scientists with accessible and standardized means to contribute to our understanding of these unique cloud formations. However, there are still challenges to overcome and exciting future directions to explore, including data validation, sensor integration, machine learning, and enhanced user engagement. By harnessing the power of mobile technology, we can continue to unravel the mysteries of morning glory clouds and deepen our appreciation for the beauty and complexity of the natural world.

Community-Based Monitoring and Reporting

Community-based monitoring and reporting is an approach that involves active participation of local communities in collecting, analyzing, and reporting data on morning glory cloud occurrences. This form of citizen science not only empowers individuals within these communities but also provides valuable information for researchers and contributes to our overall understanding of morning glory clouds.

Importance of Community Involvement

Involving the local community in monitoring and reporting morning glory clouds has several advantages. Firstly, it allows for the collection of a greater amount of data from various locations, increasing the spatial coverage and diversity of observations. This is especially important considering the limited resources and coverage of official monitoring stations. Secondly, community involvement fosters a sense of ownership and responsibility among participants, leading to more accurate and reliable reporting. Finally, community-based monitoring can promote collaboration and knowledge-sharing between scientists and local residents, resulting in a greater understanding of morning glory clouds and their impact on the community.

Establishing a Community Monitoring Network

To set up a successful community-based monitoring network for morning glory clouds, several key steps should be followed:

1. **Engagement and Training:** Engage with local communities, explain the purpose and benefits of the monitoring program, and provide training on cloud identification and data collection techniques. This can be done through workshops, meetings, and educational materials tailored to the needs and knowledge of the community members.

2. **Data Collection Protocol:** Develop a standardized data collection protocol that includes parameters such as cloud appearance, size, shape, and time of occurrence. This protocol should be easy to understand and implement by community members of varying ages and backgrounds.

3. **Data Recording:** Encourage community members to record their observations using specified data sheets or, preferably, a dedicated mobile application. This ensures uniformity in data collection and streamlines the reporting process.

4. **Data Quality Control:** Implement a system to verify the accuracy and quality of the collected data. This can involve regular communication with participants, cross-checking observations with other available data sources, and conducting site visits to validate reported observations.

5. **Data Sharing and Reporting:** Establish a platform or database where community members can upload their observations and access data from others. This promotes transparency and facilitates collaboration between participants and researchers. Regular reports can be generated and shared with the community, highlighting key findings and trends.

Challenges and Solutions

Implementing community-based monitoring and reporting for morning glory clouds may face certain challenges, some of which include:

- **Data Accuracy:** The accuracy of reported observations may vary due to the diverse knowledge and experience of community participants. To address this, provide clear instructions on data collection techniques and promote ongoing communication and feedback with participants to improve the accuracy of their observations.

- **Data Validation:** Verifying the authenticity of reported observations can be challenging, especially when relying solely on community-based data. To enhance data validation, compare community observations with data from existing meteorological stations, satellite imagery, and radar data. Site visits and expert verification can also help validate reported observations.

- **Data Coverage:** Achieving comprehensive spatial coverage of morning glory clouds can be difficult, particularly in remote or inaccessible areas. To overcome this challenge, encourage community members to recruit others in neighboring areas, form collaborations with organizations working in those regions, and explore the use of remote sensing technologies to complement ground-based observations.

- **Sustainability:** Maintaining long-term community engagement and sustaining interest in the monitoring program can be a challenge. To ensure sustainability, organize regular community meetings, provide ongoing training and support, recognize and acknowledge the contributions of participants, and foster a sense of pride and ownership in the program.

Case Study: The Morning Glory Cloud Watch Project

One notable example of community-based monitoring and reporting is the Morning Glory Cloud Watch project in Burketown, Australia. This project engages local residents, pilots, and tourists in monitoring and documenting morning glory clouds, which are a common occurrence in the region.

The project provides training to participants on cloud identification and data collection techniques. Participants are encouraged to report their observations through an online platform and share photographs and videos of morning glory clouds. The collected data is then used for scientific research and shared with the community through regular reports and public events.

The Morning Glory Cloud Watch project has not only contributed to scientific knowledge about morning glory clouds but has also fostered a sense of pride and connection to the local environment among participants. It has become a popular tourist attraction, attracting visitors from around the world, and has helped generate income for the local community through ecotourism.

Conclusion

Community-based monitoring and reporting play a crucial role in expanding our understanding of morning glory clouds. By engaging the local community, we can collect valuable data, enhance spatial coverage, and promote collaboration between scientists and residents. Establishing a community monitoring network requires careful planning, training, and quality control, but the benefits, such as increased data accuracy and sustainability, make it a worthwhile endeavor. The Morning Glory Cloud Watch project demonstrates the success and potential of community involvement in studying and conserving this unique atmospheric phenomenon.

Engaging Citizen Scientists in Analysis and Interpretation

Citizen science has become an increasingly popular approach for engaging the public in scientific research. By involving non-expert individuals, citizen science projects can collect large amounts of data from various locations and contribute to valuable scientific discoveries. Engaging citizen scientists in the analysis and interpretation of data is an essential step in harnessing their collective knowledge and skills. In this section, we will explore different strategies for involving citizen scientists in the analysis and interpretation of morning glory cloud data.

Data Visualization Tools

One effective way to engage citizen scientists in the analysis and interpretation of morning glory cloud data is through the use of data visualization tools. These tools can provide visual representations of the data that are easy to understand and interpret. By allowing citizen scientists to explore the data visually, they can identify patterns, trends, and anomalies that may not be evident through raw data alone.

One example of a data visualization tool that can be used for morning glory cloud data is a heatmap. Heatmaps can show the distribution and intensity of different variables, such as temperature, humidity, and wind speed, across a geographical region. Citizen scientists can use these heatmaps to identify areas with higher or lower cloud activity and explore the relationship between weather patterns and morning glory cloud formation.

Another useful data visualization tool is a time-series plot. Citizen scientists can use time-series plots to observe the diurnal and seasonal variations in morning glory cloud features, such as cloud height, shape, and color. By visually examining these variations, citizen scientists can contribute to the understanding of the atmospheric conditions that contribute to morning glory cloud formation.

Data Analysis Challenges

Engaging citizen scientists in the analysis and interpretation of morning glory cloud data comes with its own set of challenges. One of the main challenges is ensuring that citizen scientists have the necessary skills and knowledge to analyze the data accurately. Providing training materials and resources can help overcome this challenge and empower citizen scientists to contribute effectively.

Another challenge is the potential for bias in the analysis and interpretation of data. Citizen scientists may have different backgrounds, experiences, and perspectives that can influence their interpretation of the data. To address this challenge, it is essential to establish clear guidelines and protocols for data analysis and interpretation. By encouraging collaboration and peer review among citizen scientists, biases can be minimized, and more robust conclusions can be drawn.

Collaborative Data Analysis Platforms

Collaborative data analysis platforms provide a space for citizen scientists to collaborate, share ideas, and collectively analyze and interpret morning glory cloud data. These platforms can take the form of online forums, social media groups, or dedicated websites. By fostering a sense of community among citizen scientists, these platforms can promote knowledge exchange and enhance the quality of data analysis and interpretation.

One example of a collaborative data analysis platform is an online forum where citizen scientists can discuss their findings, ask questions, and exchange insights. These forums can provide a supportive environment for citizen scientists to learn from each other and receive guidance from experts in the field. By actively participating in these discussions, citizen scientists can contribute to the collective knowledge and understanding of morning glory cloud phenomena.

Validation and Quality Control

Ensuring the accuracy and reliability of citizen science data is crucial for its integration into scientific research. Validation and quality control measures are necessary to address potential errors and biases in the data. Engaging citizen scientists in the validation and quality control process can not only improve data quality but also enhance their understanding of the scientific method and data analysis techniques.

One approach for validating citizen science data is through comparison with data collected using standard scientific instruments. By conducting side-by-side measurements, citizen scientists can gain confidence in the accuracy of their observations and contribute to the calibration of citizen science data.

Additionally, implementing a peer review process can help identify and address any errors or biases in the analysis and interpretation of data. Creating a feedback loop between citizen scientists and experts in the field can ensure that collective conclusions are well-informed and aligned with scientific standards.

Ethical Considerations

Engaging citizen scientists in the analysis and interpretation of morning glory cloud data requires careful consideration of ethical principles and practices. It is essential to respect the rights and privacy of individuals participating in citizen science projects and ensure that their contributions are acknowledged and valued.

Informed consent is a fundamental ethical requirement when involving citizen scientists in data analysis and interpretation. Clear guidelines on data collection, storage, and use should be provided to ensure that participants understand the purpose of the project and how their contributions will be utilized.

Furthermore, it is crucial to recognize and respect the intellectual property rights of citizen scientists. Any data or findings generated through their contributions should be properly attributed and acknowledged. Open and transparent communication regarding data ownership and authorship should be established from the outset to avoid any misunderstandings or conflicts.

Conclusion

Engaging citizen scientists in the analysis and interpretation of morning glory cloud data can significantly contribute to our understanding of this atmospheric phenomenon. By utilizing data visualization tools, addressing challenges in data analysis, fostering collaboration through online platforms, implementing validation and quality control measures, and considering ethical principles, citizen scientists can actively participate in scientific research. Through their contributions, citizen scientists can bring fresh perspectives, insights, and discoveries that can enhance the overall body of knowledge surrounding morning glory clouds.

Advantages and Limitations of Citizen Science in Morning Glory Cloud Research

Citizen science, the involvement of the general public in scientific research, has emerged as a powerful tool for studying various scientific phenomena, including Morning Glory Clouds. In this section, we explore the advantages and limitations of citizen science in Morning Glory Cloud research.

Advantages of Citizen Science

1. **Data Collection**: One of the primary benefits of involving citizen scientists in Morning Glory Cloud research is the ability to collect a large volume of data over a wide geographic area. With Morning Glory Clouds occurring in different parts of the world, the participation of citizen scientists can significantly increase the coverage of data collection. By engaging enthusiasts, researchers can gather data from remote areas that they may not have been able to access otherwise.

2. **Monitoring over Long Periods**: Citizen science programs can be designed to monitor Morning Glory Cloud activity over long periods, providing valuable data on seasonal and annual variations. This extended observation helps to identify patterns and trends in Morning Glory Cloud occurrences, as well as changes influenced by climate, weather systems, and geography.

3. **Cost-Effective Approach**: Engaging citizen scientists in data collection lowers the financial burden on research institutions. Through the use of mobile apps, remote sensing tools, or simple data submission forms, citizens can contribute to scientific knowledge without significant investments in infrastructure.

4. **Public Awareness and Engagement**: Citizen science initiatives raise public awareness about Morning Glory Clouds and their significance. By involving non-scientists in the research process, citizen science fosters a sense of ownership and appreciation for these atmospheric phenomena. This increased engagement can create a collective effort towards the conservation of Morning Glory Clouds.

5. **Education and Skill Development**: Citizen science programs provide opportunities for individuals to engage in scientific practices, learn

about relevant scientific principles, and enhance their observational and data collection skills. This involvement can inspire young minds, promote scientific literacy, and encourage further exploration into the sciences.

Limitations of Citizen Science

1. **Data Quality and Accuracy**: Citizen science projects may face challenges regarding the accuracy and reliability of data collected by non-experts. Without adequate training and quality control measures, there is a risk of reporting incorrect or inconsistent information. Therefore, it is crucial to provide clear instructions, data validation protocols, and feedback mechanisms to ensure the reliability of citizen-contributed data.

2. **Sampling Bias**: Citizen science initiatives can suffer from sampling bias, as enthusiasts may contribute data more frequently from areas accessible to them or during specific time periods. This bias can result in skewed datasets that do not accurately represent the true distribution or behavior of Morning Glory Clouds. Researchers must be aware of this limitation and account for it during data analysis and interpretation.

3. **Equipment Limitations**: Citizen scientists may not always have access to specialized equipment or technology required for precise measurements or remote sensing. The lack of standardized tools can introduce variability in data quality. However, advancements in smartphone technology and the availability of inexpensive but reliable sensors are reducing this limitation.

4. **Limited Expertise**: Non-experts may lack the specialized knowledge or understanding of meteorological concepts required for detailed analysis and interpretation of Morning Glory Cloud data. While citizen science programs can provide training materials and resources, it is crucial to evaluate the suitability of citizen scientists' skills for specific tasks and to establish collaborations with experts for data verification and analysis.

5. **Ethical Considerations**: Citizen science initiatives involving atmospheric research must address ethical concerns related to the potential impact of data collection on privacy, intellectual property rights, and participant safety. Clear guidelines and informed consent procedures are essential to protect the rights and interests of both citizen scientists and the communities they interact with.

In conclusion, citizen science presents several advantages and limitations in Morning Glory Cloud research. While it offers opportunities for large-scale data collection, public engagement, and cost-effective research, concerns regarding data quality, expertise, and ethical

CITIZEN SCIENCE AND CROWDSOURCING   371

considerations must be addressed. Careful planning, training, and collaboration between researchers and citizen scientists can harness the potential of citizen science, enabling a deeper understanding of Morning Glory Clouds and their complex dynamics.

Index

-effectiveness, 294

a, 1–9, 11, 13–37, 39–49, 51, 53, 54, 56–63, 65–78, 81–108, 110, 111, 113–116, 118–129, 131–133, 135, 137–157, 159, 163–176, 178, 180, 182–184, 186–199, 201, 203–216, 219–224, 227–230, 232, 234–237, 239–242, 245, 247–251, 253, 254, 256–259, 261–263, 265–278, 280–289, 292–298, 300, 301, 305–310, 313–315, 318–323, 326–328, 330–332, 334–346, 348–357, 359–363, 365–369, 371
ability, 7, 60, 63, 83, 104, 108, 157, 161, 174, 179, 182, 199, 209, 215, 217, 218, 232, 234, 248, 251, 256, 268, 270, 294, 311, 318, 320, 328, 335, 351
absence, 150
absorption, 96, 97, 99, 101, 102, 104–106, 302
abundance, 125, 152, 153, 253, 256–258
accelerometer, 362
access, 285, 286, 296, 310, 312, 331, 352
accessibility, 151, 327, 337, 354, 360
accommodation, 123
account, 102, 124, 156, 181, 189, 229, 231, 240, 330, 342
accuracy, 42, 80, 81, 106, 156, 158, 173, 179, 185, 196, 217, 218, 225, 241, 247, 251, 294, 304, 306, 307, 317, 320, 322, 323, 327–330, 338, 340, 342–344, 351, 353, 362, 365, 367, 368
acknowledgment, 5
act, 24, 34, 59, 70, 74, 84, 95, 126, 190
action, 36, 228, 247, 252, 276, 297
activity, 1, 6, 47, 83, 113–115, 123, 125, 133, 134, 140, 150, 210, 215, 219–221, 223, 240, 251, 284, 343, 366
Adams, 275
adaptation, 8, 135, 210, 238, 240,

241, 243, 247, 248, 250–252, 345, 351
addition, 60, 92, 96, 121, 151, 171, 225–227, 229, 232, 256, 270, 331, 333, 353
adhering, 287
admiration, 268
adrenaline, 6
advance, 179, 243, 251, 283, 351
advancement, 21, 270, 361
advantage, 121, 332
advection, 84, 85, 207
advent, 310
adventure, 6, 285
advocacy, 296
advocate, 288
aerosol, 35, 104, 293, 307, 308, 329
aesthetic, 137, 150, 270, 271, 275, 279
age, 278
agility, 80
agriculture, 39, 243
aid, 105, 225, 340
aim, 5, 9, 173, 226, 228, 262, 277, 294, 343, 344, 348
air, 1, 7, 9, 17–25, 27–29, 32–37, 43–63, 66, 67, 70–76, 81–93, 99, 104, 113–115, 118, 120, 124, 133, 140, 141, 144, 145, 147, 148, 150, 163, 164, 166–170, 187–195, 198, 200, 203–207, 209, 210, 212–214, 217, 224, 226–228, 230–232, 234, 236, 241, 245, 250, 251, 293, 298, 321, 327, 329
aircraft, 36, 90, 92, 134, 156, 165, 171, 200, 218–222, 224, 225, 228, 229, 231, 250, 293, 310, 323, 324, 326, 327, 331, 336, 337
airflow, 18, 21, 125, 148, 188, 199, 220, 342, 343
airline, 222
airport, 222
airspace, 218
airspeed, 219, 220
alarm, 146
alert, 229
alignment, 59, 88, 169
allure, 133, 146, 270, 271, 278
alteration, 239, 246, 247
alternative, 107, 229, 295, 331
altitude, 38, 40, 43, 44, 77, 82, 84, 87, 89, 92, 104, 165, 166, 204, 216, 219, 220, 231, 249, 250, 321, 331, 332
amount, 18, 30, 31, 34, 71, 80, 209, 213, 301, 363
amplification, 167, 175, 181
amplitude, 70, 82, 164, 167, 175, 178, 181, 183, 185
analysis, 8, 41, 42, 68, 79–81, 93, 96, 102–105, 107–111, 128, 201, 202, 207, 211, 221, 232, 293, 294, 301–303, 308, 314, 317, 318, 320, 332, 333, 348, 350–354, 356, 360, 362, 366–368
anemometer, 92
angle, 95, 96, 98, 101, 277
Ansel Adams, 275
Antarctic, 23
antenna, 306
Anthony Doerr, 273

Index

appeal, 150, 275, 295
appearance, 1, 2, 35, 38–40, 42, 43, 55, 65, 66, 75, 77, 81, 90, 95, 96, 99–101, 121, 144, 145, 149, 151–153, 209, 240, 256, 262
applause, 153
applicability, 351
application, 93, 102, 179, 182, 183, 244, 308, 309, 313, 314, 328, 329, 344, 350, 351, 362
appreciation, 45, 69, 97, 106, 132, 146, 151, 165, 227, 254, 255, 259, 263, 265, 270, 277, 281, 283, 285, 288, 295, 356, 362
approach, 41–43, 77, 88, 91, 94, 97, 107, 108, 127, 154, 201, 220, 226, 235, 247, 248, 263, 266, 294, 300, 337, 348, 363, 366, 368
appropriation, 254
archive, 262
Arctic, 46–48
Arctic, 47
area, 17, 18, 22, 28, 29, 45, 49, 56, 61, 74, 75, 91, 128, 134, 135, 176, 180, 181, 189, 198, 199, 203–205, 210, 211, 222, 296, 298, 323
array, 133, 277
arrival, 115, 153, 195, 256, 278
art, 3, 5, 107, 109, 124, 154, 256, 269–271, 277–280, 282, 352
artist, 11, 269, 275
artistry, 135
artwork, 269, 281

ascending, 66, 164, 189
ascent, 17, 67, 84, 92, 321, 322
Asia, 114
aspect, 37, 44, 57, 81, 94, 105, 175
assessment, 215, 231, 232, 234, 238
assimilation, 81, 157, 238, 337, 342, 344, 353
association, 65, 68, 209, 213
Atlantic, 121
atmosphere, 9, 17–19, 23, 24, 27, 28, 30–35, 37, 40, 43–45, 48, 49, 56, 59, 60, 66, 69, 71, 73–75, 77, 78, 81, 83, 87–89, 91–93, 95–97, 99–101, 103, 105, 126, 129, 132, 134, 135, 141, 144, 146–150, 152, 155, 156, 159, 163–174, 180, 186–188, 190, 191, 193, 204, 206–209, 212–214, 216–218, 239, 241, 245, 248, 249, 273, 293, 301, 306–309, 321–323, 327–330, 334, 341, 342, 344, 353, 355
attention, 4, 107, 123, 126, 135, 166, 198, 238, 270, 285
attenuation, 175
attitude, 219
attraction, 22, 122, 126, 282, 297, 365
attribute, 174
audience, 111, 270, 281, 287, 361
audio, 154, 266
Augustus Gregory, 2, 4
Australia, 2, 3, 5, 6, 9, 21, 28, 29, 68, 71, 93, 113–115, 118, 120, 122, 124–126, 131,

134, 145, 148, 153, 195,
197–199, 208, 210, 214,
222, 258, 267, 276, 284,
288, 296, 333, 342, 365
authenticity, 255
authorship, 368
autonomy, 266
autumn, 29, 125, 128–132, 145
availability, 56, 71, 72, 80, 114,
115, 122, 133, 135, 151,
155, 157, 170, 171, 189,
194, 200, 210, 213–215,
240, 270, 294, 322, 359
avenue, 143, 182
aviation, 3, 6, 8, 9, 37, 39, 44, 45,
55, 61, 63, 90, 91, 121,
127, 133–135, 149, 155,
165, 193, 200, 202, 208,
216, 218, 219, 221–224,
226–237, 250, 251, 340,
341, 351
aviator, 3, 222
avoidance, 8, 218, 220, 221, 227
awareness, 5, 6, 151, 218, 227,
228, 247, 251, 252, 254,
263, 270, 271, 275, 284,
288, 295–297, 300, 356
awe, 5, 6, 25, 26, 72, 76, 108, 144,
155, 187, 253, 255, 257,
262, 268, 269, 272, 274,
277, 278, 281, 283, 285,
289, 297, 298
axis, 65, 67, 81, 87, 129

backdrop, 278
background, 164
backscatter, 307, 309
balance, 8, 30, 36, 39, 41, 124,
134, 149, 174, 182, 245,
247, 269, 276, 294, 295,
297, 298, 300, 344
balloon, 83, 92–94, 124, 207, 308,
321, 322, 327, 328,
330–332, 353
bank, 21
barrier, 84, 93, 118, 164, 169, 180,
213
base, 2, 76, 79, 352, 356
basic, 141, 306
basis, 127, 242, 338
battery, 313
Baudelaire, 273
beauty, 5, 11, 40, 72, 76, 97, 107,
108, 111, 121, 122, 132,
134, 137, 144–146, 151,
153, 244, 256, 262,
268–282, 285, 298, 300,
356, 362
behavior, 2, 3, 5, 7, 8, 13, 15,
23–26, 30, 43, 48, 51, 53,
55, 57, 60–63, 68, 75, 76,
78, 80, 81, 83, 84, 86–89,
91, 116–118, 127, 128,
130, 131, 135, 137, 140,
141, 143, 144, 147–150,
152, 155, 159, 161,
166–168, 171–174,
176–187, 189, 194–196,
205–208, 212, 213, 215,
226, 229, 239–241,
244–246, 250, 251, 265,
278, 286, 292, 293, 295,
304, 307, 309, 315, 323,
327, 331–333, 338, 339,
345, 353, 356, 360
being, 6, 8, 80, 123, 129, 142, 152,
154, 157, 235, 239, 256,
257, 259, 262, 270, 273,

Index

287, 288, 320, 323, 330, 350
belief, 153
belt, 26
bending, 106
benefit, 287, 288, 295, 353
bias, 367
biodiversity, 240
biology, 241, 344
blend, 121, 279
blessing, 256
blue, 95, 96, 98, 100, 105, 149
boat, 6
body, 49, 53, 369
book, 2, 9, 11, 267
boost, 283
Borneo, 126
boundary, 22, 44, 87, 148, 156, 163, 164, 167, 168, 172, 174, 175, 183, 190, 191, 198, 217, 328, 329, 339, 349
Brazil, 153, 222
break, 148, 283
breakdown, 167
breaking, 73, 75, 164, 167, 186
breeze, 61–63, 68, 71, 120, 138, 140, 141, 145, 148, 164, 170, 171, 180, 189, 195, 208, 328
bridge, 258, 335
briefing, 222
brightness, 101
British Columbia, 267
budget, 243
building, 287, 296
buoyancy, 44, 66, 67, 147, 163, 167–169, 190, 321, 327

Burketown, 5, 21, 115, 120–122, 124–126, 131, 134, 145, 153, 197, 198, 210, 222, 284, 285, 288, 296, 365
Burketown, 288
burst, 328

calendar, 265, 267
calibration, 320, 323, 327, 337, 368
California, 121, 122, 126
camera, 97, 275, 277
campaign, 4, 313
Canada, 1, 46, 267
Canary Islands, 195
canvas, 280
capacity, 82, 287, 296
Cape York Peninsula, 2, 120, 213
capturing, 78, 80, 97, 98, 108, 151, 196, 197, 269, 270, 274–277, 280, 302, 337, 339, 350, 360
carbon, 141, 243
case, 17, 35, 56, 57, 61–63, 66, 68, 69, 71, 72, 90, 93–95, 100, 103, 131–134, 140, 148, 164, 174, 180, 182, 188, 191, 195–200, 206, 211, 213, 222–224, 227, 232, 233, 267, 296, 297, 336, 337, 343, 348, 354, 355
catalyst, 70
catch, 153
category, 2
cause, 18, 44, 66, 84, 134, 203, 210, 216, 220
caution, 221, 227
Ceará, 222, 223

celebration, 5, 127
cell, 241, 338
center, 169, 178, 262, 284
Central Australia, 152
century, 2, 3, 269, 271, 272, 274
challenge, 6, 42, 104, 151, 157, 173, 189, 233, 272, 277, 283, 284, 308, 320, 328, 339, 349, 350, 354, 362, 367
championship, 222
chance, 287
change, 7, 8, 34, 36, 37, 39, 42, 66, 70, 73, 82, 89, 99, 128, 135, 141–145, 148, 149, 175, 184, 189, 193, 214, 215, 219, 232, 238–252, 267, 270, 276, 278, 281, 293, 298, 329, 330, 344–346, 348, 351
chaos, 269, 273
chapter, 9
characteristic, 21, 24, 28, 43, 44, 59–62, 66, 67, 70, 81, 88, 90, 91, 100, 120, 132, 133, 144, 149, 150, 164, 167, 170, 180, 183, 191, 194, 195, 213, 214, 224, 226, 339, 340
characterization, 41, 86
chemical, 241, 331
chemistry, 241, 344
China, 114
choice, 172, 302
cinematography, 276
circulation, 18, 47, 54, 58, 59, 71, 81, 138, 141, 170, 344
citizen, 57, 165, 251, 292, 294, 356, 357, 360, 362, 363, 366–371
city, 121
Ciudad Bolivar, 114
classification, 2, 37, 40–43
Claude Monet, 269, 277
cleanup, 285
Clement Wragge, 3, 4
Clement Wragge's, 3
climate, 2, 7, 8, 23, 33, 36, 37, 39, 42, 51, 62, 63, 116, 118, 119, 124, 125, 128, 130, 132, 134, 135, 137, 139–144, 162, 184, 189, 193, 199, 211, 212, 214, 215, 232, 238–252, 267, 270, 276, 281, 293, 298, 329, 330, 340, 344–348, 351, 353
climatology, 9
cloud, 1, 2, 5–9, 13, 19, 21–26, 28–45, 48, 55–57, 59–62, 65–97, 99–106, 108, 110, 111, 113–115, 120, 121, 124–127, 129–135, 140–142, 144–153, 155–157, 159, 161, 163–171, 175, 176, 178–184, 186–204, 208–218, 221, 223–225, 228–230, 240, 241, 248–251, 253, 254, 257, 258, 261, 269, 270, 276, 277, 280, 281, 284–287, 289–298, 301, 303–310, 313, 317–320, 323, 327, 329–332, 334, 337, 339–344, 348–356, 360–363, 365–368
cluster, 35, 43

Index

coalescence, 188, 339
coarser, 156
coast, 118, 120, 121, 138, 140, 170
coastline, 6, 145, 208
cockpit, 220
coexistence, 253
coherence, 182
collaborate, 228, 230–232, 270, 282, 286, 288, 360, 367
collaboration, 158, 224, 226, 229–231, 233, 234, 244, 248, 252, 257, 259, 263, 268, 282, 285–287, 290, 300, 313, 326, 352, 360–363, 365, 367, 368, 371
collection, 93, 200, 203, 247, 251, 294, 308, 310, 319, 325, 326, 330, 333, 334, 352, 356, 359–363, 365, 368, 370
collide, 18, 35, 51, 55, 88, 188, 209
collision, 218, 339
color, 96, 97, 99–101, 103–108, 110, 111, 184, 269, 275, 277, 366
coloration, 26, 94–96, 98, 101, 102, 104, 135
combination, 1, 29, 32, 44, 71, 96, 107, 111, 113, 118, 120, 129, 133, 144, 204, 205, 213, 215, 224, 227, 228, 240
comfort, 233
commentary, 281
communication, 174, 224, 226, 227, 229, 230, 236, 238, 295, 327, 355, 368
community, 4, 5, 121–124, 131, 141, 142, 144, 153, 210, 247, 251, 252, 256, 262, 263, 266, 268, 270, 284–289, 294, 361–365, 367
comparison, 76, 332, 354, 360, 368
compatibility, 337, 349, 351, 354
complex, 7, 19, 21, 26, 34, 42, 62, 66, 72, 73, 75, 79–83, 86, 88, 89, 91, 93, 102, 103, 106, 108, 110, 127, 132, 135, 136, 141, 148, 149, 155, 157, 160, 162, 166, 168, 172, 173, 180, 181, 183, 187, 189, 190, 194–196, 198, 200, 202, 208, 211, 214, 215, 233, 239, 241, 246, 247, 268, 270, 276, 300, 304, 308, 309, 333, 335, 338–340, 344, 345, 348, 350–353, 355, 361, 371
complexity, 60, 97, 104, 146, 194, 196, 333, 346, 350, 362
component, 13, 30
composition, 35, 41–43, 79, 99, 101, 102, 104, 105, 110, 111, 166, 269, 275, 277, 293, 301, 302, 306, 319
compression, 67, 84
computer, 102, 332
computing, 157, 294, 339, 342–344, 350, 352, 354, 355
concentration, 68, 101, 104, 110, 239, 307, 309
concept, 23, 144, 174
concern, 250, 329, 349

conclusion, 11, 45, 81, 86, 91, 108, 128, 135, 137, 152, 155, 165, 173, 176, 179, 184, 189, 192, 196, 202, 208, 215, 218, 248, 252, 257, 315, 317, 320, 323, 326, 362, 370
condensation, 7, 17, 29, 32–37, 56, 66, 67, 71, 72, 74, 76, 82, 84, 85, 91, 133, 167, 168, 170, 189, 190, 198, 204, 209, 339, 350
condense, 34, 37, 43, 82, 210
conduct, 297
confidence, 368
conjunction, 216, 331
connection, 5, 43, 151, 184, 253, 258, 261, 266, 270, 274, 275, 365
consciousness, 273, 283
consent, 266, 368
conservation, 8, 124, 155, 227, 240, 241, 247, 252, 263, 265, 267, 275–277, 281, 282, 284–287, 289, 291, 292, 295–300, 338, 346, 350
consideration, 174, 339, 349, 368
consistency, 320, 349, 360
constituent, 103, 105
construction, 286
consumption, 220
contact, 19, 59, 164, 187
contemplation, 274
content, 7, 18, 34, 37, 45, 47, 56, 74–76, 78, 79, 82, 84–86, 88, 92, 94, 103, 107, 133, 141, 147, 164, 189, 204, 213, 225, 245, 249, 293, 301, 304, 308, 322
context, 21, 35, 63, 78, 87, 90, 98, 135, 167, 174, 182, 185–187, 193, 199, 202, 258, 306, 309, 344, 348, 349, 351, 352, 360, 362
contingency, 235
continuity, 250, 252, 349
contour, 111
contract, 247
contrail, 37
contrast, 22, 61, 73–75, 77, 115, 189, 194, 207, 304
contribute, 8, 19, 21, 26–29, 35–37, 44, 45, 47, 48, 55–57, 60, 63, 66, 70–76, 84–88, 95, 96, 98, 101, 103, 113–115, 119–121, 127–129, 131–133, 137, 140, 141, 149–152, 161, 163–165, 168, 170, 171, 176, 184, 186, 187, 189, 194, 197, 199, 204, 212, 215, 221, 225, 226, 231, 232, 240, 243, 246, 252, 255, 262, 265, 270, 276, 287, 288, 291, 294, 327, 331, 342, 343, 348, 351, 352, 356, 362, 366–368
contribution, 36, 103, 104, 217, 261, 267, 323
control, 90, 219, 220, 222, 223, 227, 229, 256, 266, 268, 286, 288, 294, 319, 328, 338, 354, 359, 360, 362, 365, 367, 368
convection, 34, 59, 85, 137, 157, 184, 187–190, 193–202, 209, 214, 215, 224, 245,

Index

329
convergence, 17–19, 21, 29, 44, 54, 55, 57, 60, 118, 120, 121, 140, 145, 169, 189, 194, 195, 199
conversation, 270
conveyor, 26
cool, 1, 22, 36, 61, 62, 67, 81, 84, 85, 120, 121, 194, 195, 208
cooling, 32, 34, 54, 66, 67, 84, 150, 156, 167–169, 204, 208, 210, 213
cools, 29, 32–34, 37, 43, 67, 82, 85, 90, 133, 144, 149, 168, 170, 193, 194, 209, 213, 217
cooperation, 252
coordinate, 349
core, 241
Coriolis, 17, 23
correction, 303, 305, 317
corridor, 120
cosine, 103
cost, 156, 173, 294, 331, 370
cotton, 76, 145
country, 115
couple, 68, 349
coupling, 184, 293, 340, 348–351
cover, 39, 76, 80, 127, 221, 225, 233, 301, 304, 305
coverage, 71, 304, 308, 312, 327, 356, 363, 365
craft, 275, 280
crash, 91
creation, 2, 4, 21, 56, 61, 87, 90, 213, 258
creativity, 279, 355
creature, 256

credibility, 294, 346, 362
crest, 66
crew, 222, 233
crop, 256
crowd, 108
cryosphere, 241
crystal, 319
cuisine, 123
culture, 76, 134, 153, 227, 287
cumulonimbus, 33, 188, 210
cumulus, 33, 39, 43, 76, 77, 188, 198, 210
curiosity, 238
current, 37, 61, 86, 186, 197, 230
Curtis, 275
cutting, 232
cycle, 30, 62, 74, 144–146, 149, 153, 155–157, 247, 333, 337
cyclone, 169

D. Michael Warren, 267
Dallas, 90
damage, 222, 319, 322
dance, 151, 154
darkness, 149
data, 3–5, 78–81, 83, 88, 91–94, 104, 107, 108, 110, 111, 120, 127, 128, 156–158, 165, 171–173, 176, 183, 186, 187, 189, 196, 198, 200, 202, 203, 207, 208, 215–218, 225, 226, 230–233, 241, 243–245, 247, 249, 251, 252, 266, 292–294, 296, 301, 303–306, 308, 310–312, 314, 315, 317–323, 325–337, 340, 342–344,

346, 349, 351–356,
359–363, 365–368, 370
dataset, 102, 356
date, 221, 229, 233, 237, 284
David Crook, 115
dawn, 144–146, 167, 256
day, 17, 39, 62, 74–76, 95, 96, 98,
134, 144, 151, 155, 158,
170, 197, 208, 247, 256
daylight, 100, 129, 149
daytime, 62, 134, 149, 150
decision, 124, 227, 229, 232, 233,
236, 247, 266, 297
decomposition, 103
decrease, 66, 84, 166, 203, 245
definition, 13, 159
deflection, 23
degradation, 266, 283
degree, 42, 71
demand, 123
density, 19, 28, 44, 61, 187, 190,
193, 200, 213, 342
deployment, 172, 320
depth, 79, 279, 281, 310, 362
descent, 222, 322, 331
desert, 214
design, 276, 350
designation, 49
desire, 151
destination, 121, 123, 124, 231
destruction, 286, 298
detail, 342, 350
detection, 91, 224, 225, 228, 232
development, 18–20, 22, 27–29,
32, 33, 39, 44–48, 51, 56,
57, 59, 60, 62, 66, 69–72,
74, 75, 83, 84, 88–91, 94,
116, 121, 124–129, 133,
134, 137, 140, 161, 165,
167, 169, 170, 181, 186,
187, 190, 193, 194,
197–199, 204, 211–214,
217, 222, 223, 225, 226,
228–233, 244, 248, 255,
286–288, 296–298, 308,
318, 320, 323, 328, 333,
337, 340, 342, 343, 348,
351, 353, 362
device, 330, 362
dew, 31–33, 67, 85, 170, 193, 209
diagram, 22, 25
dialogue, 254
difference, 37, 71, 77, 88, 98, 145,
170, 172, 338, 346
difficulty, 218, 304
dimension, 177
dining, 123
dioxide, 141
direction, 18, 24–26, 68, 70, 71,
73, 75, 77, 79, 82, 83,
87–90, 92, 99, 100, 128,
164, 175, 207, 214, 216,
219, 220, 307, 321, 330,
334, 340, 341
disappearance, 99, 240
disappointment, 283
disaster, 90
discomfort, 219
discovery, 3, 6, 356, 359
dispersion, 106, 147, 175, 182, 186
displacement, 66, 67, 164, 167,
188, 266
display, 26, 94, 122, 135, 149–151,
295, 298
disregard, 265
disruption, 298
dissemination, 108, 222, 226
dissipation, 62, 75, 84, 85,

Index 383

147–149, 155–157, 186, 201, 207, 214, 225, 240, 245, 251, 337
distance, 82, 89, 101, 163, 180, 219, 306, 334
distortion, 179
distribution, 2, 3, 7, 9, 23, 30, 78, 84, 86, 95, 97, 100, 102, 109, 124, 125, 127–129, 135, 165, 167, 204, 209, 225, 240, 246, 247, 250, 251, 292, 294, 302, 307, 309, 318, 332, 356, 366
disturbance, 164, 283, 286
divergence, 21, 169
diversity, 72, 108, 126, 263, 268, 363
Dobrowner, 270
document, 41, 43, 108, 111, 149, 151, 154, 254, 257, 259, 262, 294
documentary, 274, 276
documentation, 3, 154, 266, 362
Doerr, 273
domain, 103, 338, 339
dome, 76
downdraft, 219, 220
drag, 220
driver, 193
driving, 82, 86, 195, 196, 337, 344
drizzle, 76
drone, 6, 276
drop, 198
droplet, 79, 95, 102, 307, 319
duration, 33, 75, 247, 249, 342
dust, 34, 47, 96, 99, 103, 170
dwelling, 258
dynamic, 21, 26, 29, 40, 42, 53, 55, 62, 77, 82, 86, 92, 97, 135, 150, 168, 194, 201, 218, 225, 232, 234, 245, 270, 277, 281, 333, 355, 360
dynamism, 67

e, 182
Earth, 7–9, 17, 23, 24, 30, 33, 34, 36, 37, 39, 41, 75, 78, 82, 87, 99–101, 129, 141, 144, 147, 148, 162, 164, 166–168, 187, 193, 239, 241, 242, 247, 248, 301, 302, 322, 329, 330, 344–348
earth, 43, 133, 150, 253, 258
Earth, 23
east, 23, 75, 118, 120
easterly, 29, 120
eco, 297
economy, 6, 122–124, 126, 151, 287, 298
ecosystem, 210, 286
ecotourism, 6, 8, 131, 285–288, 295, 365
edge, 1, 25, 26, 194, 195, 198, 232
editing, 275, 276
edition, 41
education, 222, 228, 232, 233, 238, 252, 257, 266, 277, 285, 287, 289, 291, 296
Edward Curtis, 274
effect, 17, 23, 31, 84, 134, 175, 204, 240, 306, 307
effectiveness, 226, 243, 294, 315
efficiency, 80, 173, 220, 230, 232, 234, 235, 327, 330, 340, 341, 349, 351
effort, 226, 361
element, 172, 278, 338

elevation, 66
Eloise Grills, 274
emergence, 348
emergency, 39, 222, 227
Emily Kame Kngwarreye, 275
emission, 141, 239, 241, 243, 245
emphasis, 262
employment, 123, 286, 288
encounter, 2, 46, 56, 118, 140,
    164, 169, 190, 213, 306
end, 2, 96
endeavor, 271, 359, 365
energy, 30, 36, 39, 41, 59, 66, 67,
    126, 155, 163, 164, 172,
    174, 175, 184, 187, 190,
    240–242, 338, 346, 350
engagement, 144, 228, 252, 263,
    356, 359, 361, 362, 370
engaging, 254, 283, 287, 291, 365,
    366
engine, 220
enhancement, 100, 211, 303, 305
enjoyment, 297
ensemble, 343, 346
ensuring, 45, 124, 154, 185, 208,
    224, 227, 232–234, 241,
    252, 254, 257, 262, 265,
    266, 271, 282, 286–288,
    297, 352, 354, 360, 367
enthusiast, 11
entity, 258
environment, 6, 18, 19, 26, 29, 42,
    62, 66, 67, 75, 85, 87, 89,
    93, 102, 114, 116, 118,
    123–125, 131, 133, 145,
    169, 181, 204, 212, 227,
    232, 250, 252, 261, 262,
    267, 270, 271, 276, 283,
    285–289, 292, 295, 297,
    298, 330, 340, 355, 365,
    367
equation, 177–179, 182
equator, 23, 24
equilibrium, 163
equipment, 92, 322, 352, 360
era, 279, 359
erasure, 265
erosion, 262
error, 338, 346
essence, 109, 149, 151, 275
establishment, 227, 251, 296
estimate, 78, 242, 243, 343, 346
Europe, 48
evaluation, 244, 346
evaporation, 30, 37, 74, 85, 147,
    204, 249, 339, 350
evening, 149–152
event, 17, 27, 69, 71, 72, 121, 131,
    151, 198, 211
evidence, 258
evolution, 25, 62, 68, 78, 81, 84,
    86, 87, 119, 155, 168,
    172, 175, 176, 178, 179,
    181, 183, 184, 194, 195,
    202, 270, 293, 331, 340,
    341, 343, 348, 350, 355
example, 17, 18, 21, 22, 33, 36, 39,
    46, 47, 54, 58, 71, 74, 90,
    93, 96, 101, 104, 105,
    118, 126, 134, 152, 153,
    172, 176, 208, 210, 214,
    220, 233, 243, 256, 262,
    270, 275, 276, 288, 293,
    307–309, 313, 314, 328,
    329, 336, 340, 351, 352,
    362, 365–367
exchange, 123, 124, 131, 151, 154,
    187, 188, 233, 252, 266,

Index 385

287, 329, 349–352, 355, 367
excursion, 220
exercise, 221
exhaust, 36
existence, 182, 276, 300
exosphere, 166
expansion, 66, 67
expedition, 4–6
experience, 6, 54, 56, 66, 67, 71, 122, 123, 141, 146, 149, 215, 273, 276, 279, 281, 283, 285, 298
experiment, 69, 94, 119, 197, 211, 277, 281
expert, 276, 278, 281, 366
expertise, 128, 157, 196, 231, 232, 244, 252, 287, 288, 297, 352, 355, 356, 359, 370
explanation, 17, 98
exploration, 2, 3, 6, 11, 128, 143, 152, 274
exposure, 124, 149, 277
expression, 152, 261, 271, 277, 280, 282
extent, 18, 71, 77, 78, 86, 99, 179, 180, 188, 207, 225, 240, 335, 337
eye, 356

fabric, 154, 159, 253, 257
face, 157, 211, 248, 250, 252, 254, 262, 308, 315, 364
facility, 197
fact, 1, 103
factor, 34, 44, 70, 71, 82, 89, 114, 133, 213, 342
fair, 77
fall, 2, 44, 121, 191, 209

family, 2
fascination, 4, 155, 269, 274
feathery, 77
feature, 15, 65, 81, 84, 132, 169, 199, 282, 304
feedback, 239, 348, 350, 351, 368
festival, 5, 115, 124, 134, 153
fidelity, 196, 340
field, 2, 3, 7, 40, 42, 43, 45, 65, 94, 105, 157, 161, 162, 165, 175, 176, 184, 187, 196, 232, 234, 235, 237, 315, 318, 327, 330, 334, 340, 351, 352, 360, 367, 368
figure, 269
film, 274–277, 282
filmmaker, 274, 276
filmmaking, 274
firsthand, 122, 281
fish, 153
fishing, 153, 208
fission, 182, 184
flammability, 322
flexibility, 80, 311
flight, 134, 208, 216, 218–222, 226–231, 233, 234, 251, 313, 319, 325
flock, 6, 126, 297
flooding, 19
flow, 18, 21, 62, 66, 68, 87, 89, 338, 339
fluid, 87, 119, 155, 174, 176, 187, 193, 200, 202, 338, 339, 341, 353
fluidity, 277, 278
flying, 6, 171, 218, 220, 222, 229
focus, 102, 104, 134, 202, 211, 227, 230, 302, 320
folklore, 255–257, 265

footage, 275, 276
footprint, 285
force, 86, 163, 321
forecast, 25, 32, 33, 48, 60, 63, 81, 105, 108, 179, 202, 207, 215, 218, 252, 343
forecasting, 7, 9, 15, 19, 43–45, 55, 57, 62, 83, 86, 90, 91, 104, 105, 107, 108, 127, 128, 132, 141, 161, 165, 184, 187, 189, 195, 196, 199, 200, 202, 205, 207, 208, 211, 214–218, 221, 224–226, 229, 230, 232–236, 238, 246, 247, 251, 252, 293, 321–323, 328, 329, 331, 340, 341, 343, 351
form, 1, 14, 16–18, 20, 25, 26, 30, 32–37, 43, 44, 47, 49, 54, 58, 59, 65, 67, 71, 77, 85, 88, 92, 95, 109, 120, 133, 142, 145, 147, 168, 170, 174, 178, 187, 209, 212, 213, 285, 338, 363, 367
format, 236
formation, 1–3, 5, 7–9, 13, 14, 17–38, 40, 43–48, 50, 53, 55–57, 59–63, 65–71, 73–76, 78, 81–94, 105, 109–111, 113–122, 125–129, 131–137, 140–146, 149–153, 155–157, 159–161, 163–176, 178–202, 204–210, 212–218, 220, 224–226, 228, 238–241, 244–251, 276, 278, 285–287, 292, 293, 301, 302, 305–310, 312, 315, 318, 323, 326, 327, 329, 331–334, 337, 339–344, 348–356, 360, 366
fortune, 152, 253, 256, 258
forum, 367
foster, 154, 173, 186, 247, 248, 252, 257, 263, 268, 270, 295, 352
foundation, 3, 159, 232, 275, 315
fragility, 281, 295
fragmentation, 148, 265
framework, 42, 183
France, 273
frequency, 79, 103, 128, 183, 239, 244, 246, 249, 250, 306, 309
front, 17, 21, 22, 28, 39, 77, 114, 120, 190, 194, 195, 198, 213
frontal, 7, 22, 25, 34, 46, 55, 59, 129, 134, 137, 163, 170, 184, 190–199, 207, 209, 212, 213, 215, 218, 224, 244, 245
fuel, 33, 84, 220
function, 306
fundamental, 31, 41, 99, 165, 174, 176, 179, 233, 241, 338, 341, 344, 346, 368
funding, 267, 287
fusion, 303, 305
future, 94, 102, 104, 105, 128, 144, 154, 155, 173, 181, 184, 185, 187, 199, 233, 234, 236, 241–243, 245, 246, 248, 250, 254, 256, 257, 261, 275, 276, 285, 286, 291, 300, 313, 315, 322,

334, 344, 348, 352, 356, 362

gamut, 97
gap, 335
gas, 30, 239–243, 245, 321, 322, 327, 329, 344
gathering, 267, 321, 328, 331
genera, 40
generating, 118, 145, 175
generation, 26, 55, 129, 163, 165, 171–173, 180, 186, 209, 265, 339, 342
genus, 42
geography, 29
Georgia O'Keeffe, 277
glazing, 277
glider, 3, 22, 222, 223
gliding, 6, 222, 223
glimpse, 155, 270, 278
globalization, 154, 266
globe, 120, 124
glory, 17–27, 53, 55–57, 86–89, 91–94, 137–144, 159–162, 171, 173, 182–184, 190–193, 199–202, 215–218, 248–250, 265–268, 289–297, 306–310, 315, 317, 334–337, 341–344, 352–356, 359–369
goal, 298, 313
governing, 62, 172, 174, 338, 339, 348
government, 285, 287, 290, 296, 297
gradient, 21, 61, 62, 71, 140, 141, 170, 180, 191, 194, 195
grandeur, 269, 274, 278

gratitude, 153
gravity, 1, 21, 22, 24, 26, 34, 37, 44, 55, 57, 60–62, 66, 68–71, 77, 82, 83, 93, 94, 121, 125, 129, 132, 137, 140, 150, 157, 159–165, 167–175, 180, 183, 188, 190, 191, 193, 200, 206–209, 214, 224, 239, 307, 309, 339, 340, 342–344, 355
Great Plains, 114
greenhouse, 30, 31, 141, 239–243, 245, 329, 344
Gregory, 2
grid, 156, 173, 241, 338, 339, 341, 345
ground, 5, 40, 41, 80, 92, 104, 144, 152, 156, 172, 173, 193, 196, 203, 209, 218, 222, 226, 249, 310, 321, 327, 331, 353
growth, 33, 84–86, 89, 126, 164, 168, 175, 187, 188, 190, 194, 201, 212, 214, 242, 288
guidance, 222, 254, 267, 288, 310, 367
guide, 9, 25, 27, 251
Gulf, 120, 195, 208
gulf, 113, 118
Gulf of Carpentaria, 210, 267

habitat, 283, 286, 298
haboob, 47
haboobs, 47
Hainan Island, 114
hallmark, 175

hand, 24, 28, 39, 54, 70, 73, 87, 92, 95, 99, 156, 163, 169, 170, 204, 272, 275, 280, 307, 327, 332, 349, 351
handling, 322
hang, 6
Hans-Heinrich Purner, 3
Hans-Heinrich Purner's, 4
harmonization, 320
harmony, 6, 258, 269
harvest, 256
harvesting, 265
health, 210, 243
healthcare, 124, 287
heart, 81, 276
heat, 30, 39, 47, 67, 84, 85, 132–134, 141, 144, 147, 187, 188, 193, 199, 214, 239, 329, 341, 344
heating, 17, 23, 75, 140, 144, 145, 147, 148, 150, 156, 188, 207, 208
heatmap, 366
height, 70, 73, 74, 79, 82, 84, 85, 88, 127, 149, 175, 180, 214, 301, 305, 366
Heinrich Purner's, 4
helium, 321, 322, 327, 330
help, 7, 8, 17, 23, 33, 61, 79, 87, 104, 105, 110, 111, 113, 115, 127, 128, 135, 171, 172, 175, 176, 181, 184, 204, 225–227, 233, 239, 240, 243, 245, 248, 251, 254, 263, 284, 285, 287, 292, 293, 295, 296, 305, 319, 322, 332, 333, 343, 344, 348, 361, 362, 367, 368

hemisphere, 129
heritage, 123, 155, 254, 257, 263, 265, 267, 268, 287, 288
highway, 25
history, 2, 255, 268, 269, 271
home, 5
hope, 253, 273
horizon, 2, 96, 101, 144, 153
hospitality, 123
hotspot, 113, 118, 125, 298
hub, 284
hue, 101
human, 6, 141, 154, 155, 242, 271, 273, 281, 298, 329
humidity, 5, 14, 19, 28–34, 36, 39, 44, 47, 49, 51, 53, 54, 56–58, 67, 68, 71, 75, 79, 82–84, 89, 91, 92, 94, 104, 125, 127, 141, 155, 170, 171, 186, 190, 193, 203–205, 215, 216, 220, 226, 240, 245, 249, 251, 292, 293, 301, 305, 310, 318, 319, 321, 323, 327–331, 334, 337, 339–343, 350, 366
hundred, 1, 65, 70, 125, 198
hunting, 267
hurricane, 235
hydrogen, 321, 322, 327, 330

ice, 35–37, 43, 77, 96, 99, 100, 106, 145, 147, 188, 209, 220, 319, 331, 344, 350
icing, 220, 221, 231
identification, 42, 104, 105, 186, 230, 309, 317, 365
identity, 123, 126, 180
image, 303–305

imagery, 41, 68, 78, 80, 108, 157, 189, 199, 207, 215, 216, 232, 251, 272, 273, 278, 281, 301–305, 308, 318, 336, 337, 353
imagination, 4, 94, 152, 255, 271, 274, 275, 279, 282
imaging, 78, 225, 292
impact, 4, 6, 8, 18, 19, 22–24, 26–28, 37, 39, 44, 45, 47, 48, 55–57, 59, 60, 62, 69, 71, 74, 84, 86, 89, 91, 96, 101, 104, 116, 119, 122–124, 126, 128, 131, 134, 135, 142, 143, 148, 154, 169, 171, 172, 175, 182, 184, 186, 187, 189, 193, 203–205, 207–211, 213, 215, 216, 220, 222, 224, 227, 230–232, 234, 239, 245–247, 249, 251, 254, 271, 273, 276, 280, 281, 283–287, 289, 293, 298, 304, 329, 330, 340, 342, 363
impermanence, 258, 281
impetus, 25
implement, 296, 297, 338
implementation, 221, 232, 233, 236, 349, 350
importance, 2, 5–9, 30, 33, 72, 89, 90, 155, 195, 199, 222, 224, 230, 233, 252, 258, 259, 261, 263, 267, 268, 271, 276, 277, 281, 285, 286, 295, 296, 298, 337, 352, 354
improvement, 230, 243, 346
inability, 322

incentive, 295
incident, 91, 106, 222, 223
inclusivity, 155, 265
income, 124, 283, 295, 365
incorporation, 340
increase, 34, 39, 67, 84, 127, 129, 140, 141, 145, 147, 153, 166, 167, 173, 198, 204, 245, 249, 250, 354
indicator, 7, 208, 246–249
individual, 67, 104, 133, 178, 305, 332, 348
industry, 55, 115, 122–124, 142, 227, 229–234, 236, 238, 251, 287, 288, 298
influence, 2, 7, 13–17, 19, 22, 24–30, 34, 35, 39, 47, 51, 57, 59–62, 69–75, 84–87, 91, 92, 96, 98, 99, 104, 106, 114, 116–119, 121, 124, 127, 128, 133, 138, 139, 141, 147–150, 168, 169, 172, 173, 184, 186, 190–194, 198, 199, 203, 205–207, 209, 211, 213–216, 239, 240, 242, 245, 246, 248, 277, 279, 280, 296, 308, 332, 339, 340, 342, 344, 350, 367
influx, 6, 122, 123, 286, 298
information, 16, 30, 32, 45, 52, 55, 62, 78–80, 83, 85, 92, 102, 104, 108, 157, 174, 176, 181, 183, 203–205, 208, 216, 217, 220–222, 225, 226, 228–231, 233, 236, 243, 247, 248, 250, 251, 265, 284, 292, 296, 301, 303–310, 314,

318–323, 327, 329, 330, 332, 334, 335, 340, 341, 349, 353–355, 359, 362, 363
infrastructure, 122–124, 243, 285–288, 298
ingenuity, 6
ingredient, 205
initialization, 343
initiation, 33, 60, 70, 82, 170, 194
initiative, 288, 296, 297
innovation, 186, 243, 352
input, 156, 172, 349, 351
insight, 154, 301
inspiration, 258, 269, 271, 274, 275, 280, 282
instability, 28, 59, 60, 85, 114, 133, 167, 170, 188, 191, 194, 195, 200, 205, 212, 213
instance, 198, 257, 269, 333, 343
instrument, 268, 302, 321, 326, 327
instrumentation, 324, 326, 337
integration, 93, 157, 173, 218, 233, 320, 326, 334, 336, 337, 348–350, 354, 356, 360, 362, 367
integrity, 182, 219, 286
intelligence, 80, 233, 320, 340, 362
intensification, 21, 214, 218, 246
intensity, 18, 19, 26, 33, 48, 79, 99, 101, 103, 110, 128, 147, 174, 180, 209, 211, 215, 217, 226–228, 239, 245, 246, 250, 301, 302, 307, 308, 342, 366
interaction, 2, 18, 19, 24, 26, 28, 29, 34, 39, 44, 46–48, 53–57, 59–62, 68, 70, 71, 73, 77, 82, 85, 88, 89, 99, 102, 103, 106, 114, 119, 121, 123, 131, 134, 140, 145, 147, 148, 150, 155, 157, 161, 164, 166–176, 178, 180, 181, 183, 186, 188–199, 206, 207, 210, 214, 215, 224, 244, 281, 289, 308, 309, 335, 337, 339, 342, 344, 355
interconnectedness, 261, 262, 267, 273
interconnection, 261
interest, 3, 65, 92, 126, 128, 135, 166, 175, 182, 238, 341, 357
interface, 360
interoperability, 354, 356
interplay, 21, 36, 45, 60, 66, 69, 72, 82, 83, 85, 86, 94, 106, 118, 132, 135, 137, 140, 146, 166, 175, 198, 208, 214, 215, 241, 245, 270, 271, 273, 280
interpretation, 80, 107, 108, 229, 232, 294, 304, 333, 366–368
intersection, 300
intertwining, 152
interval, 304
interview, 4
intrigue, 126, 278, 280
introduction, 308
intrusion, 46, 47
inverse, 183
inversion, 93, 94, 164, 166, 167, 180, 183
investigation, 60, 127, 181, 183, 184, 222

Index

involvement, 251, 287, 297, 356, 363, 366, 369
iron, 101
island, 126, 199
issue, 141, 207, 216, 218, 226

Jane Marie, 276
Jennifer Bartlett, 269
jet, 7, 18, 23–29, 87, 93, 114, 115
job, 124, 287
John Best, 115
John Constable, 269
John Keats, 271
journal, 237
journey, 2, 9, 27, 273, 276
judgment, 221

Kansas, 113
Karumba, 125, 126
Kelvin-Helmholtz, 88
kite, 327, 328
Kngwarreye, 275
knowledge, 5–9, 11, 33, 37, 45, 57, 60–63, 69, 83, 91, 94, 97, 122, 123, 127, 128, 146, 152, 154, 162, 185–187, 190, 198, 205, 208, 211, 218, 227, 229, 232, 233, 241, 248, 250, 252, 254, 257, 259, 261–263, 265–268, 270, 271, 282, 287, 296, 297, 310, 323, 326, 327, 330, 337, 351, 352, 354–356, 361, 363, 365–367, 369

laboratory, 69, 88, 102, 161, 175, 176, 185
lack, 203, 222

land, 21, 22, 29, 54, 58, 61, 62, 74, 115, 117, 120, 121, 138, 140, 141, 145, 148, 170, 171, 189, 195, 208, 239–242, 251, 261, 265, 266, 275, 342, 344
landfall, 235, 238
landing, 90, 91, 220, 222
landmass, 61
landscape, 113, 269, 272, 275, 280
language, 2, 266, 267, 278
lapse, 92, 94, 166, 281
laser, 79, 307, 309
latex, 321, 327
layer, 67, 79, 81, 93, 118, 121, 126, 148, 149, 164, 166–168, 174, 175, 180, 183, 309, 328, 329
layering, 67
learning, 33, 42, 43, 80, 105, 233, 320, 333, 340, 362
lee, 213
lee waves, 70
left, 23, 278
legend, 258
length, 1, 70, 96, 101, 150, 180
lens, 77, 277, 280, 281
letter, 49
level, 1, 17, 18, 40, 65, 76, 77, 89, 92, 93, 114, 188, 189, 220, 243, 275, 340, 342
Lidar, 79
lidar, 79–81, 97, 172, 176, 181, 204, 225, 251, 292, 305–310, 318, 320, 332, 351
life, 48, 69, 149, 258, 272, 274, 281, 283, 286, 313, 337
lifespan, 26, 207

lift, 90, 113, 115, 190, 220, 327
lifting, 34, 56, 90, 91, 115, 134, 163, 169, 188, 189, 209, 210
light, 27, 76, 79, 91, 95–97, 99–101, 103–107, 110, 128, 145, 149, 183, 196, 222, 269, 270, 276, 277, 279–281, 302, 307, 309
lighting, 149, 151, 277
Lik, 275
likelihood, 25, 32, 125, 129, 216, 242, 243, 343
limitation, 42, 308
line, 194
link, 238–240
liquid, 32, 34, 35, 67, 103
literacy, 359
literature, 43, 69, 197, 256, 271, 272, 274, 277–280, 282
local, 2, 5, 6, 8, 15, 16, 21, 22, 29, 36, 55, 59, 60, 62, 63, 68, 71, 72, 74, 75, 87, 90, 104, 107, 108, 114–116, 118, 119, 121–126, 131, 133, 134, 137, 139–142, 144, 148, 151–155, 197–199, 203, 204, 206, 208–213, 215–218, 227, 244–247, 251, 283–290, 292, 294–298, 300, 302, 340, 341, 344, 363, 365
location, 75, 78, 108, 115, 120, 126, 145, 172, 221, 226, 228, 297, 306, 321, 360, 362
longevity, 82, 113, 114, 123, 164, 184
loop, 368

Lorraine Muller, 267
loss, 90, 219, 220, 222, 254, 262, 266, 322, 328
lower, 17, 18, 24, 28, 35, 68, 77, 82, 84, 89, 93, 144, 147, 188, 193, 203, 204, 213, 328, 366
luck, 257
luminosity, 277

machine, 42, 43, 104, 233, 319, 333, 340, 362
magic, 146, 279
magnificence, 270, 278
magnitude, 21, 243
maintenance, 29, 68, 69, 74, 82, 84, 86, 87, 126, 140, 166, 168, 170, 186, 193, 194, 199, 206, 207, 214
making, 42, 70, 78, 124, 145, 157, 189, 196, 211, 222, 223, 227, 229, 232, 233, 236, 246, 262, 266, 280, 283, 297, 304, 330, 333, 340, 349, 360
Malaysia, 126
man, 330
management, 39, 123, 210, 211, 251, 284, 287, 288, 296
maneuver, 220
maneuverability, 80, 220, 311
manifest, 55, 70, 150, 182
manifestation, 256
manner, 67, 73, 175, 229, 286, 288, 297, 346, 349, 360, 361
map, 9, 295, 307
mapping, 111
margin, 220

Marie-Laure, 273
maritime, 46, 47, 58
mark, 278
marketing, 287
marvel, 121, 145, 284
Mary Oliver, 274
masking, 303, 305
mass, 28, 45–49, 51, 53, 55–59, 61, 73, 75, 84, 113, 155, 190, 193, 198, 241, 283, 284, 338, 346
mastery, 275
material, 65, 67, 68
meaning, 32, 178, 258
means, 2, 174, 200, 258, 261, 362
measure, 31, 32, 78, 79, 83, 87, 91, 110, 165, 171, 172, 181, 211, 248, 292, 293, 302, 306, 307, 319, 323, 328, 329, 331, 332
measurement, 30, 33, 78, 79, 183, 186, 308
mechanism, 1, 17, 18, 24, 43, 88, 95, 100, 129, 163, 167, 175, 184, 193
media, 108, 226, 270, 274–277, 281, 367
medium, 108, 174, 180, 200, 269
meet, 44, 59, 213
merging, 159, 180, 184
mesh, 338, 339
meso, 348, 350
mesoscale, 1, 22, 197, 201, 293, 341–344, 349
mesosphere, 166
metadata, 108, 362
meteorologist, 3, 4, 11, 235
meteorology, 3, 7, 9, 16, 26, 30, 31, 40, 42, 43, 45, 57, 63, 76, 97, 130, 176, 187, 232, 234, 235, 237, 270, 306, 318, 321, 323, 328, 352–354
methane, 141
method, 172, 338, 339, 367
Mexico, 258
Michael Adams, 115
microburst, 219, 220
microwave, 78, 301, 305, 332
mid, 2
midday, 101, 147–149
migration, 267
milestone, 3
mind, 272
mindfulness, 274
mindset, 94, 357
mischief, 256
mission, 3, 318, 328
Mitch Dobrowner, 269
mitigation, 8, 222, 224–226, 228, 231, 240, 241, 243, 345
mix, 19, 36, 46, 59, 147, 208
mixing, 21, 26, 31, 56, 70, 85, 92, 147, 148, 150, 156, 166, 167, 194, 277
mobile, 294, 359–362
model, 156, 157, 172, 173, 197, 241, 243, 244, 285, 288, 293, 341–343, 346, 349–351, 354
modeling, 2, 36, 57, 86, 88, 91, 102, 107, 127, 128, 157, 162, 173, 176, 181, 183, 189, 196, 199, 202, 211, 228, 243, 244, 294, 339, 340, 342–344, 346, 347, 351–354, 356
modernization, 154, 254, 262, 266

modification, 54, 57
moisture, 18, 31, 33, 34, 37, 45, 47, 49, 50, 54, 56–61, 71, 72, 74–78, 81–86, 88, 92, 94, 107, 114, 115, 125, 126, 129, 132, 133, 135, 141, 144, 145, 147, 155, 156, 164, 166, 167, 170, 171, 188, 189, 193, 194, 204, 213–215, 225, 240, 245, 249, 250, 301, 322, 329
moment, 40, 42, 153
momentum, 66, 155, 163, 241, 329, 338
Monet, 269
monitoring, 86, 90, 108, 217, 218, 221, 224, 228–230, 232, 246, 247, 250–252, 287, 294, 296, 359, 363–365
monotony, 283
monsoon, 141
morning, 1, 17–27, 53, 55–57, 61, 86–89, 91–94, 120, 132, 137–145, 149, 153, 159–162, 171, 173, 180, 182–184, 190–193, 199–203, 208, 215–218, 240, 248–250, 265–268, 289–297, 306–310, 315, 317, 333–337, 341–344, 352–356, 359–369
Morning Glory Clouds, 124, 144, 145, 283
morphology, 207, 307
motion, 3, 4, 18, 21, 34, 44, 56, 59, 61, 65–67, 69, 70, 73, 74, 77, 79, 81, 82, 89–91, 113, 150, 156, 160, 163, 164, 167–169, 172, 182, 190, 191, 194, 195, 198, 207, 209, 212, 214, 216, 217, 281, 306, 307, 338, 346, 350
mountain, 61, 62, 73, 118, 119, 140, 169
move, 1, 47, 54, 67, 73, 75, 89, 150, 163, 182
movement, 7, 14, 17, 18, 21–26, 52–55, 57, 59, 65, 73, 76, 78, 79, 84, 92, 147, 155, 157, 169, 181, 187, 189, 190, 199, 206, 209, 214, 216, 217, 219, 221, 225–228, 251, 269, 271, 272, 277, 279, 292, 301, 302, 305, 307, 309, 318, 319, 332, 335, 337, 362
mP air, 46
multi, 201, 202, 247, 340, 348, 350, 351
muse, 277, 280–282
music, 154
myriad, 258
mystery, 97, 153, 244, 269, 279, 281
mystique, 151, 256, 273, 274, 278
mythology, 119, 255–258, 262

name, 1, 81
narration, 278
narrative, 278
naturalist, 2
nature, 6, 8, 9, 23, 42, 51, 54, 55, 69, 73, 76, 77, 82, 96, 99, 120, 121, 132, 135, 144–146, 150, 154, 157, 173, 178, 181, 195, 198,

Index   395

    200, 211, 215, 218, 219, 252–255, 258, 262, 267–275, 277, 278, 280, 281, 283, 285, 295, 296, 298, 308, 309, 333, 338, 348, 352, 355, 356, 360
navigation, 218
Nebraska, 113
need, 42, 91, 94, 123, 132, 134, 154, 156, 202, 222, 233, 236, 270, 276, 283, 294, 295, 302, 315, 326, 354, 360
network, 3, 294, 363, 365
night, 61, 84, 170
nighttime, 62, 134, 149–152
nitrogen, 100
non, 267, 366
nonlinearity, 175, 182
north, 18
North America, 46, 48
northeast, 23, 24
Northern Australia, 1, 125, 213
note, 70, 74, 107, 145
novel, 273, 314
nucleation, 7, 34–37, 339
number, 79, 122, 284, 286, 288, 289

obscuration, 304, 306
observation, 3, 4, 19, 43, 122, 145, 151, 226, 249, 304, 313
observer, 42
obstacle, 66, 199
occurrence, 7, 8, 25, 29, 57, 60, 81, 84, 105–107, 113–116, 118–122, 125, 126, 128–132, 134–138, 140, 141, 145, 155, 161, 165, 171, 179, 186, 199, 202, 207, 211, 212, 214, 215, 229, 238, 239, 244, 245, 248, 251, 256, 267, 292, 298, 340, 341, 362, 365
ocean, 1, 3, 47, 58, 61, 71, 72, 118, 121, 126, 140, 145, 170, 198
offering, 137, 153, 173, 262, 270, 284
onboard, 219, 302
one, 6, 7, 17, 77, 81, 89, 124, 151, 177, 187, 199, 235, 239, 284, 349
onset, 68, 262, 342
opening, 208
operating, 222, 223, 231
opportunity, 5–7, 22, 122, 123, 146, 154, 263, 282, 319, 351
optimization, 185
orange, 95, 96, 101, 105, 149
order, 83, 163, 166, 228, 330
organization, 59, 168, 267
orientation, 65, 67, 199, 308
origin, 103, 200
other, 1, 7, 18, 19, 21, 22, 24, 28, 30, 39, 44, 48, 51, 54, 55, 60, 62, 70, 73, 74, 76, 77, 85–88, 91–96, 98, 99, 101, 103, 107, 110, 114, 121, 124, 126, 127, 132, 133, 135, 137, 156, 159, 161, 163, 165, 169, 170, 172, 174, 176, 178, 180, 184, 188, 202, 204, 208, 209, 216–218, 220, 221, 226, 228, 230–232, 236, 241, 243, 244, 250, 251,

261, 267, 272, 275, 278, 280, 281, 284, 285, 288, 296, 302, 304, 306–308, 310, 315, 321, 327, 330–332, 334, 335, 338–344, 349, 351, 352, 355, 367
outback, 275
Outback Australia, 269
outlook, 107
output, 216, 239, 341, 349
outreach, 263, 295
outset, 368
overview, 9, 205
ownership, 247, 251, 266, 287, 294, 356, 363, 368
oxide, 101
oxygen, 100
ozone, 307

pace, 266
package, 321, 330
painter, 269
painting, 269, 280
palette, 275
Papua New Guinea, 125
parachute, 321, 330
parallel, 65, 67, 179, 198, 309, 350
parameter, 31, 92, 346
parameterization, 157, 349–351
parcel, 66, 67, 87
part, 6, 30, 153, 222, 236, 295, 329
participation, 266, 362, 363
particle, 41, 78, 79, 102, 182, 292, 318, 319, 331
partnership, 230
pass, 96, 101, 209
passage, 71, 114, 120, 204, 262, 281

past, 204, 241, 310, 328, 344
path, 23, 75, 96, 101, 209, 228, 248
pathway, 26
pattern, 1, 2, 6, 18, 22, 65, 70, 87, 90, 91, 145, 147, 148, 164, 354
payload, 330
peak, 123, 147, 285
peer, 367, 368
people, 2, 4, 5, 94, 122, 132, 151–153, 255, 258, 267, 282
perception, 154
performance, 156, 157, 219, 220, 231, 294, 323, 328, 339, 342, 346, 350, 352, 354, 355
period, 150, 163
permission, 255
perpetuation, 188
persistence, 56, 82, 84, 85, 93, 113, 114, 125, 127, 150, 164, 175, 176, 199, 245, 342
personnel, 226, 230, 322
perspective, 127, 261, 269, 283, 301, 305, 318, 334
Peter Lik, 275
phase, 34, 35, 41
phenomena, 3, 5, 6, 9, 11, 13, 17–19, 22, 23, 25, 30, 36, 40, 45, 48, 55, 57, 59–61, 63, 70, 72, 73, 75, 78, 86, 89, 96, 105–108, 110, 122, 124, 126, 128, 129, 137, 139–141, 149, 151–154, 157, 166, 171, 173–177, 179, 182, 184–187, 200, 208, 213,

215, 218, 222, 224, 225, 227, 228, 231, 240, 241, 245–248, 250–253, 255, 257, 261, 265, 268–271, 273–278, 285, 288, 289, 292, 293, 296, 297, 301, 302, 304, 306, 314, 315, 318, 320, 322, 323, 326–329, 333, 337, 338, 340–342, 344, 348, 350, 351, 354–356, 361, 367, 369
phenomenon, 1, 2, 5, 7, 21, 25–27, 30, 43, 68, 76, 83, 94, 99, 102, 106–108, 113, 116, 118, 120, 122, 131, 132, 135, 151, 159, 165, 166, 171, 174, 182, 196, 197, 199, 205, 210, 212, 239, 244, 261, 263, 280, 282, 285, 286, 288, 298, 352, 356, 366, 368
Phetchabun Province, 297
photograph, 42
photographer, 109, 269
photography, 96, 109, 146, 151, 275, 277, 280
physics, 2, 7, 130, 155, 186, 195, 241, 344, 352, 354
picture, 172, 244, 308, 320, 335
pilot, 222, 223, 230
place, 2, 3, 34, 166, 228, 258, 269, 274, 286
plain, 121
plan, 134, 145, 235, 236, 251, 277, 283, 284, 297
plane, 89
planet, 8, 39, 45, 141, 241, 348
planning, 52, 106, 151, 221, 227, 229, 287, 288, 300, 313, 319, 365, 371
planting, 265
platform, 5, 41, 123, 365, 367
play, 6, 13, 17–19, 21, 23, 24, 27, 29, 39, 40, 43, 45, 48, 54, 57, 60, 61, 66, 73, 76, 81, 83, 85, 86, 100, 109, 114, 116, 119, 125, 127, 129, 137, 138, 140, 144, 145, 149, 155, 159, 163, 164, 167, 170–172, 176, 188, 190, 192, 193, 196, 205, 207, 209, 224, 229, 230, 234, 241, 245, 251, 254, 261, 269, 270, 277, 280, 287, 289, 292, 296, 301, 315, 322, 323, 326, 330–332, 337, 339, 341, 344, 348, 352, 360, 365
plot, 278, 366
poem, 273
poet, 272
poetry, 271, 274, 278
point, 31–34, 42, 67, 82, 85, 170, 193, 209, 278, 354
polarimetry, 102
polarization, 308, 309
policy, 242, 296
pollutant, 329
pollution, 103, 283, 286, 298, 304, 329
pooling, 352, 356
popularity, 297
popularization, 3, 4
population, 123, 242
portion, 96, 118, 167, 241
position, 87, 88, 150
possibility, 39

potential, 32, 39, 42, 43, 45, 48, 51, 52, 55, 90, 94, 124, 128, 131, 134, 135, 141–144, 155, 184, 187, 210, 215–218, 220, 221, 225, 227, 229–231, 235, 236, 238–248, 250–252, 276, 278, 279, 285–288, 293, 294, 297, 298, 309, 319, 322, 328, 340, 349, 352, 365, 367, 371
power, 6, 154, 157, 196, 266, 274, 275, 294, 320, 323, 327, 339, 342, 344, 347, 353–355, 359, 362
practice, 153
precipitation, 7, 14, 17, 18, 28, 31–33, 39, 43, 45–48, 51, 92, 116, 133–135, 193, 204, 209–211, 216–218, 243, 246, 306, 350
precision, 6, 202, 323
predictability, 205, 215, 218, 246
prediction, 33, 63, 72, 91, 165, 184, 202, 211, 215, 246, 321, 322, 329, 350, 353, 362
preparedness, 227
presence, 1, 6, 17, 21, 26, 29, 34–36, 39, 44, 45, 50, 66, 68, 70, 72, 74, 77–79, 82, 84, 87, 88, 90, 93, 95, 96, 98, 100, 101, 106–108, 114, 115, 123–127, 133, 134, 139, 145, 150, 151, 155, 159, 166, 169–172, 175, 178, 183, 193, 198–200, 204, 206, 208, 211–213, 215, 217, 219, 220, 222, 226–229, 253, 256–258, 271, 282, 304, 306, 307, 309, 318
preservation, 5, 45, 123, 124, 127, 128, 141, 144, 152, 154, 155, 166, 240, 241, 250, 252, 255, 257, 259, 262, 263, 265, 267, 277, 281, 282, 284, 289, 297, 298
pressure, 3, 7, 13–19, 21–24, 27, 28, 31, 34, 35, 44, 46, 59–62, 66, 67, 71, 80, 86, 87, 91, 141, 145, 146, 169–171, 180, 181, 189, 190, 198, 206, 212, 283, 293, 301, 305, 310, 321, 327, 328, 330, 341, 342, 354
prevalence, 113
pride, 126, 266, 287, 365
prime, 47, 120, 145
principle, 103, 306, 327
print, 41
prism, 106
privacy, 359, 368
probability, 242
probe, 79
problem, 25
process, 34–37, 67, 72, 89, 95, 99, 102, 105, 106, 110, 147, 158, 167–170, 173, 187–189, 194, 210, 214, 266, 322, 328, 343, 349, 353, 355, 360, 367, 368
processing, 104, 301, 303–306, 308, 333, 354, 360
processor, 306
professional, 6, 270
profile, 83, 85, 92, 110, 142, 164,

Index 399

206, 322
profiling, 186
program, 296
progress, 91, 102, 134, 162, 184, 195, 242, 243, 339, 353
project, 344, 354, 355, 365, 368
prominence, 262
promise, 80, 184, 211, 323
promotion, 123
propagation, 26, 55, 126, 127, 150, 163–165, 171–175, 178–181, 185, 214, 239, 342
property, 174, 255, 266, 320, 368
prose, 281
prosperity, 253, 256
protagonist, 273
protection, 124, 240, 254, 256, 262, 267, 288, 296
proximity, 30, 71, 126, 138, 140, 141
public, 4, 45, 52, 55, 108, 218, 226–228, 243, 251, 294, 329, 330, 356–359, 361, 365, 366, 369, 370
Purner, 3
purple, 95, 105
purpose, 368
push, 29

quality, 157, 270, 275, 285, 286, 294, 303–305, 329, 354, 359, 360, 362, 365, 367, 368, 370
Queensland, 5, 120, 122, 124, 125, 131, 134, 145, 153, 197, 210, 222, 284, 288

radar, 56, 79–81, 93, 108, 157, 165, 172, 176, 181, 183, 189, 198, 203, 219, 221, 225, 226, 230, 232, 251, 292, 306–310, 320, 332, 334
radiation, 39, 62, 78, 84, 102, 129, 134, 144, 148, 156, 193, 203, 301, 302, 304, 341
radio, 306, 307, 309, 321
radiosonde, 92–94, 321, 322, 330
rain, 19, 46, 47, 76, 188, 198, 209, 217, 258
rainbow, 106
rainfall, 134, 209–211, 216, 217, 239, 246
range, 39, 71, 72, 74, 80, 95, 97, 98, 104, 105, 107, 126, 134, 169, 173, 181, 241, 242, 258, 292, 294, 307, 308, 310, 328, 339, 346, 352
ranging, 65, 70, 88, 239, 341
rarity, 356
rate, 33, 34, 92, 94, 166
ratio, 31
Rayleigh, 95, 100
reader, 9, 281
realism, 339
reality, 272, 279, 348
realm, 27, 153, 253, 262, 272
rebirth, 256
recapture, 328
receiver, 306, 307, 321
recognition, 5, 91, 227, 259, 267
record, 3, 111, 154, 183, 257, 259, 268
recording, 360
recovery, 220
recycling, 285

red, 95, 96, 100, 105, 149
reduction, 99, 243
reference, 41, 273
refinement, 173, 185, 187, 230, 348
refining, 185, 189, 225
reflection, 78, 164, 274
reflectivity, 226
refraction, 105, 164
regeneration, 82
regime, 214
region, 1, 3, 5, 6, 17, 18, 22, 25, 27, 29, 45, 46, 51, 53, 58, 62, 68, 72, 75, 84, 85, 92, 93, 113, 114, 116, 118–121, 123–126, 131, 134, 138, 140, 145, 148, 153, 194, 195, 197, 198, 204, 208, 210, 211, 214, 221, 235, 267, 296, 297, 309, 341, 343, 354, 365, 366
reinvestment, 287
relation, 199
relationship, 8, 19, 24, 35, 43–45, 56, 72, 122, 190, 191, 205, 211, 230, 245, 246, 265, 273, 274, 276, 302, 366
relative, 31, 32, 47, 67, 87
release, 59, 67, 209
relevance, 129, 174, 310
reliability, 158, 173, 185, 220, 241, 247, 315, 317, 323, 328, 342, 346, 354, 362, 367
reliance, 267
remote, 3, 5, 37, 41, 43, 56, 57, 63, 78, 80, 81, 86, 91, 93, 96, 97, 102, 104, 107, 120, 127, 128, 151, 156, 157, 165, 173, 176, 181, 186, 187, 189, 196, 197, 203, 204, 225, 228, 230, 232, 249, 292, 293, 306–309, 315, 317, 319, 320, 324, 326, 332–337, 351–355
renewal, 253, 256, 257
report, 4, 236, 294, 365
reporting, 226, 228, 230, 247, 251, 294, 363–365
repository, 261
representation, 97, 103, 156–158, 173, 189, 273, 337, 346, 348–350
requirement, 368
research, 3–5, 8, 17, 19, 22, 35, 36, 39, 43, 45, 56, 57, 60, 63, 72, 86, 91, 93, 102, 104, 105, 119, 120, 127, 128, 131, 132, 134, 135, 141, 143, 144, 151, 152, 154, 155, 158, 162, 165, 171, 173, 176, 179, 181, 182, 184–187, 189, 190, 192, 195–197, 199–202, 205, 210, 211, 221, 224, 229, 230, 232, 234, 235, 237, 240, 247, 248, 252, 255, 257, 263, 266–268, 287, 292–294, 296, 297, 300, 301, 309–315, 317, 320, 322, 323, 326–328, 330, 331, 333, 334, 336, 340, 343, 344, 348, 349, 351–362, 365–370
resilience, 247, 248, 252, 273
resistance, 92, 212, 216
resolution, 78–80, 92, 104, 128, 156, 157, 165, 173, 176,

Index 401

181, 186, 187, 196, 199,
216, 294, 302, 304,
306–308, 312, 317, 318,
332, 333, 339, 342, 344,
355
resonance, 186, 271
resource, 11, 41, 154, 210, 211,
237, 359
respect, 17, 154, 255, 262, 263,
265, 266, 268, 282, 368
response, 278, 345
responsibility, 251, 363
restaurant, 286
restoration, 240, 252
result, 18, 20, 46, 60, 61, 66, 67,
70, 71, 73–75, 83, 84, 88,
90, 100, 101, 105, 106,
134, 147–150, 163, 168,
182, 198, 199, 210, 219,
256, 258, 283
retrieval, 319, 328
return, 163, 306
revenue, 122, 123, 126, 286, 287
reverence, 257, 261
reversal, 214
review, 43, 69, 197, 367, 368
revitalization, 266, 267
revival, 124
richness, 97, 362
ride, 6, 22, 121
right, 23, 145, 169, 268
rise, 18, 19, 33, 44, 46, 56, 61, 66,
69, 74, 93, 100, 105, 118,
121, 133, 140, 141, 144,
147, 150, 163, 191, 193,
194, 198, 207, 210, 239,
243, 276, 351
risk, 90, 217, 219, 220, 231, 232,
234, 265, 322, 328

ritual, 153
road, 9
role, 3, 6, 13, 17–19, 21, 23, 24,
26, 27, 29–32, 34, 35, 37,
39–41, 43, 45, 48, 54, 56,
57, 60, 61, 66, 68–70,
72–74, 76, 78, 81–83,
85–87, 89–92, 94, 99,
100, 105, 114, 116, 119,
121, 125, 127, 129, 133,
135, 137, 138, 140, 144,
145, 147, 148, 153, 155,
159, 161, 163, 164,
166–168, 170–173, 175,
176, 183, 184, 186–190,
192, 193, 195–197, 199,
201, 205–209, 213, 214,
224, 229, 230, 234, 241,
245, 251, 254, 261, 270,
274, 275, 277, 287, 289,
292, 296, 297, 301, 315,
322, 323, 326, 330–332,
337, 339, 341, 343, 344,
348, 352, 353, 359, 360,
365
roll, 1, 2, 9, 21, 28, 29, 44, 55,
59–62, 65–71, 73,
75–78, 81–83, 87, 88, 90,
91, 115, 120, 121, 132,
134, 147–150, 164, 167,
168, 170, 175, 178, 194,
195, 198, 224, 277, 339,
340, 342, 355
rolling, 1, 3, 4, 6, 22, 24, 44,
65–67, 69, 70, 72, 73, 77,
81, 82, 90, 121, 122, 160,
169, 182, 183, 191, 214,
218, 276, 278
rotation, 23, 65–67, 87, 169

route, 220, 221, 227, 231
row, 120
runway, 220

safety, 8, 9, 44, 45, 52, 63, 90, 91, 127, 151, 165, 193, 200, 202, 208, 216, 220, 221, 224, 227–236, 238, 251, 322, 329, 330, 341, 351
Saint-Malo, 273
salt, 103
sampling, 293, 327–331
Samuel Taylor Coleridge, 271
Santa Cruz, 121, 122
Sarah Thompson, 115
satellite, 41, 56, 68, 78, 80, 81, 93, 97, 104, 108, 127, 157, 165, 173, 189, 199, 203, 207, 208, 215, 216, 230, 232, 248, 251, 292, 301–306, 308, 318, 320, 334, 336, 337, 346, 351, 353
saturation, 31, 34–37, 67, 85, 170
scalability, 354, 356
scale, 2, 23, 47, 54, 80, 88, 102, 156, 157, 165, 167, 169, 171, 189, 196, 201, 202, 207, 245, 269, 275, 278, 292, 293, 308, 332, 335, 337, 339–344, 348–351, 370
scatter, 95, 96, 99–101, 105, 304, 306
scattering, 94–105, 172, 183, 302, 307
scenario, 61, 75, 204, 211
scene, 132, 281
scheme, 338

science, 7–9, 16, 30, 33, 36, 57, 63, 65, 69, 97, 130, 141, 162, 165, 251, 265, 270, 271, 276, 281, 292, 294, 353, 363, 366–371
scientist, 11
scope, 310
screen, 278
sea, 21, 22, 29, 44, 61, 62, 68, 71, 74, 75, 103, 114, 115, 120, 121, 138, 140, 141, 145, 148, 149, 164, 170, 171, 180, 189, 195, 208, 213, 241, 243, 249, 328, 342, 343
seaboard, 121
season, 46, 126, 129, 132, 133, 135, 140, 210, 262, 267
seasonality, 249, 250
seat, 120
second, 199
section, 2, 4, 13, 17, 19, 23, 27, 30, 34, 37, 40, 43, 45, 48, 57, 60–62, 65, 70, 72, 73, 76, 78, 81, 86, 89, 94, 99, 102, 105, 108, 113, 116, 120, 122, 124, 128, 135, 137, 141, 144, 147, 152, 159, 163, 168, 174, 179, 185, 187, 190, 193, 196, 197, 199, 200, 205, 212, 215, 218, 222, 224, 230, 238, 244, 248, 250, 255, 257, 261, 268, 271, 274, 277, 282, 289, 292, 298, 301, 306, 315, 321, 323, 327, 330, 334, 337, 341, 344, 348, 351, 352, 359, 366, 369

Index 403

seed, 34
selection, 221
self, 182
sense, 67, 153, 247, 251, 262,
    272–274, 281, 283, 287,
    294, 296, 338, 356, 360,
    362, 363, 365, 367
sensing, 3, 37, 41, 43, 56, 57, 63,
    78, 80, 81, 86, 91, 93, 96,
    97, 102, 104, 107, 127,
    128, 156, 157, 165, 173,
    176, 181, 186, 187, 189,
    196, 197, 203, 204, 225,
    228, 230, 232, 238, 249,
    292, 293, 306–309, 315,
    317, 319, 320, 324, 326,
    332–337, 351–355
sensitivity, 8, 295, 304, 340
sensor, 304, 306, 323, 362
series, 21, 70, 118, 121, 269, 275,
    318, 366
service, 233
set, 77, 94, 129, 132, 137, 146,
    200, 281, 309, 328, 338,
    353, 363, 367
setting, 62, 119, 126, 209
setup, 69, 71
severity, 228, 250
shading, 134
shadow, 273
shape, 1, 2, 21, 23, 35, 40–44, 55,
    62, 65, 67, 69–72, 74–78,
    86, 87, 89–91, 94, 99,
    119, 133–135, 147, 160,
    162, 164, 169, 174, 175,
    178–184, 199, 201, 203,
    216, 224, 240, 272, 277,
    279, 280, 308, 332, 346,
    348, 366

shaping, 6, 23, 24, 45, 48, 60, 66,
    70, 72–74, 76, 87, 90, 99,
    175, 184, 207, 262
share, 108, 111, 132, 263, 287,
    365, 367
sharing, 173, 226, 233, 244, 248,
    263, 267, 270, 296, 352,
    354, 356, 360, 363
shear, 8, 59, 60, 62, 66, 68–74, 82,
    83, 87–91, 142, 150, 167,
    174, 175, 180, 200, 214,
    215, 219–221, 328, 342
shearing, 89, 91
shift, 129, 220, 240, 247, 249, 306
shopping, 123
show, 242, 366
side, 210, 213, 368
sight, 145
sign, 39, 256, 257, 262, 267
signage, 286
signal, 77, 102, 103, 306, 308
signature, 104
significance, 2, 4–6, 8, 9, 11, 17,
    37, 39, 40, 45, 57, 63, 76,
    78, 89, 94, 119, 121, 124,
    127, 130, 132, 134, 135,
    137, 152, 154, 155, 197,
    211, 227, 253, 255–257,
    259, 261–263, 265, 274,
    277, 278, 281–283, 287,
    289, 292, 294–296, 317,
    327, 341
silver, 278
simulation, 91, 176, 197, 211, 350
sine, 103
singing, 151
sinking, 170, 200
situ, 56, 63, 79, 86, 91, 173, 183,
    184, 204, 225, 292, 293,

319, 332–337
size, 1, 35, 41, 42, 70–72, 78, 86, 94, 95, 99, 100, 102, 105, 132, 135, 164, 203, 216, 228, 240, 292, 302, 304, 306, 307, 309, 318, 332
skill, 6
sky, 1, 36, 40, 65, 76, 95, 98–100, 105, 108, 122, 132, 133, 149, 178, 223, 256, 258, 277, 356
smoke, 101
snake, 2
snow, 46, 101, 209
snowfall, 19
society, 250
socio, 288
software, 313, 354, 360
soliton, 174–176, 178–187
solution, 178
soot, 36, 101
sound, 275, 276, 279
source, 45, 48, 49, 53, 54, 58, 126, 258, 274, 280, 282, 286, 293
south, 18, 118, 120, 195, 198
South America, 114
southeast, 23, 24, 118
Southern Hemisphere, 28
souvenir, 123, 283
space, 18, 36, 39, 216, 350, 367
spacetime, 159
Spain, 195
speaker, 273
specific, 17, 19, 22, 23, 29, 31, 35, 44, 45, 48, 56–58, 60, 71, 74, 95, 97–101, 103–106, 113, 114, 119, 120, 122, 132, 133, 135, 139, 141, 145, 148, 149, 157, 166, 167, 172, 174, 200, 211, 214, 215, 221–223, 227, 229, 231, 249, 265, 267, 286, 293, 296, 298, 302, 304, 308–311, 325, 331–333, 340, 342, 349, 354
spectacle, 6, 22, 67, 105, 115, 122, 125, 127, 133, 144, 269, 280–282
spectrum, 96, 103–106, 109, 302
speed, 18, 24–26, 68, 70, 71, 73, 79, 82, 83, 87–92, 148, 164, 171, 175, 179–182, 185, 198, 214, 219, 292, 306, 307, 321, 327, 328, 330, 334, 340, 341, 366
spirit, 253, 263
splendor, 146
splitting, 180
spread, 36, 346
spring, 21, 29, 120, 125, 128–132, 134, 145, 210
squall, 22
stability, 1, 32, 33, 36, 39, 48, 51, 53, 54, 56, 57, 60, 68, 70, 72–76, 78, 82–88, 91, 92, 107, 114, 116, 125, 127, 129, 132–135, 142, 147, 149–152, 155, 164, 166, 168, 171, 174–176, 179, 180, 182, 184, 189, 205–208, 212, 214–219, 240, 245, 247, 248, 293, 322, 328, 338, 342, 343
staff, 123
stage, 22, 166, 209
standard, 123, 222, 368

Index 405

standardization, 354, 360
start, 39, 144, 256
starting, 9, 354
state, 3, 5, 85, 167, 172, 222, 339, 352
station, 92, 197, 321, 331
step, 225, 366
stewardship, 263, 296
storage, 368
storm, 1, 19, 47, 119
storyteller, 4
storytelling, 5, 124, 151, 261, 263, 270, 274, 279, 281, 282
strain, 123
strategy, 219, 251
stratocumulus, 2
stratosphere, 28, 166
stratus, 43, 77, 80
stream, 23–26, 29, 93, 114
strength, 21, 68, 71, 170, 207, 244, 306
stretching, 71, 89–91, 178
string, 327
strip, 121
structure, 3–5, 21, 24, 29, 51, 55, 61, 65–70, 73, 77–79, 81–83, 85–92, 94, 104, 115, 120, 132, 134, 135, 143, 147–152, 156, 160, 164–166, 168, 169, 172, 174–176, 180, 186, 194, 203, 214, 217, 225, 301, 305, 307–309, 313, 315, 318, 319, 323, 327, 334, 335, 337, 339, 340, 353, 355
study, 2–5, 7, 8, 17, 45, 55, 61, 62, 68, 71, 72, 78, 83, 88, 89, 93, 94, 96, 102, 111, 122, 130–132, 134, 144, 148, 149, 152, 159, 162, 165, 171, 172, 174–176, 178, 179, 181, 182, 187, 189, 198, 199, 211, 215–218, 245, 248–250, 257, 263, 265, 267, 285, 292, 293, 296, 302, 306, 309, 315, 317, 320, 329, 330, 333–337, 339, 340, 343, 348, 350–352, 354, 355, 359, 360
style, 109
sub, 241
subgrid, 350
subject, 60, 65, 148, 174, 221, 268, 276, 328, 356
success, 243, 285, 288, 296, 362, 365
sum, 103
summary, 8, 26, 39, 75, 89, 171, 230
summer, 21, 33, 120, 132–135, 140
sun, 61, 62, 96, 133, 144, 145, 147, 149, 167, 189, 203
sunlight, 36, 39, 78, 94–96, 98–103, 105, 106, 108, 133, 280
sunrise, 1, 75, 96, 101
sunset, 96, 101
superstition, 265
supply, 213
support, 21, 154, 251, 287, 294, 295
suppression, 100
surface, 17, 18, 23, 24, 33, 34, 37, 39, 43, 49, 50, 58, 61, 75, 82, 87, 101, 102, 133,

134, 144, 147, 148, 150, 163, 164, 166, 167, 180, 187, 188, 193, 203, 204, 207, 208, 212, 213, 239, 241, 249, 301, 302, 329, 342–344
surge, 21
surrealism, 272
surrounding, 3, 14, 17, 19, 35, 36, 42, 63, 66–68, 72, 74, 78, 87, 115, 116, 118–120, 123–125, 133, 144, 147, 152, 154, 163, 186, 188, 193, 203, 204, 208, 210, 247, 254–257, 259, 262, 278, 293, 308, 369
survival, 266
sustainability, 124, 186, 234, 252, 263, 285–287, 291, 295, 297, 365
sustenance, 131, 190
symbol, 253, 256
symbolism, 119, 153, 256, 257, 262, 272, 273, 279, 280, 282
synchronization, 349
synergy, 309
system, 8, 16–19, 21, 23, 27, 33, 41, 49, 71, 93, 141, 156, 169, 174, 194, 197–199, 239, 241, 242, 247, 249, 250, 306–309, 321, 337, 341, 344–348

takeoff, 90, 220
tale, 4
target, 306, 354
task, 96, 173, 300
team, 3, 5, 236, 276

technique, 103, 171, 236
technology, 6, 93, 94, 108, 128, 221, 226, 270, 274, 276–279, 291, 304, 306, 308, 310, 321, 327, 328, 330, 356, 359, 362
television, 278
temperature, 5, 7, 14, 17, 19, 21, 24, 28–37, 39, 43–45, 49–51, 53, 54, 57–62, 67, 68, 71, 72, 74, 75, 77–79, 81–87, 89, 91, 92, 104, 115, 116, 119, 125, 127, 129, 132, 134, 140–142, 144, 145, 147, 148, 155, 161, 166–172, 181, 186–191, 193–195, 198, 200, 203–207, 213, 215–217, 220, 226, 239, 240, 243, 245, 247–251, 292, 293, 301, 305, 310, 318, 319, 321, 327–332, 334, 337, 339–343, 346, 350, 366
template, 34
ten, 40
tendency, 73, 206
term, 3, 7, 116, 124, 135, 139, 141–143, 239, 241, 246–249, 251, 285, 288, 293, 294, 297, 298, 344, 346, 348
terrain, 21, 74, 80, 88, 141, 171, 199, 210
testament, 6, 135, 146, 154
tether, 92
texture, 107, 277
Thailand, 297
the Atlantic Ocean, 18

the Canary Islands, 195
the Cape York Peninsula, 343
the Carpentaria Coastal Plain, 210
the Channel Islands, 121
the East Coast, 46
the East Coast of North America, 18
the Great Plains, 114
the Gulf Coast, 93
The Gulf of Carpentaria, 1, 6, 29, 68, 118, 120, 122, 195, 208
the Gulf of Carpentaria, 1, 3, 6, 9, 28–30, 68, 71, 72, 93, 113–115, 118–121, 125, 126, 134, 145, 148, 195, 197–199, 208, 213, 222, 288, 333, 342, 343
the Gulf of Mexico, 46, 125, 126, 145
the Gulf of Thailand, 121, 125, 126
the Gulf of Tonkin, 114
the Gulf Stream, 126
The Gulf Winds, 115
the Gulf Winds, 113, 114
the Middle East, 47
the Northern Hemisphere, 17, 18, 23, 24
the Northern United States, 46
the Owens Valley, 126
the Pacific Ocean, 121
the Santa Cruz Mountains, 121
the Southern Hemisphere, 18, 23, 24, 29
the Torres Strait, 125
the United Kingdom, 121

the United States, 1, 46, 93, 121, 125, 126, 145
theory, 179, 182, 183
thermal, 17, 75, 78, 147, 148, 167, 188
thermodynamic, 184, 313
thermosphere, 166
thickness, 26, 75, 79, 149, 188
thinking, 357
thought, 24
thrill, 6
thunderstorm, 1, 33, 47, 163, 204, 205
tilting, 89, 90
time, 36, 39, 41, 42, 74–76, 78, 96, 103, 108, 111, 115, 132, 144, 145, 172, 176, 204, 207, 208, 211, 217, 220, 226–229, 233, 238, 241, 242, 245, 248, 250, 251, 262, 265, 269, 273, 274, 280, 281, 293, 294, 304, 306, 307, 318, 321, 322, 328, 331, 332, 340–343, 349, 350, 356, 360, 366
timing, 156, 170, 216, 221, 239, 247
tint, 101
togetherness, 154
Tom Bierbaum, 276
tool, 41, 78, 93, 102, 110, 155, 157, 216, 327, 348, 360, 366, 369
top, 1, 76, 79, 82, 133
topic, 175, 205, 333, 357
topography, 2, 21, 60, 62, 68, 70, 72, 74–76, 87, 114–116, 118, 119, 121, 124, 125, 131, 138, 169, 171, 172,

184, 198, 199, 206–208, 210–213, 215, 302, 342
touch, 278
tour, 122, 127, 283, 285, 295, 297
tourism, 8, 63, 115, 122–124, 126, 142, 151, 227, 282–291, 294, 295, 297–300
tourist, 22, 122, 126, 297, 365
town, 5, 21, 115, 120, 124, 131, 222, 288
trace, 93
track, 19, 78, 165, 228, 251, 292, 307, 309, 318, 322, 337
tracking, 92, 104, 105, 232, 328
trade, 23, 24, 29, 54, 73, 118, 120, 141, 180, 195
traffic, 90, 220, 224, 226–231, 236, 238, 251
trail, 36, 256
training, 42, 222–224, 227, 229, 230, 232–234, 294, 295, 365, 367, 371
transfer, 84, 126, 175, 187, 193, 199, 214, 341, 344, 349
transformation, 67, 278, 281
transience, 278, 280
transition, 20, 59, 129, 149–151, 267
transmission, 262, 265, 266
transmitter, 306, 321
transpiration, 30
transport, 23, 24, 54, 59, 85, 188, 194, 209, 277, 278
transportation, 25, 26, 123, 286
trapping, 30, 39, 93, 141, 166
travel, 73, 118, 174, 183, 224, 230, 232, 234, 285, 306
trigger, 20, 24, 28, 59–62, 66, 68, 93, 121, 129, 167, 169, 170, 188, 198, 199, 213
tropopause, 164
troposphere, 28, 166–168
trust, 266
trustworthiness, 346
tube, 9, 81, 144, 180
turbulence, 56, 73, 74, 79, 134, 147, 148, 157, 164, 188, 219, 221, 222, 231, 250, 328, 329, 339–341
turn, 60, 91, 147, 149, 210, 244, 245, 247
type, 1, 43, 44, 47, 100, 104, 199, 212

ultraviolet, 307
uncertainty, 242, 346
undercut, 21
understanding, 2–9, 11, 15–17, 25, 27, 30, 32–34, 37, 41–45, 48, 56, 57, 60, 62, 63, 68, 72, 76, 80, 81, 83, 86, 88, 89, 91, 93, 94, 97, 102, 105, 107, 108, 110, 111, 119, 123, 127–129, 132, 134, 135, 137, 141, 143, 144, 149, 152, 154, 157, 159, 161, 162, 165, 168, 171–173, 176–179, 181, 182, 184–187, 189, 190, 193, 195–199, 201, 202, 204, 206, 211, 215, 217, 218, 224, 225, 228–230, 232, 238, 240, 241, 243, 245–248, 251, 252, 254, 257, 259, 261, 263, 265, 267, 268, 270, 271, 275, 277, 280, 285, 287, 292, 293, 296, 297,

307, 308, 312, 314, 315, 317, 319, 320, 323, 326–332, 334, 337, 340–345, 347–349, 351–356, 359, 361–363, 365–368, 371
uniform, 2, 45, 53, 61, 73, 124
United States, 47, 113, 114
unity, 153
universe, 274
unpredictability, 283, 284
uplift, 19, 29, 33, 55, 145, 199, 209, 210
upslope, 118
urbanization, 254
use, 41, 42, 62, 79, 97, 127, 151, 155, 156, 171–173, 181, 207, 215, 217, 219, 226, 230, 232, 240, 242, 245, 251, 266, 270, 275, 276, 307, 308, 310, 312, 321–323, 327, 329, 330, 336, 342, 344–346, 349, 352, 354, 355, 362, 366, 368
user, 360, 362
utility, 303, 304, 306
utilization, 297, 307

validation, 157, 173, 187, 320, 337, 343, 362, 367, 368
valley, 61, 62
value, 142, 164, 240, 265, 267, 270, 296
vantage, 6, 153
vapor, 7, 17, 30–37, 43, 56, 67, 71, 74, 76, 78, 82, 84, 85, 90, 133, 170, 193, 209, 213, 307, 308, 339, 350

variability, 7, 42, 60, 132, 157, 210, 211, 239, 241, 244–246, 248, 319, 337, 340, 344
variable, 145
variation, 70–73, 76, 82, 86–89, 140, 155
variety, 42, 43, 61, 73, 95, 137, 167, 256, 330
velocity, 28, 88, 174, 178, 182, 183, 306, 307, 321
Venezuela, 114
venture, 146
version, 41, 42
viability, 124, 288
vice, 201
vicinity, 309
video, 266, 332
videography, 332
Vietnam, 114
view, 42, 78, 80, 151, 201, 218, 308, 309, 337
viewer, 278
viewing, 283, 284
violet, 95, 100, 105
visibility, 43, 149, 151, 152, 218, 221, 231
vision, 109
visit, 120, 283
visitor, 284–286, 288, 295
visual, 36, 42, 67, 81, 105, 107, 110, 111, 150, 154, 269–271, 274, 276, 277, 280–282, 366
visualization, 110, 111, 360, 366, 368
volume, 45, 294, 338, 346, 355
vortex, 23, 47, 223
vorticity, 79, 81–83, 86–90, 194
vulnerability, 251

war, 273
warm, 1, 17, 18, 20, 22–24, 28, 29, 37, 39, 43, 46, 47, 54, 56, 58, 61, 62, 72, 76, 77, 81, 84, 93, 101, 113–115, 118, 120, 125, 126, 132, 133, 140, 144, 147, 167–169, 187–195, 198, 204, 207, 212, 213, 217, 248
warming, 17, 129, 276
warning, 8, 61, 91, 165, 218, 224–226, 228, 229, 251, 252, 257
waste, 123
water, 6, 7, 17, 30–37, 43, 56, 67, 68, 71, 74, 76, 78, 79, 82, 84–86, 90, 95, 96, 99–101, 106, 114, 133, 145, 147, 163, 170, 176, 180, 187, 193, 204, 209–211, 213, 243, 249, 307, 308, 339, 350
wave, 24, 55, 66, 75, 87, 88, 93, 132, 150, 152, 163–165, 167, 168, 172–180, 182, 186, 214
wavelength, 70, 82, 95, 100, 101, 104, 105, 163, 185, 309
way, 27, 43, 100, 146, 149, 173, 209, 268, 272, 278, 286, 313, 355, 359, 366
wealth, 280, 314
weather, 2, 3, 5–9, 13–19, 21–31, 33, 36, 37, 39–48, 50–55, 57, 59–63, 72, 77, 79, 83, 86–88, 90–94, 104, 105, 108, 116, 120, 122, 127–129, 131–133, 135, 137, 139, 141, 155, 156, 162, 165, 166, 169–171, 173, 176, 183, 186, 189, 190, 196, 198, 199, 202–205, 207, 208, 211–236, 238, 239, 241, 243–251, 256, 257, 262, 265, 284, 293, 302, 308, 318, 321–323, 327–333, 335, 337, 341–344, 348–355, 362, 366
weaving, 255, 282
web, 226, 360
welfare, 329
well, 8, 30, 45, 47, 66, 67, 74–76, 78, 113, 114, 123, 126, 133, 142, 149, 152, 174, 180–182, 197–199, 229, 239, 265, 273, 284, 287–289, 309, 322, 330, 332, 335, 337–339, 344, 346, 368
west, 23, 118, 120
westerly, 19, 113
width, 180
wildlife, 283, 285, 286
William Wordsworth, 271
wind, 3, 6–8, 14, 17, 18, 23–27, 29, 30, 44, 45, 50, 54, 59, 60, 62, 66, 68–77, 79, 81–83, 86–91, 107, 114–116, 119, 125, 127, 132, 134, 141, 142, 144–146, 148, 150, 155, 161, 164, 167, 171, 172, 174, 175, 180, 181, 186, 190, 194, 198, 205–208, 214–216, 219–221, 226, 239, 240, 245, 247, 249,

251, 278, 292, 301, 305–307, 309, 310, 321–323, 327–331, 334, 337, 339–343, 350, 354, 366
window, 271
winter, 24, 46, 47, 135, 322
wisdom, 154, 155, 254, 265
woman, 273
wonder, 152–154, 257, 262, 272–274, 277, 278, 281, 283, 297
work, 3, 67, 123, 124, 226, 227, 232, 233, 275, 296, 352
world, 1, 2, 4, 6, 9, 11, 15, 17, 22, 23, 26, 29, 102, 111, 113–115, 118–123, 125–128, 131, 132, 137, 146, 155, 156, 208, 253, 254, 257–259, 261, 262, 265, 270, 272–283, 285, 295, 297, 362, 365
Wragge, 3

year, 5, 132, 134, 215, 249, 285
yellow, 95, 96, 100
yield, 256

zone, 120, 140, 180